SURFACE ENGINEERING BY FRICTION-ASSISTED PROCESSES

Methods, Materials, and Applications

SURFACE ENGINEERING BY FRICTION-ASSISTED PROCESSES

Methods, Materials, and Applications

B. Ratna Sunil, PhD

Apple Academic Press Inc.
3333 Mistwell Crescent
Oakville, ON L6L 0A2, Canada

Apple Academic Press Inc.
1265 Goldenrod Circle NE
Palm Bay, Florida 32905, USA

First issued in paperback 2021

Exclusive worldwide distribution by CRC Press, a member of Taylor & Francis Group

ISBN 13: 978-1-77463-451-6 (pbk)
ISBN 13: 978-1-77188-769-4 (hbk)

Library and Archives Canada Cataloguing in Publication

Title: Surface engineering by friction-assisted processes : methods, materials, and applications / B. Ratna Sunil, PhD.

Names: Sunil, B. Ratna, author.

Description: Includes bibliographical references and index.

Identifiers: Canadiana (print) 20190114886 | Canadiana (ebook) 20190114908 | ISBN 9781771887694 (hardcover) | ISBN 9780429398094 (ebook)

Subjects: LCSH: Surfaces (Technology)

Classification: LCC TA418.7 .S86 2019 | DDC 620/.44—dc23

Library of Congress Cataloging-in-Publication Data

Names: Sunil, B. Ratna, author.

Title: Surface engineering by friction-assisted processes : methods, materials, and applications / B. Ratna Sunil, PhD.

Description: 1st edition. | Oakville, ON ; Palm Bay, Florida : Apple Academic Press, Inc., 2019. | Includes bibliographical references and index. | Summary: "Key Features : Explores a multitude of topics within the field of surface engineering at length Discusses three primary examples of friction-assisted processes Summarizes and explores the mechanical foundation of friction stir processing, fabrication of surface metal matrix composites, and friction surfacing Incorporates figures and tables to aid in illustrating the concepts discussed Offers potential applications and discusses future benefits of specific elements pertaining to surface engineering"-- Provided by publisher.

Identifiers: LCCN 2019026738 (print) | LCCN 2019026739 (ebook) | ISBN 9781771887694 (hardcover) | ISBN 9780429398094 (ebook)

Subjects: LCSH: Surfaces (Technology) | Finishes and finishing. | Tribology.

Classification: LCC TA418.72 .S85 2019 (print) | LCC TA418.72 (ebook) | DDC 620/.44--dc23

LC record available at https://lccn.loc.gov/2019026738

LC ebook record available at https://lccn.loc.gov/2019026739

Apple Academic Press also publishes its books in a variety of electronic formats. Some content that appears in print may not be available in electronic format. For information about Apple Academic Press products, visit our website at **www.appleacademicpress.com** and the CRC Press website at **www.crcpress.com**

About the Author

B. Ratna Sunil, PhD
Associate Professor, Department of Mechanical Engineering; Bapatla Engineering College, Bapatla, Andhra Pradesh, India

B. Ratna Sunil, PhD, is an Associate Professor in the Department of Mechanical Engineering at Bapatla Engineering College, Bapatla, Andhra Pradesh, India. Earlier, Dr. Sunil worked as an Assistant Professor in the Department of Mechanical Engineering and also served as an Associate Dean of Academics at Rajiv Gandhi University of Knowledge Technologies (RGUKT), Nuzvid, Andhra Pradesh, India. He currently teaches courses on metallurgical and materials science, manufacturing processes, welding technology, composite materials, advanced materials technology, and surface engineering. Dr. Sunil has published several articles in international peer-reviewed journals and conference proceedings. He has given conference presentations and invited talks at various scientific and technical events. From his research findings, Dr. Sunil has filed four Indian patents. He has received several awards, including the Sudarshan Bhat Memorial Best PhD Thesis Award from the Indian Institute of Technology Madras, India, and the Bajpai-SAHA Award from the Society for Biomaterials and Artificial Organs, India. He is an Associate Fellow (elected 2017) of the Andhra Pradesh Akademi of Sciences. Dr. Sunil earned his PhD in Metallurgical and Materials Engineering from IIT Madras, India.

Contents

Abbreviations

ABE	accumulative back extrusion
ARB	accumulative roll bonding
HB	Brinell Hardness number
BUE	built-up edge
CMCs	ceramic matrix composites
CVD	chemical vapor deposition
CNC	computer numerical controlled
CAD	computer-aided design
CGP	constrained groove pressing
CR	consumption rate
CDRX	continuous dynamic recrystallization
DR	deposition rate
DFSP	direct friction stir processing
DDRX	discontinuous dynamic recrystallization
DRX	dynamic recrystallization
EBSH	electron beam surface hardening
EBVD	electron beam vapor deposition
ECAP	equal channel angular pressing
FSC	friction stir channeling
FSP	friction stir processing
FSR	friction stir riveting
FSSW	friction stir spot welding
FSW	friction stir welding
FS	friction surfacing
GDRX	geometric-dynamic recrystallization
HAZ	heat affected zone
HDPE	high-density polyethylene
HPT	high-pressure torsion
HVOF	high-velocity oxy-fuel
LBSH	laser beam surface hardening
MMC	metal matrix composite
MAF	multi-axial forging
MWCNT	multi-walled carbon nanotubes

PD	penetration depth
PBS	phosphate buffer solution
PVD	physical vapor deposition
PCBN	poly cubic boron nitride
PMCs	polymer matrix composites
PFF	primary flash formation
RE	rare earth
RFF	rate of flash formation
SEM	scanning electron microscope
SFF	secondary flash formation
SPD	severe plastic deformation
SBF	simulated body fluid
SMAT	surface mechanical attrition treatment
TWI	The Welding Institute
TMAZ	thermomechanical affected zone
3D	three dimensional
TEM	Transmission electron microscope
TE	twist extrusion
UTS	ultimate tensile strength
UFG	Ultrafine-grained
HV	Vickers hardness number

Preface

Development of unconventional processing methods in the past two decades revolutionized the effective utilization of resources in the manufacturing industry and strengthened the arms of the engineers to efficiently address various issues in solving several industrial problems. Evolution of friction-assisted processes is an example of one such development. When friction stir welding (FSW) was developed and successfully demonstrated for the first time at The Welding Institute (TWI), UK, in 1991, a new area was opened in the field of welding. Later, it has been widely used to join various similar and dissimilar metals, particularly those that are difficult to be joined by conventional fusion welding techniques. Aluminum, copper, magnesium, titanium, and their alloys are the best examples of such metals that have been successfully joined by FSW. With the same principles of FSW, friction stir processing (FSP) was developed to alter the surface microstructure without melting the workpiece at the surface level.

Furthermore, a new area has been developed, known as "surface composites by FSP," in which composites are fabricated by introducing the secondary phase particles into the workpiece during FSP. This kind of composite material is similar to that of metal matrix composites (MMCs) in which secondary phase particles are dispersed in a suitable metal matrix. There is a great potential in using FSP to fabricate surface MMCs for many structural applications in the automobile and aerospace industries. There are several material systems that have been used to fabricate MMCs by FSP, and the performance of these composites was found to be excellent. However, there are a few challenges and issues that still need to be addressed in developing surface composites by FSP.

Friction surfacing (FS) is another variant of the friction-assisted processes in which a consumable tool called as "mechtrode" is used instead of non-consumable tool. The consumable rotating tool undergoes plastic deformation due to the generated heat resulted from friction between the mechtrode, and the substrate and material is transferred from the consumable tool to the substrate. FS has gained a special attention as a promising solid-state method in developing surface coatings of different alloys and composites.

One common observation that can be made in all friction-assisted processes is the state of the material. The substrate and consumable tool do not melt during the process. Hence, the processes completely avoid any complex issues associated with the liquid state of the material. Compared with other conventional manufacturing processes, which have been in use for centuries in the manufacturing industry, friction-assisted processes, are at their infant stage. However, for the past 25 years, tremendous research has been carried out in this area and garnered a significant attention of many research groups working worldwide. The number of articles being published in this area is also rapidly growing. Here arises the necessity of a single source of information where one can obtain complete information and the state of the art of the technology. Therefore, the objective of the present book is to provide a comprehensive summary of friction-assisted processes that are used to modify the surfaces of structures. As a single volume, this book gives an instant snapshot of the upcoming and potential area in the advanced manufacturing processes.

This book is intended for graduate students, researchers, and engineers working in the areas of surface engineering and composite materials. I hope this book will be really helpful for those who are working or about to start work in these related areas.

First, I would like to thank the Almighty and my parents for their blessings. I especially thank my wife Lalitha and my sweet daughter Amruta. It would be not possible to complete this book without their support. I thank my research guide of my postgraduation, Mrs. R. Subasri, Scientist-E, ARCI, Hyderabad, India, who actually inspired me to choose Materials Engineering as my research field. I extend my sincere thanks to my PhD guides, Prof. T. S. Sampath Kumar and Prof. Uday Chakkingal, Department of MME, IIT Madras, India, for their continuous encouragement in engaging in research. Thanks are due to my research colleagues at MML, IIT Madras (Mr. Kranthi, Mr. Yogesh, Dr. Hanas, Mrs. Madhumati, Mrs. Jayasree) for their encouragement. I thank my friends Dr. S. Anandkumar, Assistant Professor, Department of ME, IIT Jammu, India, Dr. Ravikumar Dumpala, Assistant Professor, Department of ME, VNIT Nagpur, India, and Dr. M. Jagan, TVS group, Chennai, India, for their continuous support. I also thank my students, friends, and colleagues at RGUKT Nuzvid who helped and encouraged me in writing this book.

— B. Ratna Sunil, PhD

PART I

Surface Tailoring of Metals by Friction Stir Processing

CHAPTER 1

Introduction to Surface Tailoring of Metals

1.1 MANUFACTURING PROCESSES

Human civilization stands on the level of understanding the materials and their properties available in nature. Utilizing the knowledge of different materials and their properties and by which manufacturing of required structures and components targeted for different applications accelerate the quality of the human life. Manufacturing field has played a crucial role in the era of the industrialization. It is impossible to imagine the development of science and technology without the manufacturing industry. All manufacturing processes can be broadly grouped into four basic categories known as casting, welding, machining, and joining. The selection of a specific manufacturing process or combination of different manufacturing processes depends on a number of factors such as the size and shape of the component, quantity, cost, precision, required properties, type of material and targeted application. A brief discussion of these basic manufacturing processes is given below.

1.1.1 Casting

Casting is the oldest process records its existence in different ancient civilizations from thousands of years around the world. Casting involves filling a cavity with molten metal and allowing solidification to develop a predefined shaped component or raw materials in the form of ingots. Overall, 80% of the castings are in the form of ingots. Cast ingots or billets are used as starting materials for all other manufacturing processes. The fluidity of the molten metal, shrinkage during solidification, porosity in the solidified structure and level of impurities are major concerns while casting a metal.

The success of the casting and obtaining a quality product depends on the aforementioned characteristics. Casting is the most economical, shortest and simple route to develop complex shapes which are not possible with other manufacturing methods. Almost all pure metals and alloys can be used to develop structures and end products through casting. In the iron-carbon system, cast irons such as gray cast iron, nodular cast iron, chilled cast iron, etc. are a group of alloys named after this process as casting is the only suitable process to produce components of these alloys.

From a small component weighing a few grams to a large structure of tons of weight can be produced by casting. Cast components are good in compressive strength but exhibit poor tensile strength. Cast structures usually contain columnar grains and free from residual stresses. Casting reduces or minimizes the necessity of other manufacturing processes like machining, forging, and welding in developing a single object. Conventional casting processes require follow-up manufacturing process such as machining to tailor the surface of the structure before it is used for the application. Among all other casting processes, sand casting is the old and most widely used method. The basic steps involved in conventional casting are pattern making, core making, molding, melting, pouring, cleaning, and finishing. In recent days several advanced casting methods such as shell mold casting, investment casting, die casting, centrifugal casting, strip casting, and slip casting have been developed which revolutionized the production engineering in terms of quantity and quality of the produced components. Figure 1.1 shows the classification of casting processes.

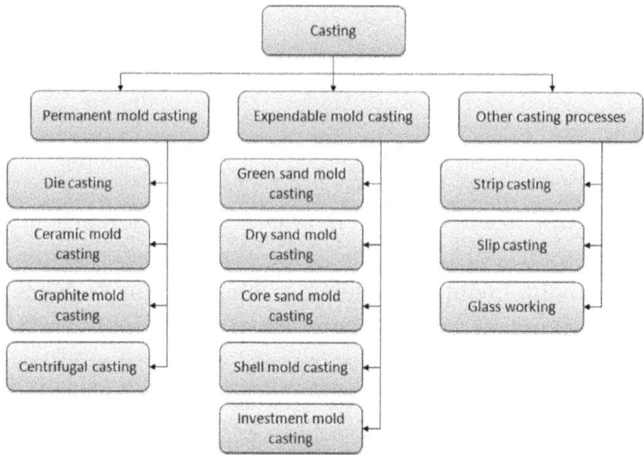

FIGURE 1.1 Flowchart showing the classification of casting processes.

Components and structures developed by casting include engine blocks and heads of automobile engines, pump housings, railway wheels, machine tool bodies and frames, jewelry, idols, statues, etc.

1.1.2 Metal Forming

Metal forming is an another important category of manufacturing processes which involves applying higher amounts of mechanical loads to deform materials to obtain designed geometric shapes plastically. The basic principle behind applying mechanical loads using specially designed dies is to make the material cross the yield stress, so the shape induced by forming operation is retained. Extrusion, forging, rolling, wire drawing, sheet metal bending, and punching are a few examples of metal forming techniques. Figure 1.2 shows the classification of metal forming techniques based on the dimensions of a material. If the material has a higher surface area to volume ratio such as sheets, the processes are called sheet metal forming. If the material has a relatively lower surface area to volume ratio, the processes can be called as bulk-forming processes. It is difficult to process brittle materials by metal forming operations. Most of

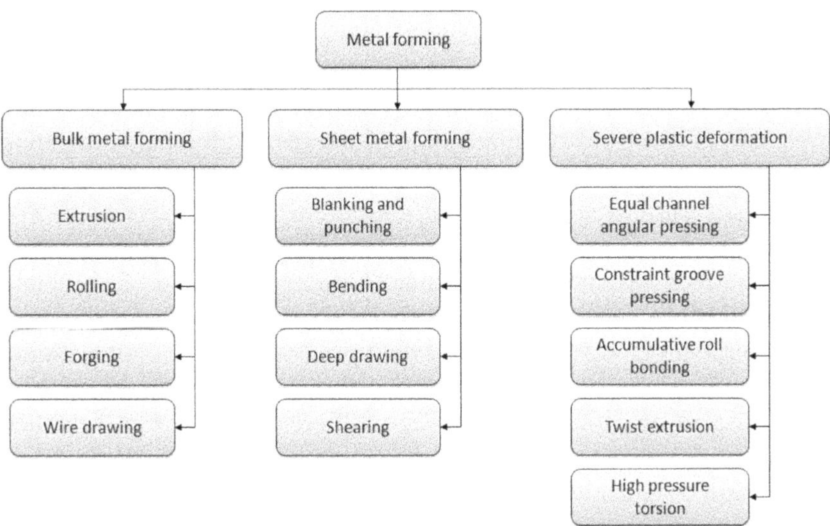

FIGURE 1.2 Flowchart showing the classification of metal forming processes.

the forming operations result in high compressive stresses in the processed material. Therefore, the ductility of the material plays an important role in metal forming operations. Brittle materials can be processed by applying a suitable amount of heat as the yield strength of a material is decreased, and the ductility is also increased with increase in temperature.

Broadly metal forming techniques can be classified as hot working, warm working and cold working processes based on the temperature at which processing is carried out. In hot working, the material is heated to a temperature above its recrystallization temperature before processing. In warm working, the material is processed at a temperature lower than recrystallization and above the room temperature. If the processing is carried out at room temperature, it is called a cold working process. Depending on the level of ductility of the processing material, a suitable range of temperatures is selected.

1.1.3 Machining

Machining processes involve removing of material from a workpiece by different means to attain required dimensions. The remaining workpiece is the desired component. Usually, in machining (conventional) a cutting tool is used which is having a sharp cutting edge against the workpiece to cut undesired material by applying a certain amount of load.

The material, which is removed against the movement of the cutting tool, is called a chip. Turning, milling, drilling, shaping, planning, etc., are a few examples of the machining processes. Additionally, grinding operations are considered as a special type of machining processes where hard and brittle abrasive particles are used to remove undesired material in the form of very fine chips. Along with grinding, other finishing processes such as honing and lapping are also used to obtain surface finish in which very fine abrasive particles are used to remove the material from the surface. Recently, another group of machining processes is gaining wide popularity in the manufacturing industry in which instead of a sharp cutting edge, different forms of energy such as mechanical, electrical, and chemical are used to remove the material and named as unconventional machining processes. Figure 1.3 shows different machining processes used in the manufacturing industry.

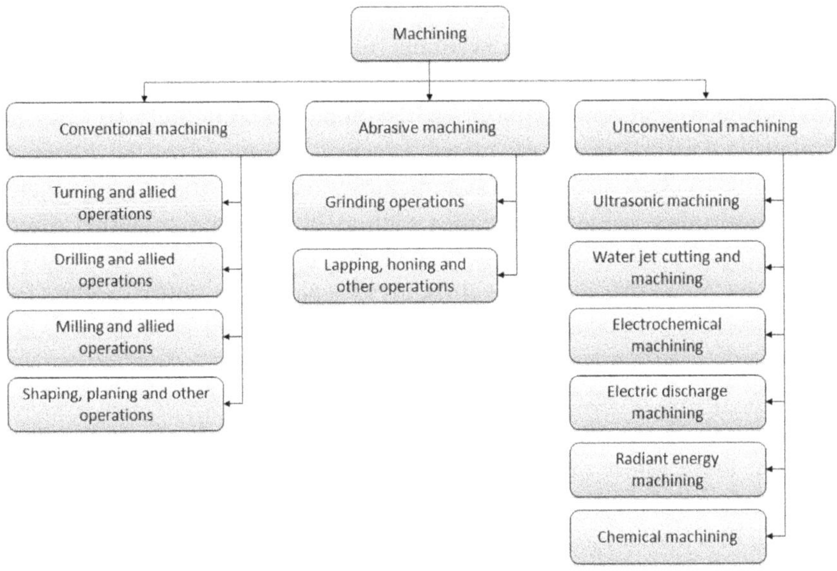

FIGURE 1.3 Flowchart showing the classification of machining processes.

1.1.4 Joining

Welding occupies the most important position in materials joining techniques. Welding is an operation in which metallurgical continuity is established between two entities using heat or pressure in some cases both heat and pressure. The basic objective in welding is to obtain equilibrium inter-atomic distance between the atoms which belong to the mating surfaces if so the two surfaces are said to have metallurgical continuity. In order to achieve this objective, a sufficient amount of heat or pressure is required. In some welding techniques, applying both pressure and heat is required to get a perfect joint. All welding processes can be grouped into two categories known as fusion welding and solid-state welding. The given heat causes the material to melt at the joint and the subsequent solidification results to establish a perfect metallurgical continuity between the two surfaces in fusion welding processes. Oxy-acetylene gas welding, manual metal arc welding, submerged arc welding, tungsten inert gas welding, gas metal arc welding, electron beam welding, laser beam welding, plasma welding, and resistance welding are a few examples of fusion welding processes. In solid-state welding processes, metallurgical continuity is established between two surfaces without melting. Explosive welding, diffusion welding, ultrasonic welding, forge welding, and friction stir welding are a few examples for solid-state welding processes. Welding is a necessary process required in the manufacturing industry and is obvious in developing any structures. Figure 1.4 summarized the different welding processes used in the manufacturing industry. Additionally, there are several other joining techniques such as brazing and soldering often used in the manufacturing industry. Welding, brazing, and soldering give permanent joints that mean it is difficult to separate the entities joined by these processes. On the other hand, several temporary joining methods known as assembling processes are also widely used in the manufacturing industry. These operations involve using fastening components such as rivets, screws, bolts, nuts, etc. to assemble structures, sheets, and plates. Some of these joints cannot be disassembled like in the case of riveted joints. Other fastening techniques allow disassembling the structures for example bolt and nut joints.

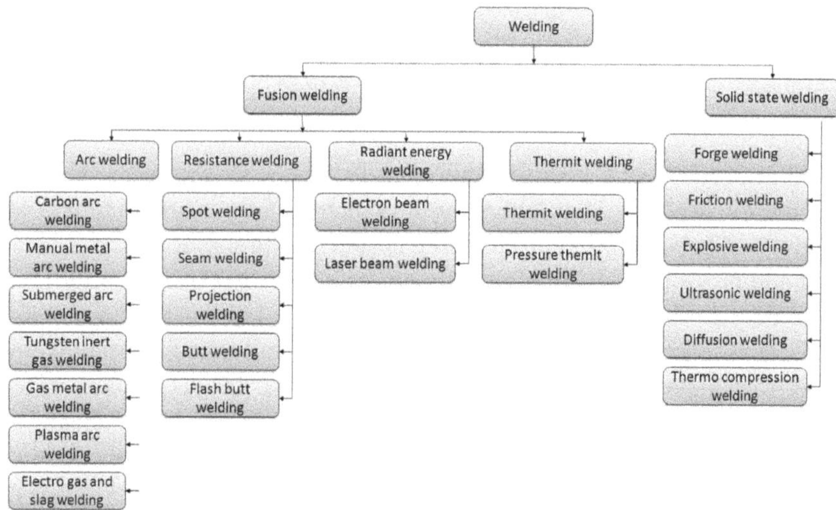

FIGURE 1.4 Flowchart showing the classification of welding processes.

1.1.5 *Advanced Manufacturing Processes*

In recent years, many advanced manufacturing processes have been developed in which traditional mechanisms behind material processing are not observed as usually found in conventional manufacturing processes. Machining of high strength alloys without applying mechanical loads, cutting of materials by chemical and electrical means, developing complex shapes which are not possible with conventional machining are a few examples of advanced manufacturing processes [1–3]. Among these methods, unconventional machining also called some times as nontraditional machining is a well-known group of processes now being widely used in various industries. Figure 1.3 shows different unconventional machining processes which involve using mechanical, chemical and electrical energies or in some processes combination of these energies to cut the material or to remove undesired material from the workpiece without using any sharp-edged cutting tools [3].

Developing near net shaped components by minimizing or completely eliminating the material wastage in the form of scrap or chips is the next important group of unconventional manufacturing processes. Rapid prototyping is one of such techniques gaining tremendous attraction in the manufacturing fields. In rapid prototyping, computer-aided design (CAD) and computer-aided manufacturing (CAM) technologies occupy a central part [4]. Computer numerical controlled (CNC) machines are used in material removal/addition in rapid prototyping methods. Layer-by-layer is added to develop a three-dimensional component in material addition rapid prototyping methods. Stereolithography [5], droplet deposition [6], fused deposition [7], selective laser sintering [8], and 3D printing technology [9] are the best examples for rapid prototyping techniques.

1.2 SURFACE ENGINEERING

Surface engineering is an interdisciplinary field that basically provides information related to properties observed at the surface of the materials which include surface roughness, topography, texture, surface energy, hardness, wear resistance, etc. Surface engineering covers aspects from basic disciplines of engineering and sciences, for example, mechanical engineering, materials science, chemistry, physics, metrology, and instrumentation.

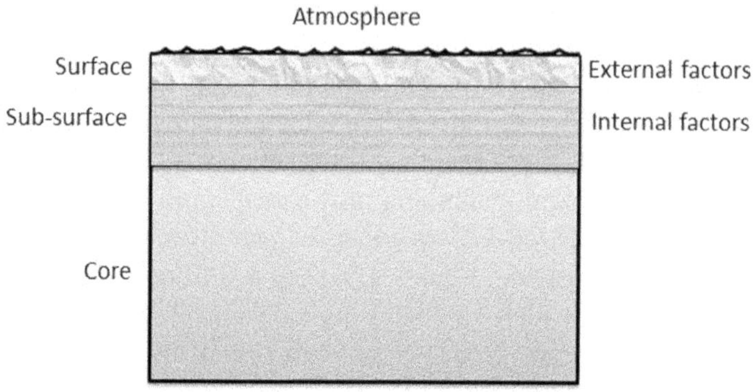

FIGURE 1.5 Schematic representation of surface and sub-surface influencing factors in a material.

All material properties observed at the surface can be grouped into two categories as external factors and internal factors as listed in Table 1.1. Eternal factors are the influencing features observed at the surface layers of thickness varying from a few nanometers to micrometers. The sub-surface factors or internal factors are the influencing features observed beneath the surface. Figure 1.5 schematically illustrates how these influencing factors are distributed across the thickness. In some applications, external factors play an important role such as appearance, corrosion behavior, wear resistance, and initiation of cracks. Internal factors along with external factors at the surface influence the mechanical behavior and failure mechanisms in the structures.

TABLE 1.1 Different Influencing Factors Usually Found At the Surface and Subsurface of a Material

Surface (External factors)	Sub-surface (Internal factors)
Chemical composition	Microstructure
Surface roughness and waviness	Hidden stresses
Topography and texture	Hardness
Impurities and inclusions	Cracks
Oxides	Chemically non-homogeneity
Pits and peaks	Presence of new phases

1.3 SURFACE PROPERTIES

Understanding different surface properties are important in the study of surface engineering. Specific surface engineering processes alter particular surface properties either by introducing microstructural modification or by altering the chemical composition at the surface. The important surface properties, which are crucial to understand in the field of surface engineering and manufacturing industry, are discussed below.

1.3.1 Surface Energy

The surface energy of solids and surface tension of liquids are identical terms which relate the molecule interaction in forming bonding at the surface. However, surface energy (γ, $J.m^{-2}$) of solids is composed of elastic and plastic components and hence slightly differs with the surface tension of liquids. Determining surface energy of solids is difficult due to the reasons such as the type of the solid (ionic, covalent and metallic), chemical non-homogeneity at the surface and measuring parameters adopted in different methods. Measuring the contact angle made by a solvent when placed a drop on the surface of a solid is a common method to calculate the surface energy. In this experiment, the surface tension of the solvent used to measure the contact angle must be known. Figure 1.6(a) shows the schematic representation of contact angles on high-energy surface and low energy surface. Figure 1.6(b) shows the three energy components as per Young's equation [10, 11].

When a drop of liquid contacts the solid surface in a medium of gas as shown in Figure 1.6(b), a relationship can be established between the measured contact angle of the liquid (θ), surface tension of the liquid (γ_{lg}), interfacial energy between solid and liquid (γ_{sl}), and the surface energy of the solid (γ_{sg}) as given below.

$$\gamma_{sg} = \gamma_{sl} + \gamma_{lg} \cos \theta \tag{1}$$

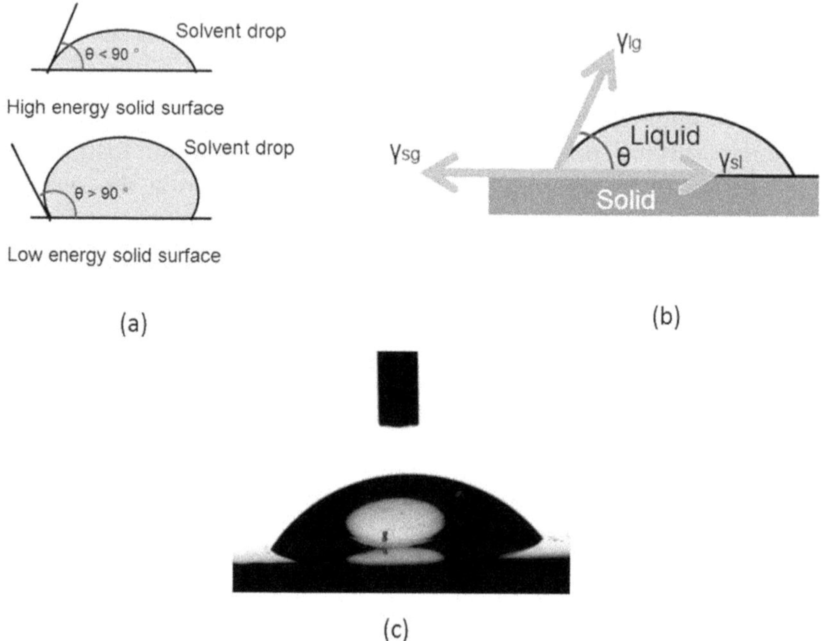

FIGURE 1.6 Schematic representation of contact angle measurement: (a) contact angle on high and low surface energy solids, (b) three force components as per Young's equation, and (c) typical water droplet on the surface of AZ31 Mg alloy.

This relation is valid if the three systems (solid, liquid and gas) are thermally equilibrium and chemically homogeneous. If any surface has contact angle (θ) < 90°, if water is used as the solvent, such surface can be called as hydrophilic surfaces and records higher wettability. If the water contact angle (θ) of any surface is higher than 90°, those surfaces show hydrophobic nature. Depending on the application, surfaces are engineered to achieve the desired level of wettability. For example, developing high hydrophobic surfaces is the prime objective behind developing corrosion resisting coatings and paintings. Similarly, developing high hydrophilic surfaces is the prime concern in developing new biomedical implants with improved bioactivity where tissue interactions are demanded.

Additional methods are also used to determine the surface energy of solids. Among them, (i) measuring the surface tension of liquid metal and from which calculating the free surface enthalpy of its solid-state, and (ii) measuring the work required to create a crack (an extra surface in a bulk volume) from which calculating surface energy are well-known methods.

1.3.2 Surface Topography

Surface topography gives the details of overall surface features mainly include roughness, waviness, and form. Understanding surface topography by studying individual surface features is crucial as the surface characteristics greatly influence the bulk behavior of the material. Figure 1.7 shows the schematic representation of typical surface parameters. Among them, roughness is the most influencing parameter which is observed at the microscopic level.

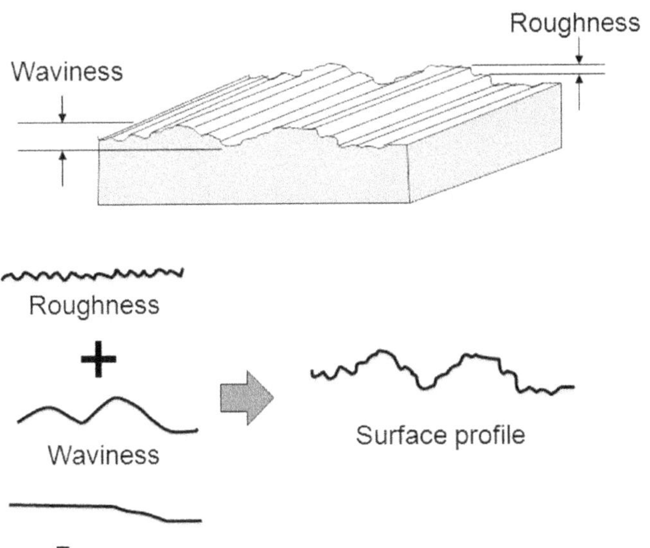

FIGURE 1.7 Schematic diagram explains roughness and waviness of a typical surface profile.

Surface undulations resulted from material removal can be observed at two different scales. The surface irregularity or undulation at the small scale as shown in Figure 1.7 can be called as *Roughness,* and at the large scale, it can be called as *waviness*. Together roughness and waviness give surface profile which is usually measured by 2D or 3D profilometers. Surface roughness at two dimensional and three-dimensional levels play an important role in engineering applications where two surfaces come into contact and friction is inevitable. Next level of the surface unevenness is *form,* which provides the macroscopic surface profile.

1.3.3 Corrosion

Corrosion is a surface degradation phenomenon resulted due to a chemical reaction in an environment. Most of the metals are unstable at room temperature and ready to participate in a chemical reaction which is further accelerated with an increase in temperature. Usually, the corrosion mechanisms are initiated from the surface of a metal. If metals with two different potentials come into contact in any medium, a galvanic cell is formed as shown in Figure 1.8 (a).

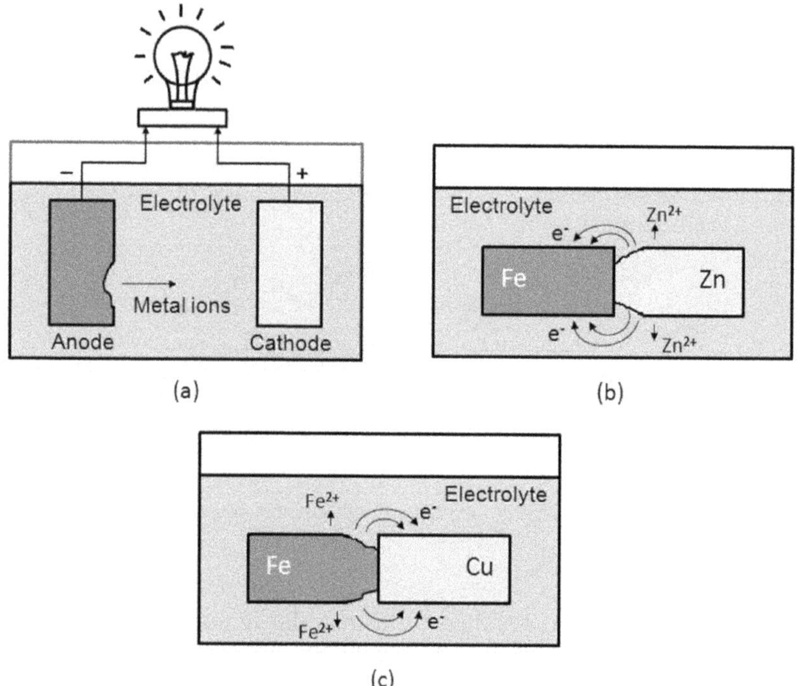

FIGURE 1.8 (a) Schematic representation of galvanic corrosion, (b) iron-zinc system, and (c) iron-copper system.

The metal, which is more active in the galvanic series, is corroded rapidly and called an anode, and the other metal is protected known as a cathode. The corrosion present in such cases is called galvanic corrosion. For example, if iron and zinc metals form a galvanic couple, being an active metal, zinc becomes the anode and is slowly degraded as shown in Figure 1.8 (b). If, iron-copper form a galvanic cell, now, iron becomes as an anode and copper acts as cathode and iron is degraded due to galvanic corrosion as schematically shown in Figure 1.8 (c). In mechanical fastening, components made of two different metals come in contact and undergo galvanic corrosion. In such cases, non-metallic washers are used as in-between elements to separate the direct contact between the components and to reduce the galvanic corrosion. In addition, general corrosion, pitting corrosion, crevice corrosion, intergranular corrosion, stress corrosion cracking, fretting corrosion are the other forms of corrosion usually found in metals. Tailoring the surface chemical composition or modifying the microstructure by various processes or providing anti-corrosive coatings and paintings at the surface are the different strategies adopted in the surface engineering to address the corrosion issues.

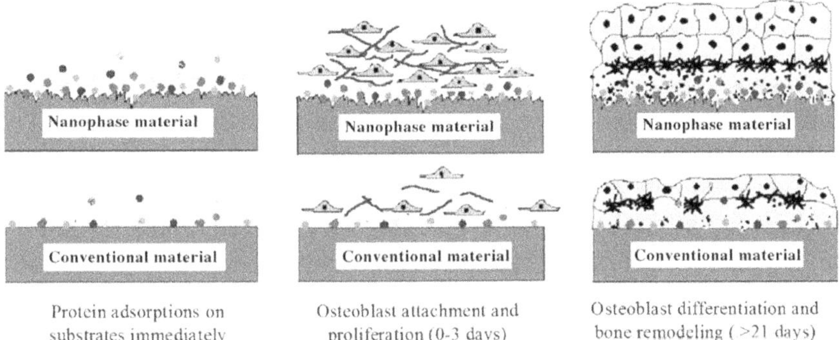

FIGURE 1.9 Advantage of nanostructuring the biomaterials: "The bioactive surfaces of nanomaterials mimic those of natural bones to promote greater amounts of protein adsorption and efficiently stimulate more new bone formation than conventional materials" (Reprinted from L. Zhang, T. J. Webster, 'Nanotechnology and nanomaterials: Promises for improved tissue regeneration.' *Nano Today*, 4, 66–80, Copyright (2009), with permission from Elsevier, USA [12]).

1.3.4 Bioactivity and Biomineralization

In biomaterials, *bioactivity* is desired for an implant which is used as an external member to support a fractured bone for example in orthopedic implant applications. Bioactivity can be defined as the ability of a biomaterial to form a strong bond with the host tissue during the healing process. Steels, Co-Cr alloy, and titanium alloys are the best-known metallic implants widely used in the medical industry. However, these metals are bio-inert in nature, i.e., do not form any bond with the tissue. Hydroxyapatite, a well-known bioceramic material is the best example for bioactive biomaterial. Enhancing the bioactivity of metallic implants to promote a higher level of healing rate in orthopedic implants is the prime objective behind many research works carried out in the biomedical field. Providing hydroxyapatite surface coatings on the surface of the aforementioned metallic systems is the widely adopted strategy by several research groups across the world to develop novel biomaterials.

1.3.4.1 Biomineralization

Biomineralization can be defined as the ability of the implant material to be deposited with mineral phases in the presence of the physiological environment. Biomineralization enhances the tissue activities during the healing process of a fractured bone. The mineral phases and growth factors which deposit on the surface of an implant in the biological environment signal the primary cells which further attract the host cells and help for higher healing rate. In order to enhance biomineralization of a biomaterial, polymer or ceramic coatings are provided on the surface of the implants by different techniques such as electrodeposition, electrospinning, dip coating, spin coating, etc. Modifying the microstructure to nono/submicrometer level also enhances the bioactivity and biomineralization as explained by Zhang et al., [12] and schematically shown in Figure 1.9.

Developing nanostructured bulk metals or nanostructuring the surface itself without altering the microstructure of the core of the metal can be achieved by several methods which include a top-down approach and bottom-up approach as schematically explained in Figure 1.10. In the top-down approach, bulk nanostructured metals are developed from coarse-grained metals by introducing enormous strain through mechanical processing. Severe plastic deformation (SPD) techniques such as equal channel angular pressing (ECAP), constrained groove pressing (CGP), accumulative roll bonding (ARB), twist extrusion (TE), high-pressure torsion (HPT), and friction stir processing (FSP) are a few examples of the top-down approach [13]. The top-down approach gives bulk nanostructured structures which can be readily used in different industrial applications where high specific strength can be achieved by using lower weight metals. In the bottom-up approach, atom by atom are added to develop clusters, and the number of clusters is used to produce nanoparticles. These nanoparticles are consolidated to develop bulk nanostructured materials. Chemical vapor deposition (CVD), physical vapor deposition (PVD) and sol-gel methods are a few examples of the bottom-up approach [14]. Consolidating the produced nano-particles is a necessary follow-up process to develop bulk nano-structured components.

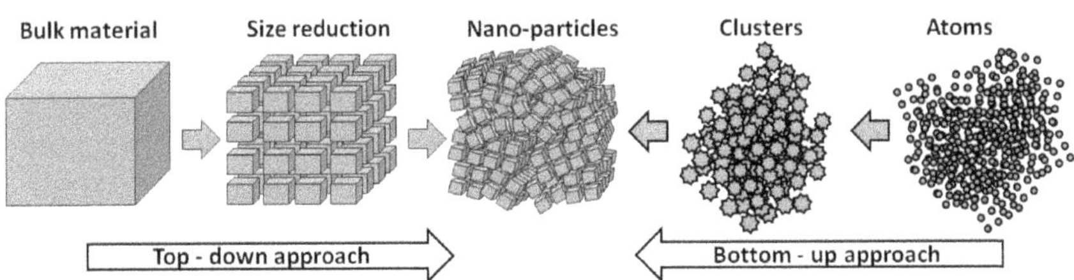

FIGURE 1.10 Schematic representation of the top-down and bottom-up approach.

1.3.5 Microstructure

Microstructure at the surface influences several properties. Basically, microstructure at the surface includes the grains, grain boundaries, twins, stacking faults and any two-dimensional lattice imperfections. Modifying these microstructural features alter the structure sensitive properties such as mechanical behavior, corrosion performance, electric conductivity and wear resistance. Altering the microstructure at the surface up to certain depth requires certain special methods which induce the effect selectively at the surface and do not introduce any effect to the core. Shot peening, sandblasting, ball/roller burnishing, mechanical attrition, and selective laser melting are a few examples used in the industry to alter the microstructure at the surface. Recently surface nanostructuring is gaining wide popularity in surface engineering. Among several methods, surface mechanical attrition treatment (SMAT) proposed by K. Lu and J. Lu [15] has shown excellent potential as a promising method to develop surface nanostructured metals. Figure 1.11 shows the schematic representation of the SMAT process used to develop nanostructured surfaces and regions subjected to different amounts of strain during SMAT.

1.3.6 Hidden Stresses

Hidden stresses sometimes called as residual stresses are common in the metals where mechanical loads are applied or thermal gradation results within the structure. When mechanical loads are applied, that is common in machining and metal forming operations, the strain that is introduced into the structure lead to lattice distortion. If these stresses and strains which are developed during the process are not relieved, hidden stresses result and may lead to failure in the form of crack or distortion during the functioning if they are not balanced properly. During solidification, thermal gradation may lead to developing thermal stresses which are either compression or tensile type and remain within the structure as hidden stresses. Intentionally, higher residual compression stresses are introduced at the surface by certain methods such as shot peening where a number of shots or balls have bombarded the surface with a high velocity to increase the fatigue resistance.

1.3.7 Hardness

Hardness has a different meaning from different perspectives. For a tribological engineer, hardness means resistance offered by a material towards wear and scratch. For a design engineer, hardness is an ability of a material to withstand against permanent deformation. For a production engineer, hardness means resistance towards material removal or being cut by a cutting tool. Overall, hardness can be defined from the materials engineering perspective as the resistance offered by a material towards plastic deformation measured as applied load per area.

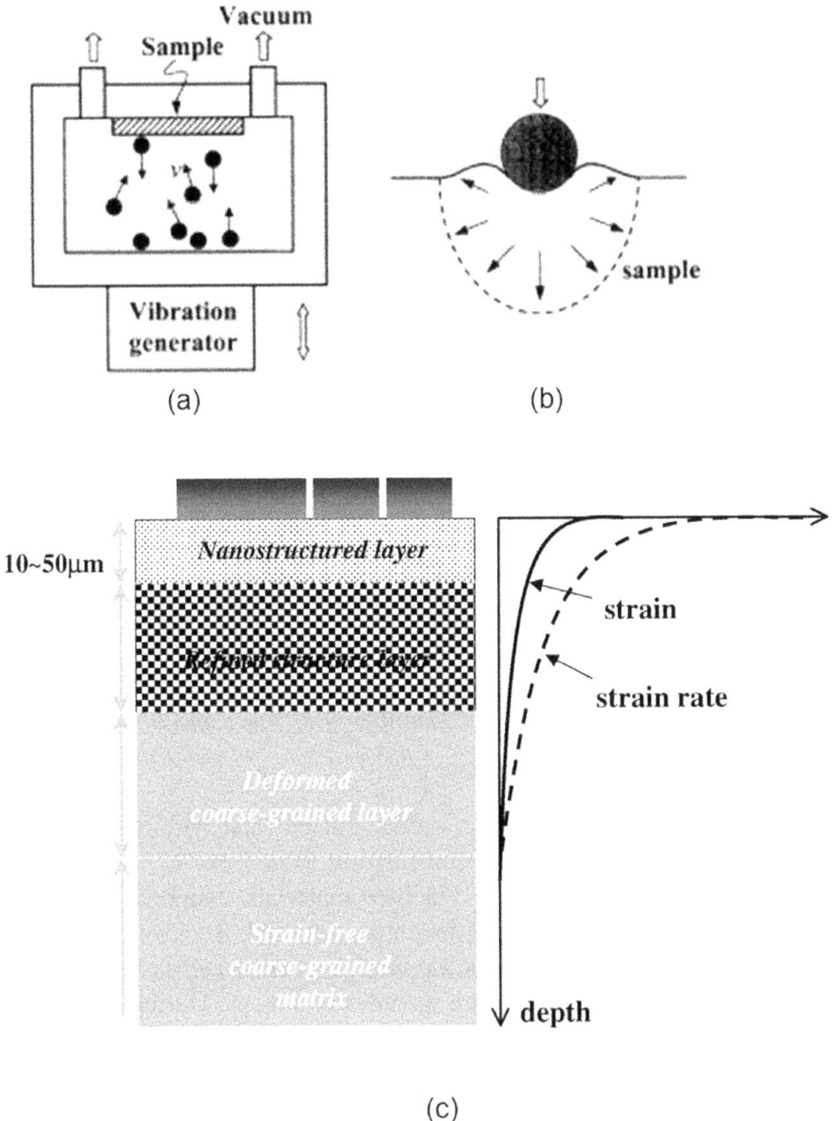

FIGURE 1.11 (a) Schematic representation of SMAT, (b) compression stress induced by a ball on the surface, and (c) strain distribution in the thickness direction during SMAT (Reproduced with kind permission from K. Lu and J. Lu [15] © 2004 Elsevier.).

On the other hand, several standard procedures were developed and widely accepted by the engineers in which an indent is placed on the surface by applying a suitable load with the help of an indenter. Brinell, Vickers, Rockwell and Knoop hardness testers are the most widely used variations. In Brinell hardness testing, ball indenters are used to generate an impression under a certain load on the surface, and the diameter of the impression is measured as shown in Figure 1.12(a). Then the Brinell Hardness (HB) number is calculated using the following equation:

$$\mathrm{HB} = \frac{2P}{\pi.\mathrm{D}\left(\mathrm{D}-\sqrt{\mathrm{D}-\mathrm{d}}\right)} \tag{2}$$

where HB = Brinell hardness number; D = diameter of the ball; d = diameter of the impression; and P = load.

In Vickers hardness testing, a diamond indenter is used as schematically shown in Figure 1.12 (b). The angle between the two opposite faces of the indenter is 136°. Then the following equation (3) can be derived to measure the Vickers hardness number (HV)

$$HV = 2P\sin\left(\frac{136}{2}\right)/d^2 = \frac{1.854P}{d^2} \tag{3}$$

where HV = Vickers hardness number; P = load; and d = arithmetic mean of the diagonals of the indent $\left(\frac{d_1+d_2}{2}\right)$.

In Rockwell hardness testing, a diamond indenter with the round tip as well as a ball indenter (diameter ranging from 1/16″ to ½") are used depending on the material type. For hard materials, diamond indenter is used, and for soft materials, ball indenters are used. Rockwell method is most reliable and directly gives the hardness value and is widely used in the industry. Testing procedure includes two steps. During measuring the hardness, the workpiece is placed below the indenter, and a preload is applied this is phase I which gives a reference position. In phase II, an additional amount of load is applied known as a major load for a dwell period to allow the elastic recovery. Then the major load is removed, and the final position with respect to the reference point is measured to get the depth of penetration and then converted to a number called Rockwell hardness number. In Knoop hardness test, an indenter similar to Vickers indenter is used which contains non-equal diagonals unlike in the case of Vickers indenter. After the applied load is removed, the major diagonal of the indent is measured, and the Knoop hardness number is calculated. Figure 1.12 (c) shows the diagonal of a Knoop indenter and the corresponding angles of the faces. Usually, for surface hardness measurement, Vickers microhardness, Rockwell and Knoop hardness methods are adopted for measuring surface hardness by considering the fact that the dimensions of the surface coatings and features which generally range in a few micrometers. Brinell hardness is not suitable to measure hardness variation in a range of few hundreds of micrometers and hence not used for surface hardness measurements.

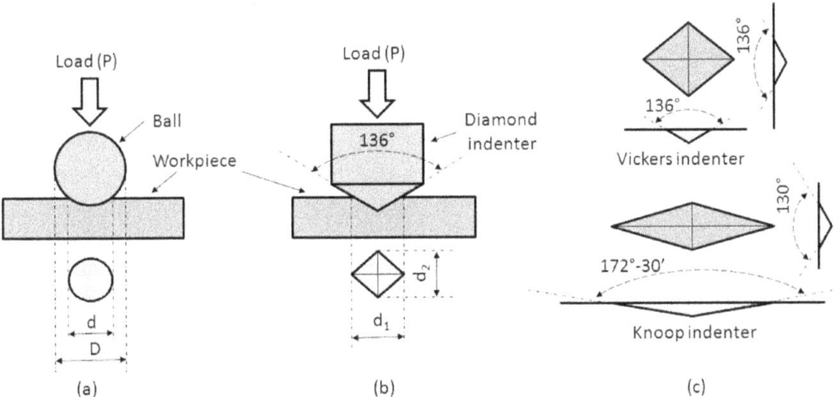

FIGURE 1.12 Schematic representation of hardness measurement by (a) Brinell indentation, (b) Vickers indentation methods, and (c) comparison of Knoop indenter and Vickers indenter.

1.3.8 Friction and Wear

When two surfaces come into contact under external loads, *friction* and *wear* prevail. Friction causes loss of energy and wear leads to loss of matter. Friction results when a surface is relatively in motion with respect to others. The level of friction between the contacting surfaces is represented with a coefficient of friction (μ). The ratio of force required to sustain the motion of the contacting body to the normal force resulted between the surfaces is a numerical value known as μ as schematically shown in Figure 1.13.

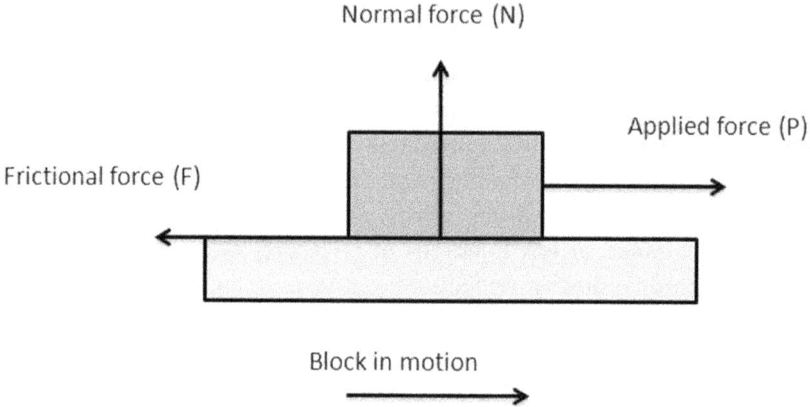

FIGURE 1.13 Schematic representation of frictional force resulting when a body is in motion on a surface.

$$\text{Frictional force (F)} = \mu.N$$

where N = normal force between the body and the counter surface; and resultant force (F_r) = P – F.

In order to sustain the motion of the block as shown in Figure 1.13, the resultant force (F_r) must not be zero. Here comes the importance of the coefficient of friction. Altering the μ value by different routes is a well-growing work in the surface engineering field. Friction can be broadly classified as sliding friction and rolling friction. In sliding friction, the counter surface may get adhered or at least at the regions of the asperities adhesion takes place and the presence of hard particles introduce plowing effect. Whereas in rolling friction, several other factors play an important role such as slip, elastic and plastic deformation, etc.

Wear can be defined as the surface deterioration phenomenon or loss of matter at the surface by mechanical means. When two surfaces possess different hardness or of same hardness to come into contact in the solid-state, due to sliding or rolling, the material is removed from one surface or both the surfaces. Continuous material loss with a steady material removal rate may lead to undesired results. Wear is an immense factor in developing structures and components when they are designed to function with relative moments. There are three important modes of wear mechanisms include wear by adhesion, abrasion, and erosion. In adhesion wear, both the asperities from the contacting surfaces cross the yield point and plastically deform. Now the contact area between the surfaces is increased, and cold-welding of the asperities demand more force to sustain the motion of the sliding surface resulting breaking of the bonded asperities and material removal. Abrasive wear occurs when a hard moving surface is in contact with the relatively soft surface, the material (mostly from the asperities) of the soft surface is removed, and the debris resulted from the material removal present at the interface and further leads to material removal. In erosion wear, a solid particle strikes the surface, and due to the impact, the material is eroded from the surface. Developing wear-resisting surfaces necessarily require making the surface harder up to certain depth or introducing high hard and brittle particles into the surface. If sudden loads arise in the structure during the functioning, surface wear resistance coupled with higher toughness in the core is required which is really a difficult task to achieve. With the advent of the recent surface engineering processes, structures with high hardness and wear-resisting surfaces along with tough core can be manufactured by adopting the optimum combination of different processes.

1.4 SURFACE ENGINEERING PROCESSES

Recent technological developments and increased level of understanding the basic science behind the material behavior and performance immensely contributed to develop several surface engineering processes. All surface engineering processes can be grouped into two categories known as surface treatments and surface coatings as shown in Figure 1.14.

Surface engineering processes can also be classified again into two groups considering the mode of modification at the surface. If the surface chemical composition is unaffected, only microstructure and stresses are altered such processes can be called as surface non-chemical treatments. If a process involves modifying the chemical composition of the surface or adding different chemical species at the surface, such a process is called surface chemical modification treatment. Furthermore, based on the state of the material (gas, liquid or solid) again all surface engineering processes can be called as vapor, liquid, and solid-state methods. All processes, which involve the formation of vapor of the coating material at the surface, can be called as vapor state methods such as CVD and PVD.

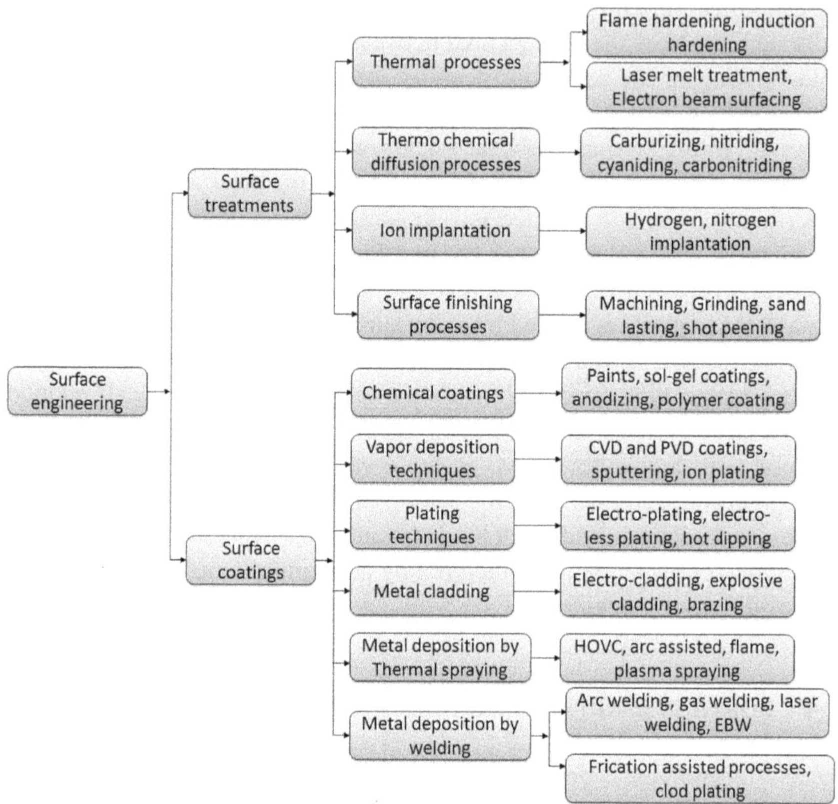

FIGURE 1.14 Flow diagram showing classification of surface engineering processes.

In liquid state methods, the material at the surface undergo liquid state and called as liquid state processes for examples, laser melt treatment. If processing is done without melting the material at the surface, such a process can be called a solid-state method. Friction-assisted processes such as friction stir processing and friction surfacing are the best examples for solid-state methods.

1.5 FRICTION-ASSISTED PROCESSES

Development of friction-assisted processes has opened a new area in the manufacturing industry. Figure 1.15 schematically shows different

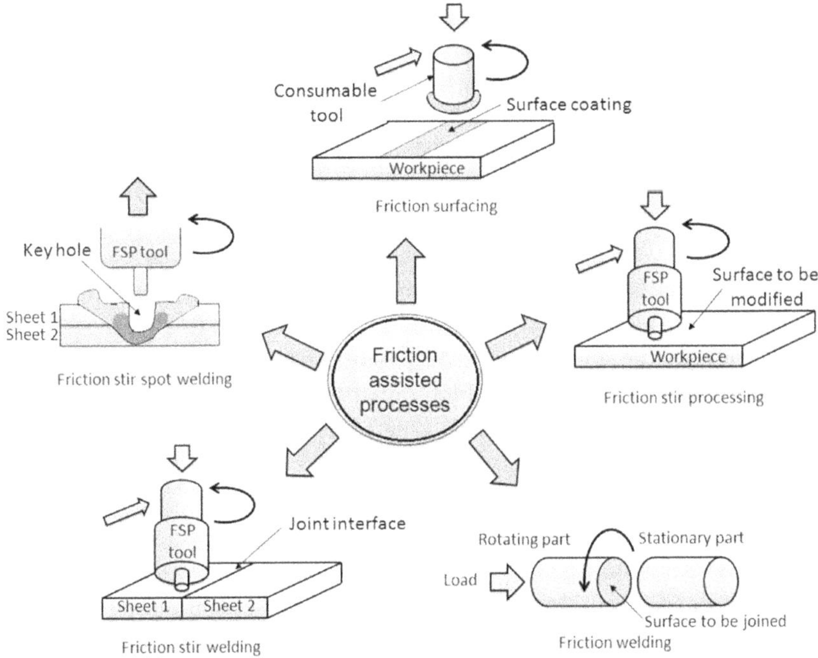

FIGURE 1.15 Schematic representation of different friction-assisted processes.

friction-assisted processes used in the manufacturing industry. Heat is generated between two surfaces if either one or both the surfaces relatively move against the other. All friction-assisted processes use the same principle to generate heat which is required to induce plastic deformation and to develop a perfect metallurgical continuity. Initially, a method was developed and demonstrated at The Welding Institute (TWI), UK in which friction is utilized to perform a welding operation. Later on, several subsequent developments opened a new solid-state processing area in the manufacturing industry for several applications. Among them, five important processes friction welding, friction stir welding, friction stir processing, friction stir spot welding, and friction surfacing are important.

In *friction welding,* the components which are joined are relatively moved against one another, and due to the friction between the counter surfaces, heat is generated. This heat is utilized to soften the material at the surface and to decrease the yield point. Then a forging load is applied so that a metallurgical continuity is established which is called as a weld joint. The impurities, oxides, and inclusions present at the interface are flushed out when the load is applied. Similarly, heat is generated due to friction between a non-consumable rotating tool that contains a pin and work material in other friction-assisted processes. Two sheets can be joined by using a rotating non-consumable tool (containing pin) as shown in Figure 1.15, and due to the stirring action of the pin at the joint, the material is plastically deformed, and a perfect weld joint can be obtained. This process is called friction stir welding (FSW). Based on the same principle of FSW, another variant which was developed to alter the surface microstructure known as FSP. In FSP, similar to that of FSW, a rotating non-consumable tool is used which consisting a pin to introduce severe plastic deformation at the surface. Due to dynamic recrystallization fine grains are evolved at the FSP regions.

The same FSP tool is used to join two sheets by overlapping them and performing joining at different intermittent distances similar to resistance spot welding and known as friction stir spot welding (FSSW) as shown in Figure 1.15. In the place of a non-consumable tool, if a rotating consumable tool is used to coat material on the surface, that process is called friction surfacing (FS). In FS, the consumable tool which is coated on the surface is also called as a mechtrode. Among these processes, friction stir processing and friction surfacing are the two process variants used for surface engineering of metals. The next chapters of the present book provide detailed information on friction stir processing, surface metal matrix composites by FSP and friction stir surfacing.

KEYWORDS

- **friction-assisted processes**
- **surface engineering**
- **surface engineering processes**
- **surface properties**

REFERENCES

1. Joao Paulo Davim, (2013). *Nontraditional Machining Processes*, Springer-Verlag London.
2. Kartal, F. (2017). A review of the current state of abrasive water-jet turning machining method, *Int J Adv Manuf Technol 88,* 495. https://doi.org/10.1007/s00170–016–8777-z.
3. Çakir, O., Yardimeden, A., & Özben, T. (2007). Chemical machining, *Archives of Materials Science and Engineering, 28*(8), 499–502.
4. Xue Yana, & Gu, P. (1996). A review of rapid prototyping technologies and systems, *Computer-Aided Design, 28*(4), 307–318.
5. Amit Joe Lopes, Eric MacDonald, & Ryan B. Wicker (2012). Integrating stereo lithography and direct print technologies for 3D structural electronics fabrication, *Rapid Prototyping Journal, 18*(2), 129–143, https://doi.org/10.1108/13552541211212113.
6. Buelens, J. J. C. (1997). Metal droplet deposition: a review of innovative joining technique, *Science and Technology of Welding and Joining, 2*(6), 239–243.
7. Komarnicki P. (2017). A Review of Fused Deposition Modeling Process Models. In: Rusiński E.,& Pietrusiak D. (eds.). *Proceedings of the 13th International Scientific Conference*. RESRB 2016. Lecture Notes in Mechanical Engineering. Springer, Cham, 241–247.
8. Olakanmi, E. O., Cochrane, R. F., & Dalgarno, K. W. (2015). A review on selective laser sintering/melting (SLS/SLM) of aluminum alloy powders: Processing, microstructure, and properties, *Progress in Materials Science, 74,* 401–477.
9. Jian-Yuan Lee, Jia An, & Chee Kai Chua, (2017). Fundamentals and applications of 3D printing for novel materials, *Applied Materials Today, 7,* 120–133.
10. Young, T. (1805). *Philos. Trans. R. Soc. Lond., 95,* 65.
11. Schwartz, L. W. & Garoff, S. (1985). *Langmuir 1,* 219.
12. Zhang, L., & Webster, T. J. (2009). Nanotechnology and nanomaterials: Promises for improved tissue regeneration, *Nano Today, 4,* 66–80.
13. Valiev, R. Z., Islamgaliev, R. K., & Alexandrov,I. V. (2000). Bulk nanostructured materials from severe plastic deformation. *Prog Mater Sci, 45*(2), 103–189.
14. Dieter Vollath (2013). *Nanomaterials: An Introduction to Synthesis, Properties, and Applications,* 2nd Edition, Wiley.
15. Lu, K. & Lu, J. (2004). Nanostructured surface layer on metallic materials induced by surface mechanical attrition treatment, *Materials Science and Engineering A, 375–377*, 38–45.

CHAPTER 2

Development of Friction-Assisted Processes

Gas welding and arc welding processes are best-known fusion welding techniques widely used in the welding industry in the early years of the 20th century. Later on, modern welding techniques such as laser beam welding, electron beam welding, and other solid-state welding processes have been developed and addressed several issues associated with conventional fusion based welding processes. Gas welding and arc welding processes are low energy density processes; therefore, results in large weld areas, heat affected zones and a higher level of distortion in the weld structures. By the year 1960, the advent of new welding techniques revolutionized the metal joining field and enabled the manufacturing industry to address several issues existing in the conventional welding processes. Solid-state welding techniques are another group of processes; completely eliminate the problems encountered during solidification in fusion welding processes [1].

2.1 FRICTION WELDING

Friction welding is one such development in the area of solid-state welding, opened a new area in the welding industry and later became a reason for the development of many friction-assisted processes. Basically, in all friction-assisted processes, heat is generated by friction and utilized to join similar and dissimilar materials or to modify the microstructure and chemical composition of metals without melting. Figure 2.1 shows the schematic representation of the friction welding process. The two surfaces which are intended to be joined by friction welding are relatively moved against each other and cause the material at the interface to reach yield point and become soft. Then a forging load is applied to flush out any unwanted matter at the interface and to develop a perfect metallurgical continuity between the two surfaces.

Friction between the components to be joined can be generated in three different ways:

1. rotational movement;
2. angular reciprocating; and
3. liner reciprocating.

Figure 2.2 illustrates the generation of friction between the parts by *rotational movement*. Either one component or both the components rotate against each other so that the surfaces intended to be join come into contact as shown in Figure 2.2 (a). After a certain period of time, a suitable amount of load is applied to establish the joint. Once the interface reached to a certain temperature then the load is applied by two different ways: (i) radially and (ii) axially. When the joining surface is parallel to the axis of the component, the load is applied in a radial direction. Whereas, when the joining surface is perpendicular to the axis of the rotating component, the load is applied axially.

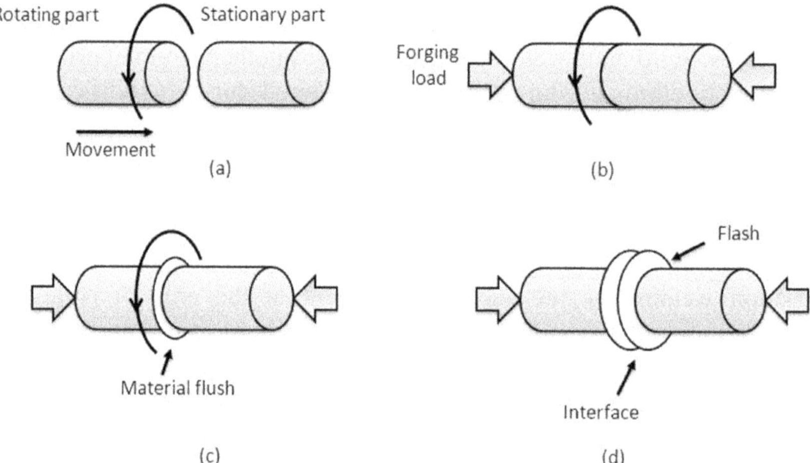

FIGURE 2.1 Schematic representation of friction welding process: (a) two parts of rotational symmetry, (b) generating friction at the interface, (c) applying upsetting load and material flushing at the interface, and (d) weld joint with material flash.

Figure 2.2 (a) and (b) illustrate the two ways of applying load during rotational friction welding. The crucial point that must be observed in rotational friction welding processes is the geometry of the components which are intended to be joined irrespective of the direction of the load. Rotational friction welding is possible with the components which possess rotational symmetry. Irregular shaped or complex shaped parts are difficult to join by rotational friction welding. When the rotating component is brought into contact with another rotating (counter direction) or stationary component as schematically shown in Figure 2.2 (a), heat is produced due to friction between the contacting surfaces. The generated heat brings the material at the interface to its plastic region, and subsequent forging load applied either along the axis of rotation or perpendicular to the axis of the rotation produces a strong bonding and perfect metallurgical continuity at the interface. Figure 2.3 shows a photograph and corresponding cross-sectional micrographs of alloy 718 weld joints achieved by rotational friction welding.

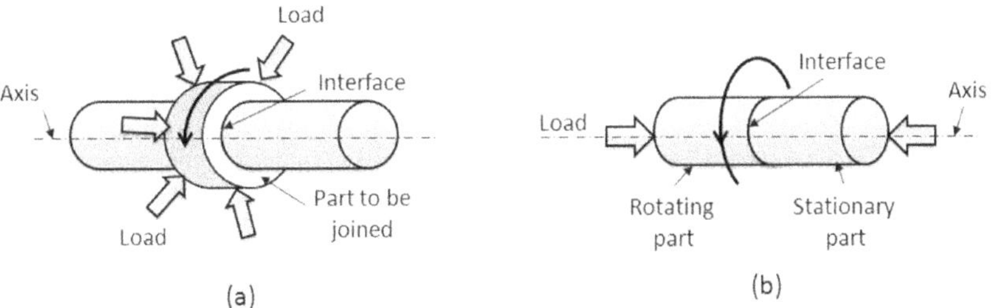

FIGURE 2.2 Schematic illustration of rotational friction welding: (a) load applied perpendicular to the axis of rotation, and (b) load applied parallel to the axis of rotation.

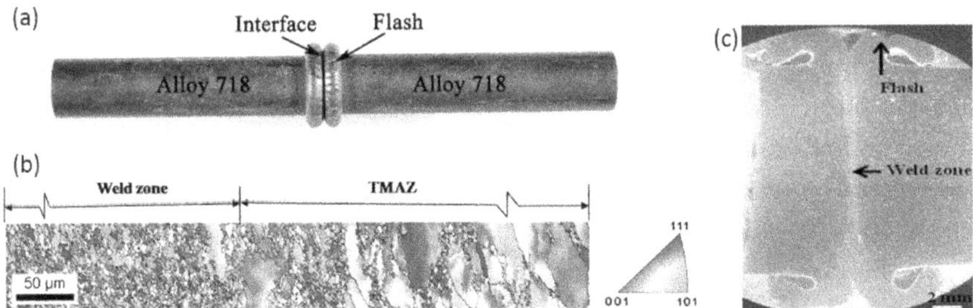

FIGURE 2.3 (See color insert.) (a) Photograph showing friction welded 718 alloy (nickel-iron based superalloy widely used in aerospace and chemical plant industries), (b) EBSD map showing the microstructure at the weld zone and thermomechanical affected zone (TMAZ), and (c) microstructure of the interface. a) *Source:* Reprinted with permission from Damodaram et al., [2]. © 2013 Elsevier. c) *Source:* Reprinted with permission from Damodaram et al., [3]. © 2014 Elsevier.).

Angular reciprocation between the surfaces to be joined also generates friction and raises the required temperature at the interface. Figure 2.4 (a) shows the schematic representation of angular reciprocation of parts to be joined by friction welding. The repeated change in the direction of rotation (clockwise and anti-clockwise for an axisymmetric part) up to certain angle with some relevant angular speed brings the material at the interface to a plastic state and then the applied forging load causes to form a weld joint. Similarly, in *linear reciprocation* also heat is generated due to the forward and backward motion of the surface against another surface and the applied forging load causes to establish a weld joint as schematically explained in Figure 2.4 (b).

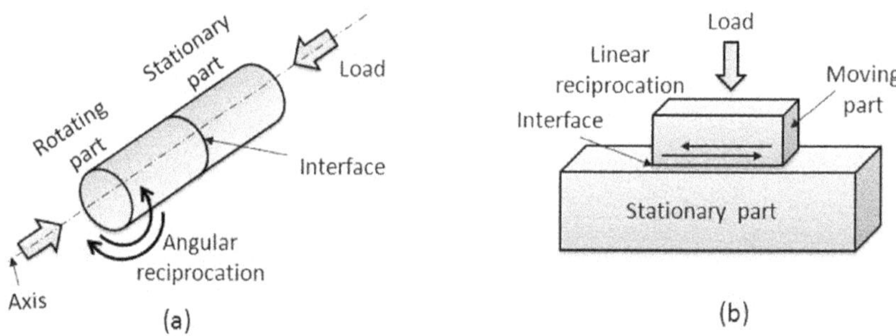

FIGURE 2.4 (a) Angular reciprocation, and (b) linear reciprocation in friction welding.

Development of friction welding technique has not gained considerable attention during its early years until the end of the 20th century. Several research groups have done a considerable amount of contribution in developing solid-state welding techniques with the evolution of friction welding. Several parts with different geometries can be successfully joined such as rods to rods, rods to plates, tubes to rods, tubes to tubes, etc. However, the space that friction welding process occupied in the manufacturing industry is unfortunately limited and not as per the expected level when the process was introduced to the welding world. Nevertheless, the friction welding process opened a new area in the welding field, and evolution of friction-based processes was accelerated. The volume of research publications and dedicated books published on friction-assisted processes testifies the same.

2.2 FRICTION STIR WELDING

Even though the friction welding process was familiar in the welding industry for a few decades, the development of friction stir welding (FSW) did not happen until the end of the 20th century. It took several years from the introduction of friction welding to see the evolution of FSW to appear in the history of the welding industry. Here, it is worthy of mentioning the significant incident that happened in the year A.D. 1991 that is Joining of aluminum alloys by FSW. It was demonstrated by The Welding Institute (TWI), UK, in particular, Wayne Thomas and his group [4, 5]. Completely eliminating the melting of material in the context of welding really offers several advantages compared with fusion based welding techniques. There is no necessity of using protective gas to shield the liquid in the weld zone. Solidification is eliminated, and hence, the shrinkage and corresponding distortion are minimized. Furthermore, no filler material is used which eliminates the formation of unwanted phase and deposition of additional material in the weld zone. Since there is no solidification in FSW joints; the microstructure of the weld zone differs from the cast microstructure as usually observed in fusion welding techniques. Figure 2.5 (a) schematically explains the FSW process. In FSW, two surfaces which are to be joined are placed as shown in Figure 2.5 (b), and a non-consumable rotating tool is plunged into the interface and moved along the interface to get a weld joint as shown in Figure 2.5 (c). A typical microstructure of AZ31–AZ91 Mg alloys weld joint obtained by FSW at the cross-section can be seen in Figure 2.5 (d).

Heat is generated at the processing zone and softens the material to plastically deform and flow within the stirring zone while the tool pin is introducing stirring action. The three important reasons behind the heat generation in FSW are:

i. Friction between the shoulder of the FSW tool and the surface of the workpieces.
ii. Friction between the FSW tool pin and the base material
iii. Due to the plastic deformation of the material during welding

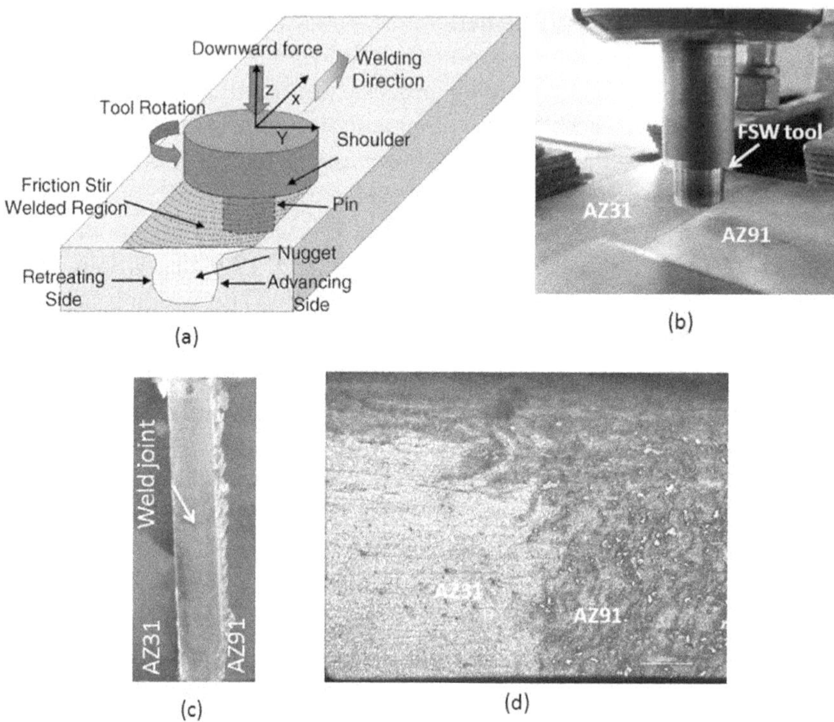

FIGURE 2.5 (a) schematic representation of friction stir welding process (*Source:* Reprinted with permission from Mishra and Ma [6]. © 2005 Elsevier.), (b) photograph showing FSW of AZ31 and AZ91 Mg alloy sheets, (c) photograph of weld joint, (d) crosssectional microstructure of the weld joint.

During FSW, while the tool is moving across the joint, due to the heat generated from the friction between the shoulder-workpiece and the pin-workpiece, and the plastic work in the base material, the material in the tool travel direction becomes soft and helps the tool to move further. FSW tool is a specially designed tool that contains two important elements as given below.

i. tool shoulder; and
ii. tool pin.

Figure 2.6 shows a typical photograph of the FSW tool and the schematic representation of the FSW tool shoulder and the pin. Shoulder of an FSW tool plays a very important role in obtaining a successful weld joint.

It prevents the escape of the softened material from the weld zone as the tool is progressing across the joint. The shoulder also applies hydrostatic pressure on the material at the weld zone; therefore, the stirred material does not flow away from the joint. Most of the shoulders of FSW tools are a flat type or concave type. Sometimes, tool shoulder contains specially designed scrolls to direct the flow of the material at the surface.

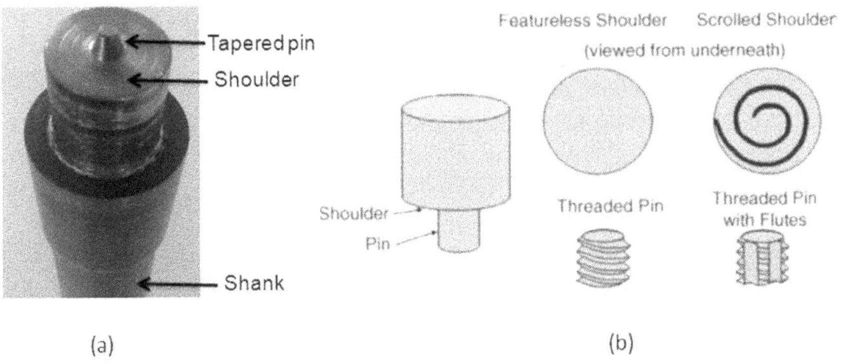

FIGURE 2.6 (a) Photograph of typical tapered pin FSW tool (unpublished) and (b) schematic diagram of FSW tool shoulder and pin with different features (*Source:* (b) Reprinted with permission from Mishra and Ma [6]. © 2005 Elsevier.)

FSW tool pin plays a very crucial role in the entire process. The material plastically flows around the pin, and the profile of the pin also influences the mechanisms of material flow during welding. Pin serves to break the material at the interface and facilitates the plastically deformed material to flow and fill the gap leftover by the pin to avoid any cavities and pores. Several pin profiles such as a plain round pin, a tapered pin, triangular pin, square pin, threaded pin, tapered threaded pin, etc. were used in the literature to weld different alloys. Usually, in FSW, the tool is tilted for 2 or 3° with respect to the axis of the tool rotation to obtain better results. The important process parameters which influence the quality of the weld joint include tool geometry (shoulder diameter, pin profile, pin length and ratio of shoulder diameter to pin diameter, etc.), load, tool rotational and travel speeds, tool tilt angle, weld material size, chemical composition, properties, etc.

FSW involves three sub-phases known as plunge and dwell, travel and retract. During the first step (plunge and dwell), the rotating FSW tool is penetrated into the joint until the shoulder sufficiently touches the surface of the workpieces to be joined and hold for some time to see that the sufficient heat is produced. In the second step (travel), the rotating tool as in penetrated condition travels in the direction of the weld joint. At the end of the process, the FSW tool is retracted as it is rotating to complete the welding process which is the third step. In most of the cases, the FSW zone contains fine microstructure with the lower possibility to form secondary phases or intermetallics. The advent of FSW has shown promising results in many potential applications in particular, difficult to weld materials such as aluminum alloys and magnesium alloys. In FSW weld joints, the microstructure contains three distinct regions named as stir (nugget) zone, TMAZ zone, and heat-affected zone (HAZ). Table 2.1 lists the key benefits of the friction stir welding process compared with other conventional welding techniques [6].

TABLE 2.1 Key Benefits of Friction Stir Welding

Metallurgical benefits	Environmental benefits	Energy benefits
• Solid phase process	• No shielding gas required	• Improved materials usage (e.g., joining different thickness) allows a reduction in weight
• Low distortion of the workpiece	• No surface cleaning required	
• Good dimensional stability and repeatability	• Eliminate grinding wastes	
	• Eliminate solvents required for degreasing•	• Only 2.5% of the energy needed for a laser weld
• No loss of alloying elements		
• Excellent metallurgical properties in the joint area	• Consumable materials saving such as rugs, wire or any other gases	• Decreased fuel consumption in lightweight aircraft, automation and ship applications
• Fine microstructure		
• Absence of cracking		
• Replace multiple parts joined by fasteners		

Source: Reprinted with permission from Mishra and Ma [6]. © 2005 Elsevier.

2.3 FRICTION STIR PROCESSING

Based on the basic principles of friction stir welding, friction stir processing (FSP) technique was developed to modify the microstructure of metals

[8]. The techniques developed for welding industry has become a source for a new method to alter the surface microstructure. In FSP, similar to that of FSW, a non-consumable tool is used which contains a pin to penetrate into the material and causes microstructural modification. The rotating FSP tool is penetrated into the surface of the metal sheet or plate until the shoulder touches the surface and travel in the desired direction under certain load. The stirring action of the pin introduces plastic deformation beneath the shoulder, and due to dynamic recrystallization (DRX), grain refinement is happened. Figure 2.7 schematically illustrates FSP and typical grain refinement achieved in AZ31 Mg alloy by FSP.

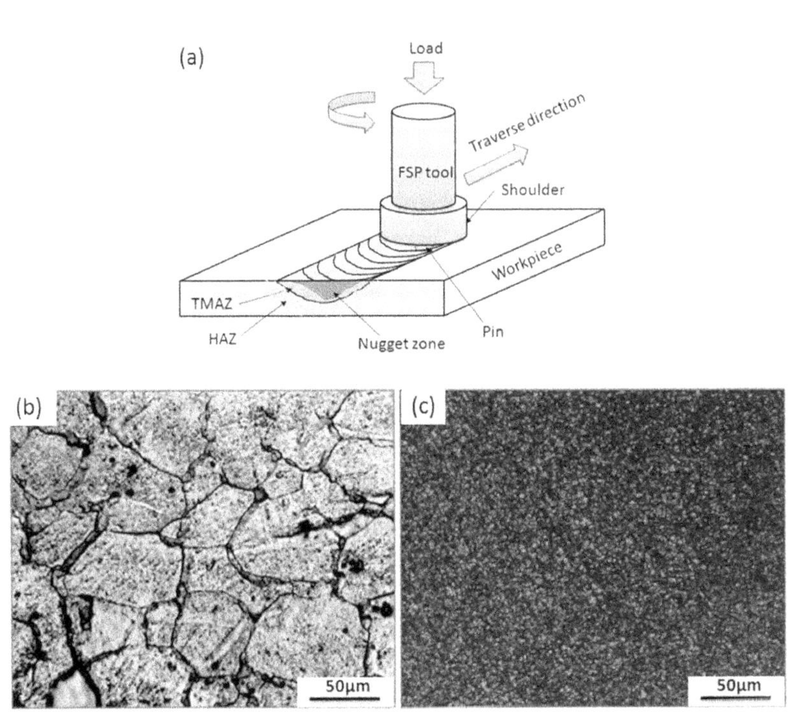

FIGURE 2.7 (a) Schematic representation of FSP and (b) microstructure of AZ31 Mg alloy and (c) FSPed AZ31 Mg alloy showing the grain refinement.

The same tool which is used in friction stir welding can be used for FSP. The mechanisms behind the material flow, influencing process parameters and material systems processed by FSP are almost of the same observations made in FSW. In addition to microstructure modification, Mishra et al., [9] demonstrated a new method to produce a surface composite by FSP. The idea behind the development of this new method is utilizing the stirring action of the FSP tool to incorporate secondary phase into the surface of the sheet or plate. Several material systems, e.g., magnesium, aluminum, copper, titanium alloys and steels were successfully processed by FSP. However, the technology has around 15 years of history; tremendous interest has shown by different research groups globally. FSP can be visualized as in infant stage compared with the other conventional surface engineering processes. However, the potential applications promised by FSP made the technology as advanced and most reliable in several applications in automobile, aerospace and marine applications. Chapters 3 to 12 are completely dedicated to discuss FSP in detail.

2.4 FRICTION SURFACING

Friction surfacing (FS) is another method developed to coat the surface with a metal with less dilution based on the friction – heat generation principle. In FS, a consumable rotating rod is used to contact the surface, and the material from the tool at the contact region become soft due to the friction-induced heat and plastically transferred to the targeted surface within the solid-state [10, 11]. Figure 2.8 schematically illustrates the FS process. The consumable rod used in FS also can be called as mechtrode. It is possible to coat any alloy on the metal substrate if the mechtrode is made of that alloy. The dilution at the interface is also eliminated or decreased to a great extent when coatings are done by the FS method. Producing coatings within the solid-state by FS offers several advantages in the manufacturing industry. The detailed coverage of the topic is provided in Chapters 13 to 18.

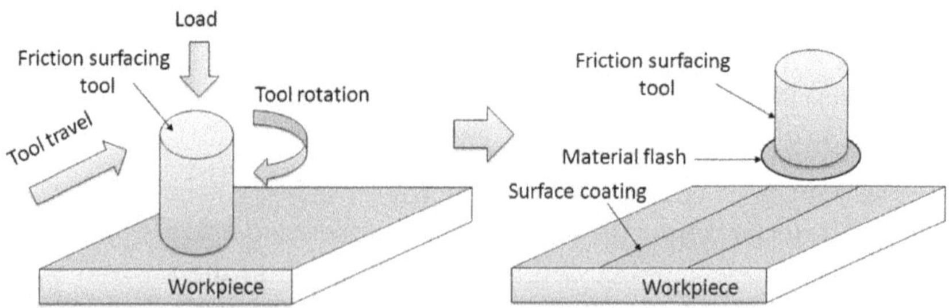

FIGURE 2.8 Schematic representation of friction surfacing process.

2.5 FRICTION-BASED ADDITIVE MANUFACTURING

Recently, additive manufacturing, a new promising route in manufacturing engineering to develop three dimensional (3D) structures has grained tremendous attention. With the development of computer technology in assisting design and manufacturing revolutionized the material processing area. Basically, in additive manufacturing, a model is generated by computer-aided design, and with the help of software, sophisticated machines deposit the material layer-by-layer as per the input of the 3D model to complete the manufacturing of the component. After successful demonstration of friction surfacing, further development to produce 3D structures by depositing a number of layers within the solid-state is a novel idea in manufacturing engineering. Several authors reported using friction surfacing principle to develop solid structures and named as "friction deposition" or "friction free form" [12, 13]. Figure 2.9 schematically shows additive manufacturing by friction surfacing.

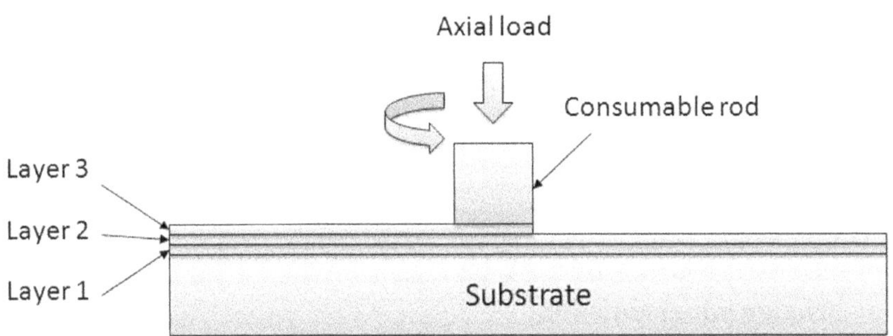

FIGURE 2.9 Schematic representation of friction based additive manufacturing.

2.6 FRICTION STIR SPOT WELDING

Development of friction stir spot welding (FSSW) has offered several advantages to join sheets and plates without melting the base materials. FSSW is completed in three steps including plunging, stirring and retreating [14]. In the initial step, the rotating non-consumable tool is plunged into the stack of the plates and penetrated until the tool shoulder is touched the surface of the top plate. In the second step, the rotating tool is allowed to stir for sufficient time to facilitate appropriate mixing of the material from both the sheets. In the third step, the rotating tool is retreated by leaving a region of spot welded which contains a sound metallurgical joint. However, after retreating, the tool leaves an unfilled hole in the center of the stir zone that may decrease the mechanical strength. Later on, several variations were proposed by different authors to close the cavity produced during the third step by modifying the tool design [15–18]. Development of FSSW addressed several issues such as joining non-ferrous metals which usually results in brittle phases in fusion-based welding processes. The technology has been already transferred and commercially used in automobile and aerospace applications.

There are other frictions assisted processes reported in the literature such as friction riveting, friction cladding and friction channeling. Heat generation due to friction between two counter surfaces to plastically mix the material or deform the material is the basic principle in all these processes. In *friction stir riveting*, two sheets are joined at specific regions from one side unlike in the case of conventional riveting. Whereas in *friction stir cladding,* a thin layer of material is deposited on a substrate by using a hollow tool. In *friction stir channeling,* a hidden continuous channel is produced by leaving the surface of the substrate is closed and similar to unprocessed regions. A brief discussion on these additional processes is given in Chapter 19.

KEYWORDS

- **friction stir channeling**
- **friction stir cladding**
- **friction stir processing**
- **friction stir riveting**
- **friction stir spot welding**
- **friction surfacing**
- **friction-based additive manufacturing**

REFERENCES

1. Robert W. Messler, Jr. (1999). *Principles of Welding: Processes, Physics, Chemistry and Metallurgy,* John Wiley & Sons, Inc.

2. Damodaram, R., Ganesh Sundara Raman S., & Prasad Rao, K. (2013). Microstructure and mechanical properties of friction welded alloy 718, *Materials Science & Engineering A 560,* 781–786.
3. Damodaram, R., Ganesh Sundara Raman S., & Prasad Rao, K. (2014). Effect of post-weld heat treatments on microstructure and mechanical properties of friction welded alloy 718 joints, *Materials and Design 53*, 954–961.
4. Thomas, W. M. (1988). Solid-phase cladding by friction surfacing, *International Welding for the Process Industries,* London, 18–16.
5. Nicholas, E. D., & Thomas W. M. (1986). Metal deposition by friction welding, *Welding Journal,* 17–27.
6. Mishra, R. S., & Ma, Z. Y. (2005). Friction stir welding and processing, *Materials Science and Engineering R 50,* 1–78.
7. Ratna Sunil, B., Pradeep Kumar Reddy, G., Mounika, A. S. N., Navya Sree, P., Rama Pinneswari, P., Ambica, I., Ajay Babu, R., & Amarnadh, P. (2015). Joining of AZ31 and AZ91 Mg alloys by friction stir welding, *Journal of Magnesium and Alloys, 3*(4), 330–334.
8. Mishra, R. S., Mahoney, M. W., McFadden, S. X., Mara, N. A., & Mukherjee, A. K. (2000). High strain rate superplasticity in a friction stir processed 7075 Al alloy. *Scr Mater. 42,* 163–168.
9. Mishra, R. S., Ma, Z. Y., & Charit, I. (2003). Friction stir processing: a novel technique for fabrication of surface composite. *Mater Sci Eng A. 341*, 307–310.
10. Gandraa, J., Krohn, H., Miranda, R. M., Vilaca, P., Quintino, L., & dos Santos, J. F., (2014). Friction surfacing—A review, *Journal of Materials Processing Technology 214,* 1062–1093.
11. Batchelor, A. W., Jana, S., Koh, C. P., & Tan, C. S., *Journal of Materials Processing Technology, 57,*172–181.
12. Dilip, J. J. S., & Janaki Ram, G. D. (2013). Microstructure evolution in aluminum alloy AA 2014 during multi-layer friction deposition, *Materials Characterization, 86,* 146–151.
13. Dilip, J. J. S., Babu, S., Varadha Rajan, S., Rafi, K. H., Janaki Ram, G. D., & Stucker, B. E. (2013). Use of Friction Surfacing for Additive Manufacturing, *Materials and Manufacturing Processes, 28*(2), 189–194.
14. Prangnell, P. B., & Bakavos, D. (2010). Novel approaches to friction spot welding thin aluminum automotive sheet,"*Materials Science Forum,* 638–642, 1237–1242.
15. Fang, Y. (2009). *The Research on the Processes and Properties of Pinless Friction Stir Spot Welding,* Jiangsu University of Science and Technology, Jiangsu, China.
16. Sun, Y. F., Fujii, H., Takaki, N., & Okitsu, Y. (2012). Microstructure and mechanical properties of mild steel joints prepared by a flat friction stir spot welding technique, *Materials Design, 37,* 384–392.
17. Mazzaferro, J. A. E., Rosendo, T. S., Mazzaferro, C. C. P., Ramos, F. D., Tier, M. A. D., Strohaecker, T. R., & dos Santos, J. F., (2009). Preliminary study on the mechanical behavior of friction spot weld. *Soldagem e Inspecao, 14,* 238–247.
18. Venukumar, S., Yalagi, S., & Muthukumaran, S., (2013). Comparison of microstructure and mechanical properties of conventional and refilled friction stir spot welds in AA 6061-T6 using filler plate. *Trans Nonferrous Metals Soc China, 23*, 2833–2842.

CHAPTER 3

Friction Stir Processing of Metals

3.1 FRICTION STIR PROCESSING (FSP)

Friction stir processing (FSP) has emerged as a promising technique and demonstrated the tremendous potential to enhance several properties in the field of surface engineering. FSP is a generic tool developed on the basic principles of friction stir welding (FSW) and enhances several properties at the surface such as mechanical, corrosion, wear, and superplastic properties. Similar to friction stir welding, FSP is a solid-state processing technique which uses a non-consumable tool. The rotating tool is plunged into the material and causes the material to undergo severe plastic deformation beneath the tool shoulder, and the pin serves to stir the material. Heat is generated due to the following three reasons:

1. friction between the tool shoulder and the surface of the workpiece;
2. friction between the tool pin and the material;
3. plastic deformation of the material during stirring.

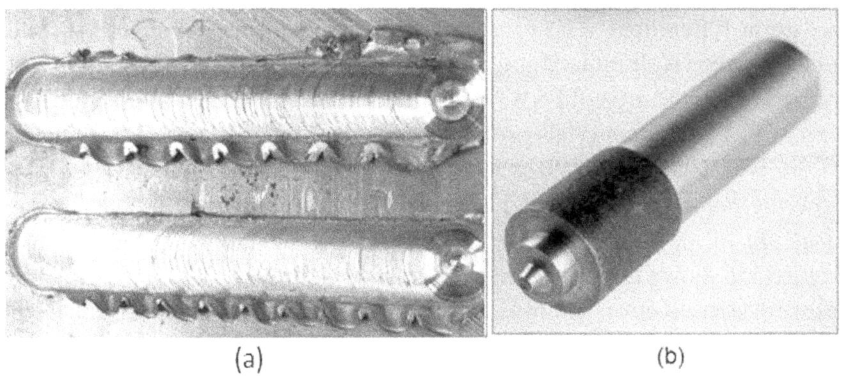

FIGURE 3.1 (a) Photograph of typical FSPed region (AZ 31 Mg alloy) and (b) photograph of a typical FSP tool.

Due to the generated heat, the yield point of the material at the vicinity of the tool pin is decreased, and the material becomes soft. The plasticized material is stirred as per the rotation of the FSP tool, and due to recrystallization, new grains and grain boundaries are evolved. Since the recrystallization has happened during the plastic deformation in FSP, the microstructural modification also happened due to the dynamic recrystallization (DRX) and resulted in very fine grains.

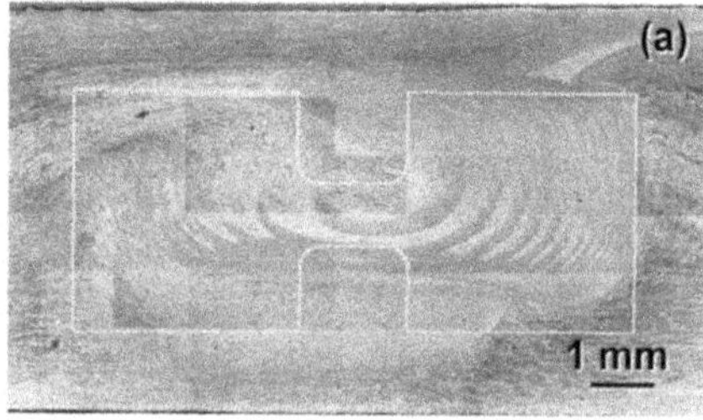

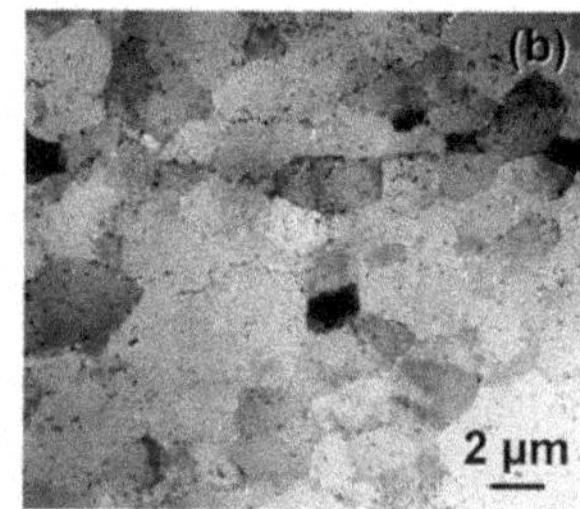

FIGURE 3.2 (a) Optical microscope image obtained at the traverse cross section of FSPed Al7075 alloy and (b) TEM bright field image at stir zone. (*Source:* Reprinted with permission from Mishra et al., [1]. © 1999 Elsevier.)

Figure 3.1 shows the photographs of FSPed region and FSP tool. The primary goal of the FSP is to modify the microstructure at the surface of the material. FSP can be categorized as one of an important severe plastic deformation (SPD) technique to process plates and sheets. Surface properties which are influenced by microstructural modifications certainly affected due to the grain refinement achieved after FSP. It is also true from the reported literature that optimizing the process parameters is crucial to get grain refinement. The principles and mechanisms behind solid-state welding were well understood, and quite a good amount of research was carried out in the area of FSW. However, it took ten years to see the evolution of FSP in the materials processing research since the FSW introduced to the welding industry. It was first demonstrated by Mishra et al., [1], how the basic principles of FSW can be applied to modify the microstructure of an aluminum 7075 alloy to achieve superplastic properties. Figure 3.2 shows the optical micrograph of the traverse cross-section and transmission electron microscope bright field image obtained in the stir zone. The typical material flow patterns usually seen in FSP were clearly seen in the microstructure, and average grain size was measured as 3.3 ± 0.4 μm. The superplastic properties assessed by conducting tensile tests show enhanced % elongation compared with the earlier published data by Xing-Gang et al., [2]. Figure 3.3 shows the photographs of tensile test samples captured before and after the tests and corresponding tensile data. The work demonstrated for the first time, high strain rate superplasticity in aluminum alloy. Furthermore, explained a new route to produce fine-grained structured monolithic plates with simple procedure later which led to open a new surface engineering method.

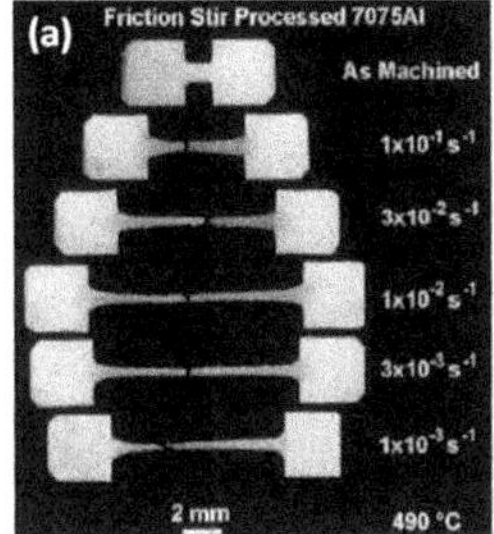

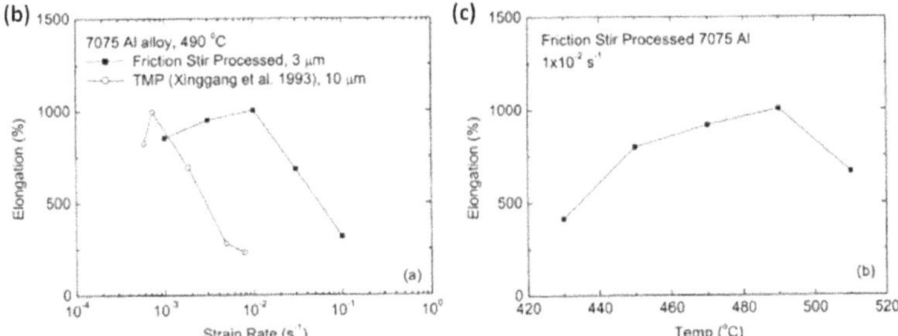

FIGURE 3.3 (a) Photograph of specimen before and after tensile test carried out at 490°C, (b) variation of % elongation with respect to strain rate (compared with the previous results of Xing-Gang et al., [2]) and (c) % elongation with respect to different temperatures at a strain rate of 1×10^{-2} s^{-1}. (*Source:* Reprinted with permission from Mishra et al., [1]. © 1999 Elsevier.)

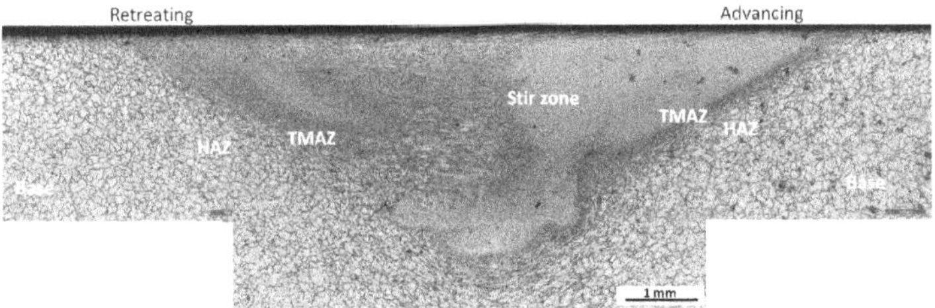

FIGURE 3.4 Optical microscope images at the cross section of FSPed ZE41 Mg alloy showing different zones.

3.1.1 Three Microstructural Zones in FSP

Three different regions are generally identified at the cross section of FSPed workpieces namely (i) stir or nugget zone, (ii) thermo-mechanically affected zone (TMAZ), and (iii) heat affected zone (HAZ) as shown in Figure 3.4.

i) Nugget Zone

Nugget zone is the region just beneath the tool shoulder where the material undergoes severe plastic deformation. The rise in temperature is higher in the nugget zone compare with the other regions usually up to 0.7 to 0.8 times the melting temperature of the base material. Additionally, high strain with dynamic recrystallization results in very fine equiaxed grains in the stir zone after FSP. Refined microstructure in the nugget zone is the main reason behind the increased mechanical and wear properties in FSPed materials. The secondary phase present at the alloys also dissolved due to FSP and lead to the formation of supersaturated grains in alloys where more amount of intermetallic phases present at the grain boundaries. For example, as shown in Figure 3.5, the nugget zone of FSPed AZ91 Mg alloy shows decreased secondary phase which suggests the formation of supersaturated grains after FSP. The depth of the nugget zone is generally not uniform from the surface of the workpiece. The shape of the nugget zone is also dependent on processing parameters such as tool geometry, rotational speed and traveling speed, type of workpiece, etc.

ii) Thermo-Mechanically Affected Zone (TMAZ)

Next, to the nugget zone, TMAZ appears after FSP. Based on the direction of tool rotation and travel two sides can be identified in FSP, e.g., advancing and retreating. As the rotating FSP tool shoulder moves in traverse direction on the workpiece, it enters the unprocessed surface at the advancing side and leaves the forefront of the moving tool at the retreating side. At the advancing side, the tool rotational and transverse directions are the same. Intense plastic deformation leads to a higher level of grain refinement with a sharp nugget interface. At the retreating side, both the tool rotation and travel are in the opposite direction and results in the diffused interface of the nugget. Due to insufficient heat compared with the nugget zone, the material flow and the level of grain refinement are lower in TMAZ. However, some fine grains may result but more elongated grains usually seen in TMAZ. If alloy contains secondary phases at the grain boundaries, the presence of this secondary phase also can be clearly observed in TMAZ. More amounts of dislocations and formation of sub-grains also is a common observation in TMAZ. For example, Figure 3.5 shows the two different TMAZ zones at the cross section of FSPed AZ91 Mg alloy at advancing and retreating side and the elongated grains with remaining secondary phase at grain boundaries.

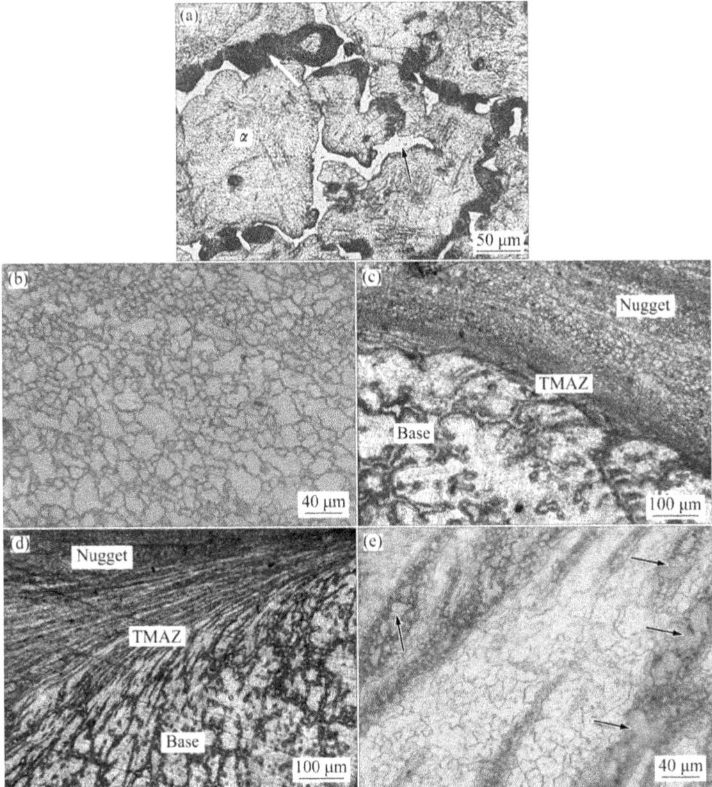

FIGURE 3.5 Optical microscope images of FSPed AZ91 Mg alloy: (a) unprocessed AZ91 (black arrow indicate Mg17Al12 secondary phase, and white arrow indicate eutectic region), (b) fine grains without secondary phase in the nugget zone, (c) TMAZ at the advancing side, (d) TMAZ at the retreating side and (e) magnified TMAZ zone at retreating side showing secondary phase (black arrows). (*Source:* Reprinted with permission from Surya Kiran et al., [3]. © 2017 Elsevier.)

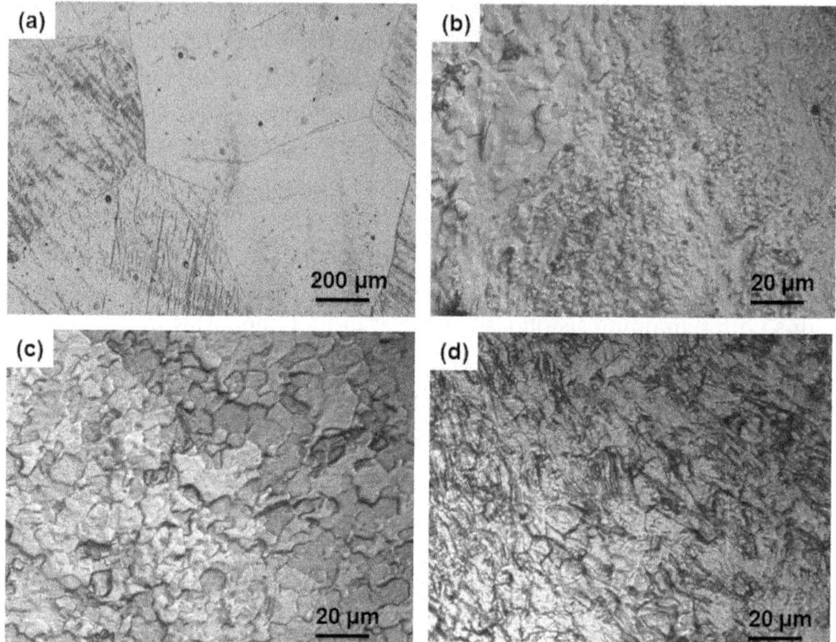

FIGURE 3.6 Microstructure observations of FSPed Mg: (a) unprocessed Mg, (b) Nugget zone and thermomechanical affected zone (TMAZ), (c) Grains in TMAZ and (d) twins appeared in heat affected zone (HAZ).

iii) Heat Affected Zone

HAZ is the third important zone in FSPed regions that involves no plastic deformation but heat dissipation from the nugget zone. Since the higher amount of heat is transmitted through HAZ, alloys may experience any phase modifications, change in the amount of precipitates, aging effect or annealing effect. However, compared with fusion based welding techniques the severity of HAZ on the material properties is lower as the thermal gradation is less in FSP. In some materials where heat is sufficient to introduce microstructure modification such as producing thermally induced twins, HAZ may contain completely different kind of microstructural features compared with the base material. For example, as shown in Figure 3.6, the appearance of thermally induced twins is common in Mg [4].

Each zone has its own type of microstructure and hence, results in a different impact on the material properties. As the variations in the microstructure are higher in the thickness direction, obtaining uniform microstructure throughout the processed regions is, of course, crucial to promote uniform material properties. Multi-pass FSP with simultaneous tracks will help to develop more area of stir zone on the surface with relatively uniform microstructure.

3.1.2 Heat Generation and Transfer During FSP

Heat is generated due to two basic reasons during FSP which are friction between FSP tool and the workpiece and plastic deformation of the material. The generated heat is dissipated through the workpiece, tool and the ambient by the three modes of heat transfer, e.g., conduction, convection, and radiation. Figure 3.7 schematically shows the heat generation and transfer during FSP. The heat is generated due to friction between the shoulder of the FSP tool and workpiece; the FSP tool pin and workpiece, and the plastic deformation of the material. The produced heat decreases the yield point of the material and helps to plasticize it to deform during the stirring action of the FSP tool.

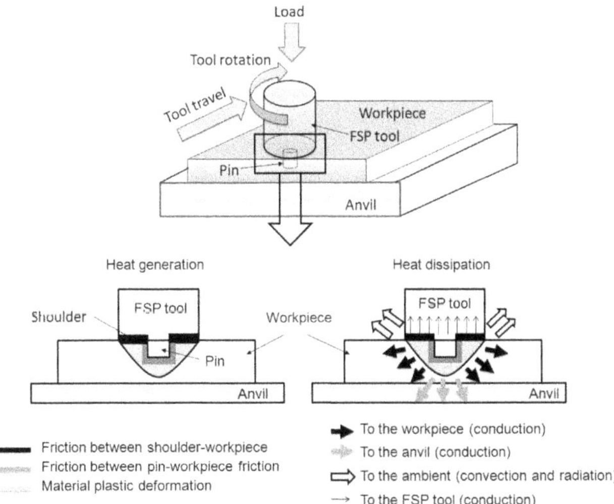

FIGURE 3.7 Schematic diagram showing (a) heat generation and (b) heat dissipation during FSP.

The produced heat in FSP is less than the heat observed in fusion-based techniques but sufficient to initiate solid-state plastic deformation. The overall heat produced during FSP is dissipated through four different channels as given below:

i) workpiece by conduction;
ii) anvil by conduction;
iii) tool by conduction; and
iv) ambient by convection and radiation.

As the amount of heat transfer is lower compared with fusion-based processes, the resulting thermally induced stresses are also lower in FSP. The heat generation and heat transfer play an important role in the success of processing. The process parameters are such as tool rotational speed, traverse speed, load, and tool penetration depth influence the heat generation. The heat carrying capacity of the tool material and work material play an important role in heat dissipation during FSP. The same process parameters may not suitable for different materials.

3.1.3 Advantages with FSP

FSP is a green process and offers several advantages compared with other surface treatment techniques. In fact, FSP is the only process of its kind compared with other surface engineering techniques and offers the following advantages.

- The process is simple and easy to be adopted in the existing industries.
- No energy sources are required such as high electrical or chemical sources as observed in other surface treatment processes.
- No harmful gases or fumes are released during the process which makes FSP as pollution free technique.
- Lower amount of heat generation during processing which further eliminates heat affected distortion in the processed materials and deceases thermally induced residual stresses.
- Excellent plastic flow of the material enables the technique suitable for a wide variety of material systems.

- Excellent mixing of the material at the surface and sub-surfaces leads to better metallurgical properties at the surface.
- Able to introduce orientation change or texture effect in the processed region.
- Results very fine grain size (in most of the cases) after FSP that helps to alter the structure-sensitive properties.
- Produces equiaxed grains by which superplastic properties can be enhanced.
- Localized processing enables engineers to repair or heal the cracks, pores, and defects of fabricated components or components which are already in functioning.
- Clean and green technology makes FSP a process of environmentally friendly.

3.2 FSP TOOL

In FSP, the tool plays a crucial role in performing three functions, which are heating the workpiece, causing material flow, and holding the plasticized material beneath the tool shoulder. Figure 3.8 shows a schematic diagram of a typical FSP tool. A specially designed pin or probe and a shoulder are the important elements of an FSP tool.

FIGURE 3.8 Schematic diagram of a typical FSP tool consisting of a tapered cylinder probe.

The shank is inserted into the tool holder, and the rotating tool is brought into contact with the workpiece. As the tool is penetrated into the workpiece, rotating pin generates heat due to friction between the pin and the work material. Additionally, pin helps the material to plastically flow at the vicinity of the pin due to the stirring action. The tool is plunged until the shoulder touches the surface of the workpiece. Due to friction between the shoulder surface and the workpiece, an additional amount of heat is generated. Fabricating FSP tool is simple and can be readily adopted by the industry.

Based on the possibility for the relative movement of shoulder and pin, FSP tools can be classified as fixed, adjustable and self-reacting. In fixed type, the entire shoulder and probe are fabricated from a single piece. The probe length is fixed and cannot be adjusted. Therefore, the affected thickness after FSP is constant for a specific tool. It is also not possible to replace the probe if it worn out or broken. The adjustable tool is made of two independent components. One is a probe, and the other is a shoulder. The probe is fabricated in such a way that is inserted within the shoulder, and the length of the probe can be adjusted. In an adjustable tool, probe material can be different from the shoulder material. Often, the probe can be replaced based on the requirement or if the probe is worn out or broken. By altering the probe length, different affected thicknesses can be achieved by using the same tool. In addition, the hole that is left at the end of processing can be eliminated by decreasing the probe length at the end of the processing. A self-retreating tool consisting of three components include top shoulder, probe, and bottom shoulder facilitate to process with different thicknesses. The requirement of the anvil is eliminated by using the self-retreating tool. However, tilting is not possible with this tool.

3.2.1 Tool Shoulder

The effect of the shoulder on the successful modification of the surface during FSP has been well studied. In addition to generating heat due to friction, the tool shoulder does the following functions:

- induces downward forging effect on the material beneath the shoulder; and
- consolidates the plasticized material in the processed zone and encapsulate to avoid escape of material under the shoulder.

Several types of shoulders were investigated to understand the effect of shoulder design on the material flow. Shoulder outer surface, the end surface, and the end profile are the three factors associated with the design of FSP tool shoulder design. Figure 3.9 shows the illustrations of different tool shoulders. The outer surface of the shoulder can be a cylindrical or conical (Figure 3.9 (a) and (b)). Using cylindrical shoulder is the most common observation from the literature. Most of the case, the type of outer surface plays an insignificant role in the process. However, in some cases where no probe is used in the process, the type of shoulder's outer surface play a crucial role as the material flow depends on the shoulder alone. Because the area is decreased for the conical outer surface compared with the cylindrical shoulders, the amount of friction and then heat generation is also decreased compared with cylindrical outer surface

There are three types of shoulder end surfaces which are flat, concave and convex type as shown in Figure 3.9 (c), (d) and (e). In FSP tool shoulder design, end surface plays a very important role. Flat end shoulder is the most widely used design. However, it is less effective in holding the plasticized material during FSP and results in the excess amount of material flash. To address this limitation, the concave end shoulder was designed. Producing concave end shoulder by simple machining is not a complex operation. During processing, the tool is tilted (1 to 3°) and plunged into the material surface. The concave end provides a reservoir of the cavity for the flowing material due to the stirring action of the pin beneath the shoulder and avoids the escape of the plasticized material in the form of flash. After plunging, when the rotating tool is moved in traversal direction, new material enters the concave cavity as the existing material in the concave cavity is moved to behind the pin. The tool end edge applies a compressive forging load on the surface and imposes a higher amount of forging load and hydrostatic pressure. Convex type shoulder end is another design demonstrated by The Welding Institute (TWI) during the early years of the evolution of friction stir welding. The results were not as expected, and the design was less seen in the literature in connection with FSP. The main limitation is the excess amount of material flash. However, a few studies demonstrated the use of convex type end particularly in friction stir welding. The other influencing factor in shoulder design is the end surface profile.

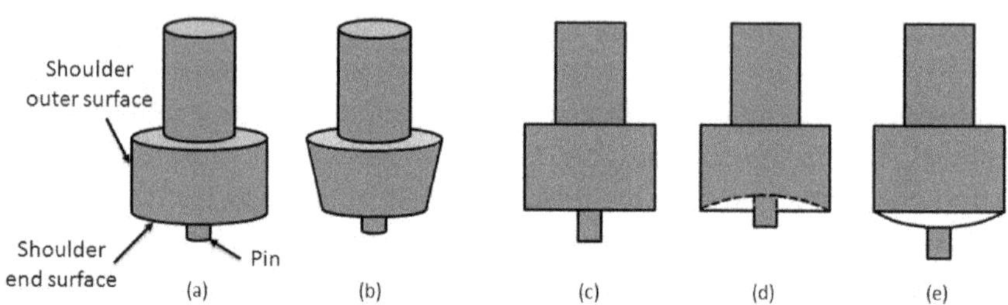

FIGURE 3.9 Illustration showing different types of tool shoulders: (a) cylindrical, (b) conical, (c) flat end, (d) concave end and (e) convex end.

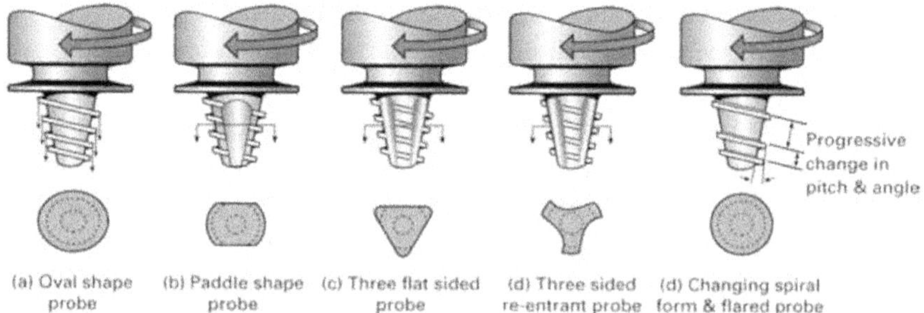

FIGURE 3.10 Schematic diagram showing different threaded profiles on Whorl pin as demonstrated by Thomas et al., (*Source:* Reprinted with permission from Thomas et al., [5]. © Springer 2003.)

The common end surface profiles used are flat, scrolls, ridges, knurling, grooves and concentric circles [5]. All these surface profiles can be developed for flat, concave or convex shoulder ends. However, some combinations may be difficult to machine. The flat profile and scroll profiles are the most commonly used end profiles. Scroll profile consisting of a flat end surface and spirals machined such a way they are directed from the outer edge of the shoulder to the center of the end surface (towards the pin). Scrolls profile helps the material to flow from the edge of the end to the center and avoids the material flash and also reduces the necessity of tool tilting. The concave end coupled with scrolls profile reduces the tool lift during processing at higher speeds. Further, scrolls profile helps to reduce the undercut defect which is usually observed in concave end surface profile. Other profiles were used a few times in the literature, and the work was insignificant [6, 7].

3.2.2 *Tool Pin*

The tool pin is the second most influencing factor in FSP and dictates the success of the process. The dimensions of the pin, shoulder diameter to pin diameter ratio and the profile of the pin are the important elements in designing an FSP tool pin. Circular pin with or without taper is the simplest form and most widely used pin profile. When the pin is penetrated into the surface of the workpiece, due to the friction between the pin surface and the work material, heat is generated, and the plasticized material is sheared and stirred from the front to the rear of the pin and fills the cavity created by the pin due to the forward tool motion. Several pin profiles are available in the literature. Figure 3.10 shows the schematic representation of threaded Whorl pin profiles as reported by Thomas et al., [5]. Figure 3.11 shows various pin designs available in the literature as reported by Zhang et al., [6]. The thickness of the affected surface mainly depends on the pin dimensions. The end surface of the probe is either flat or angled; and the manufacturing tools with flat pin end are simple. But a higher amount of plunge load is required during penetrating the pin into the work material. Further, a flat end may lead to an increase in the stresses in the pin and results in end shape change or tool wear. Producing angled or domed end shaped pins are slightly difficult but offers an advantage compared with flat endpin such as lower plunge loads, decreased stress in the pin and increased tool life [8]. The outer profile of the pin contains different shapes such as threads, flats, and flutes. Threaded probes are most widely used in FSP which promote higher material flow in the workpiece thickness direction. As the material is dragged down along the thread, the material is circulated around the tool pin surface and is deposited in the cavity generated by the pin. Threaded probes help to efficiently close the voids and break the oxides present on the surface. Pins with flat shaped outer profile completely change the material flow mechanism around the pin. Flat edges act as cutting edges and the material which is being cut by the edges is trapped behind the flats and excellent material mixing is resulted. Further, the flat profile on the pin increases the width of the nugget zone.

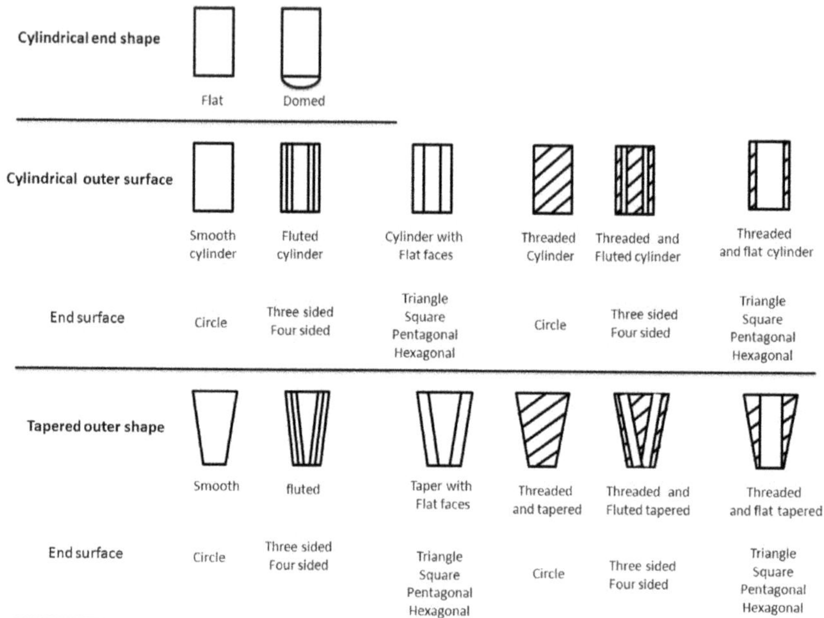

FIGURE 3.11 Schematic representation of different pin profiles and their cross-sectional geometries.

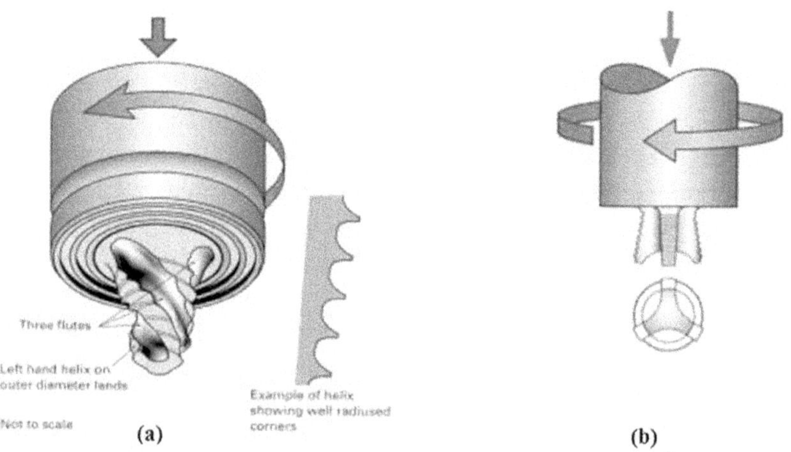

FIGURE 3.12 (a) Whorl profile pin and (b) MX Triflute profile pin. (*Source:* Reprinted with permission from Thomas et al., [5]. © Springer 2003.)

The flutes on the threaded probe increase the material flow along the thread and the material are trapped in the flutes, and better mixing can be achieved. It is interesting to learn that several special designs were proposed for tool geometries to achieve better material flow mechanisms. By adding complex-shaped profiles and features on a pin, better material flow and mixing, quality processed regions and decreased processing loads can be achieved. Whorl and MX Triflute pin profiles developed by TWI are examples for such complex profiled tools as schematically shown in Figure 3.11. A higher amount of material flow around the pin is the advantage with tools having complex profiles such as Whorl and MX Triflute [9]. The swept rate is known as the ratio of the volume of the material affected or stirred to the pin volume. Higher swept volume is an important common observation in the case of complex profiles that result in the quality-processed zone.

3.2.3 Tool Materials

Different materials were used to fabricate FSP tools. Among them, tool steels are the most widely used material. Depending on the work material, several alloys and advanced tool materials were used for FSP tool as explained below. Aluminum, magnesium, and copper alloys can be easily processed by tool steels. Tungsten (W)-based tools are another group commonly used in the FSP tool fabrication. Usually, W based tools are stronger even at the higher temperatures best suited to process steels and titanium alloys. Tungsten carbide (WC)-based materials are another important tool material used to fabricate FSP tools. However, W or WC based tool suffers from lower toughness at room temperature. In order to increase the toughness in W and WC based materials, elements such as rhenium, cobalt, lanthanum, etc. are used to develop tools for FSP. Polycrystalline cubic boron nitride (pcBN) tools are another group of tool materials used in FSP. These materials possess high strength and high hardness at elevated temperature. These kinds of materials are best suitable to treat steels and titanium alloys. Silicon nitride-based tools and molybdenum-based alloys are other types of materials used in FSP particularly to process high hard materials.

3.2.4 Tool Wear

Wear is the degradation phenomena by mechanical means. Usually, hard materials undergo lower wear compared with soft materials when they come in contact and relatively in motion. However, due to the removed metal particles present at the interface, it happens that the hard material also undergoes wear. The rotational and translational moments of FSP tool during the process lead to tool wear. Furthermore, the generation of higher temperature during FSP also lead to tool wear as the hardness and yield strength of the tool material is decreased at elevated temperature and may lead to pin failure. Oxidation of the tool material at high temperature also leads to tool material wear. Tool pin undergoes a higher rate of wear compared with the shoulder. More wear is usually seen for the pin compared with the shoulder due to the following four important reasons.

i) During FSP, tool pin is completely penetrated into the workpiece compared with the shoulder. Therefore, more resistance is faced by the pin to its forward motion compared with the shoulder.
ii) The fraction of the total heat generated during FSP is concentrated near the shoulder and therefore, shoulder experience lower resistance to motion compared with the pin.
iii) A higher amount of loads on the pin compared with the shoulder.
iv) The combined effect of bending and torsion stresses induced in the pin.

Material loss due to wear leads to change in the pin shape, dimensions and profile which are crucial factors play an important role in successful processing. Figure 3.12 shows the photographs of tool pin affected due to wear as reported by Prado et al., [10]. In the context of tool wear, FSP tools made from tungsten based alloys and molybdenum alloys are promising. Additionally, a few strategies such as providing wear-resisting coatings and preheating the workpiece are also adopted

3.3 MATERIAL FLOW DURING FSP

Understanding the material flow is of immense importance to choose appropriate tool design and to avoid any defects in the processed regions. Several studies have been conducted to investigate the material flow during

FSW. The process parameters and tool that is used for both the FSW and FSP processes are the same, and hence, the material flow behavior during FSP can be understood by knowing the material flow during FSW.

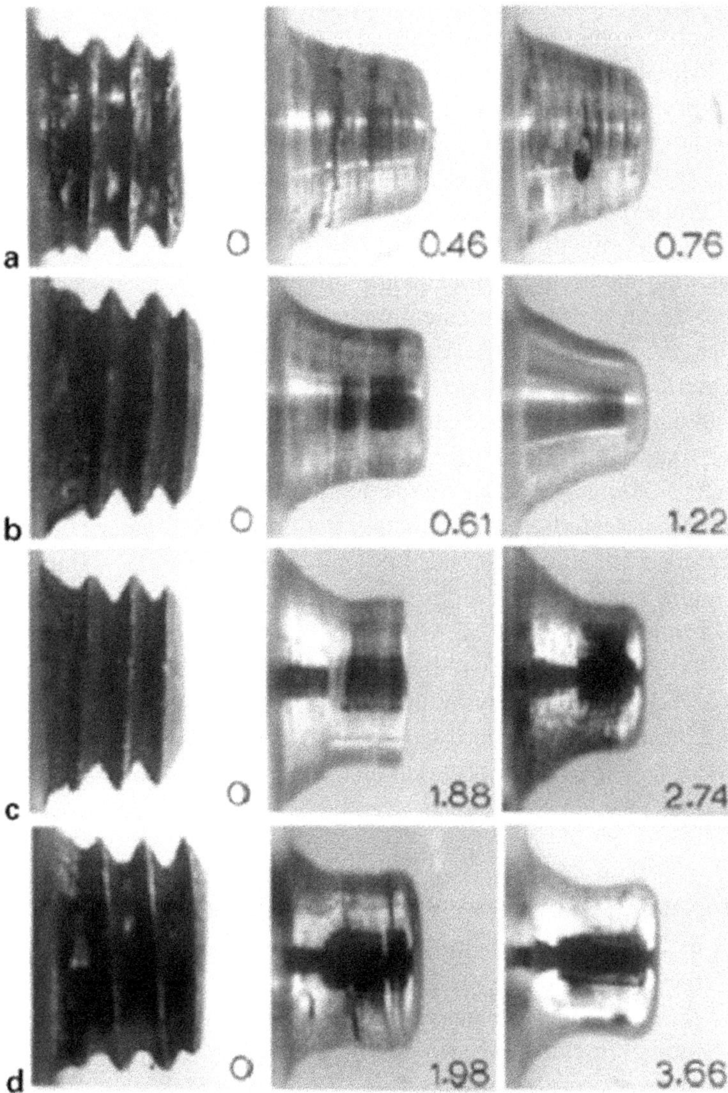

FIGURE 3.13 Photographs showing worn out threaded probes after FSP (*Source:* Reprinted with permission from Pardo et al., [10]. © 2003 Elsevier.)

The material flow during FSW/P depends on several factors as given below.

i) Process parameters which include tool rotational speed, travel speed, penetration depth, and load.
ii) Tool geometry, e.g., pin and shoulder dimensions, the ratio of shoulder diameter to pin diameter, type of pin and shoulder profiles.
iii) Material characteristics and preheating of the material.

Several experimental and computational methods have used to know the material flow during FSP. Among them, marker tracer technique and welding of dissimilar metals are the most widely used experimental methods along with finite element methods are the approaches adopted to visualize the flow of material during FSP.

In marker tracer technique, a material which is different from the base material and can be easily distinguished by etching is used. Initially, the marker material is embedded into a base material, and FSP is carried out. After FSP, the pattern of the marker is distributed in the FSPed region which is studied to understand the flow of the material. Different marker materials have been used to study the material flow during FSP. Aluminum, copper, steel, aluminum composite and tungsten wire are a few examples for the marker materials [11–18]. By using markers made of 5454Al-H32 embedded in the matrix of 2195Al-T8 (base material) in the processing path, Reynolds and coworkers [11–13] investigated the flow of the marker by microstructural observations after FSW, and the following five important observations were made in understanding the material flow.

i) The flow of the material was non-symmetric with respect to the center line of the processed region, and the backward movement of the material was limited to a distance equal to the diameter of the pin behind its initial position.
ii) A clear interface was observed between the advancing and retreating side, and also the material across the processed region was not completely stirred.
iii) Within the pin diameter, the direction of the material flow was observed as downward at the advancing side and was upward at

the retreating side. Most of the material flow was noticed beneath the shoulder.

iv) The rate of material flow in the vertical direction at the retreating side was inversely proportional to the rate of tool advancement per rotation.

v) At the same process parameters, by increasing the pin diameter, increased material flow across the center line of the weld zone was observed.

Further, Reynolds and coworkers [12, 13] suggested visualizing FSW process as a localized in situ extrusion around the pin. Guerra et al. [14] studied the flow of the material during FSW by using a thin copper foil (0.1 mm) placed along the faying surface of the weld. At the end of the process, the tool was kept penetrated by stopping the tool rotation and travel movement to get the frozen pin in the stir zone. From the microstructural observations, they concluded that the material flow happened due to two processes:

(i) Highly deformed material in front of the weld at the advancing side is flushed to the regions behind the pin with arc-shaped flow pattern.

(ii) Extrusion of the material on the front side of the retreating side between the rotational zone and the base material.

Additionally, they observed the higher rate of material flow at the top of the processed zone due to the influence of the shoulder rather than the pin. Colligan [15, 16]used small steel balls of diameter 0.38 mm to disperse in an aluminum alloy matrix in the welding direction at different positions to understand the material flow during FSW. After each welding, radiography was carried out to see the distribution of the marker material. "Stop-action" method was adopted in which the FSW tool is suddenly stopped and retreated. The distribution of the steel shots was observed as broader at the top regions of welding and extrusion of material through the threads of the pin was observed. Further, Colligan suggested that the material in the stir may not completely stir but is extruded through the pin threads.

London et al., [17] used Al606–30%SiC composite and Al-20%W composite as markers during welding of 7050Al-T7451 to investigate the material behavior. From their observations, the material flow was observed as happened in three steps. Initially, the uplifting of the material was observed in front of the pin due to the tool tilt followed by material shear around the pin and downward movement due to the threads on the pin. Further, they also noticed that the wide distribution of the marker at the advancing side compared with the centerline of the weld zone.

A few studies can be found on welding of dissimilar materials to understand the material flow in FSW. Midling [18] joined two dissimilar aluminum alloys and showed the interface shapes from the microstructural studies. Ouyang and Kovacevic [19] investigated the material flow during welding of 2024Al to 6061Al dissimilar alloy plates and reported mechanically mixed regions with vortex-like structure and alternative lamellae due to the stirring action and travel motion of the FSW tool and the localized in situ extrusion. However, other material flow patterns were also reported by several authors. Among them, onion ring patterns are one typical observation in many alloy systems as reported by Krishnan [20], Biallas et al. [21–23]. Form the work of Ma et al., [24, 25] during FSP of cast A356 to modify the microstructure, it is suggested that not to consider the material flow as a simple extrusion.

In addition to the experimental works, several modeling and simulation works have been done in understanding the mechanisms behind material flow during FSW/P [26–31]. These simulation works enhanced the level of understanding the material flow behavior. As per the explanation given by Arbegast [32], the material flow in FSP contains combinations of processes as schematically illustrated in Figure 3.14.

In FSP, five conventional zones can be identified as given below:

i) preheat;
ii) initial deformation;
iii) extrusion;
iv) forging; and
v) postheat and cooling.

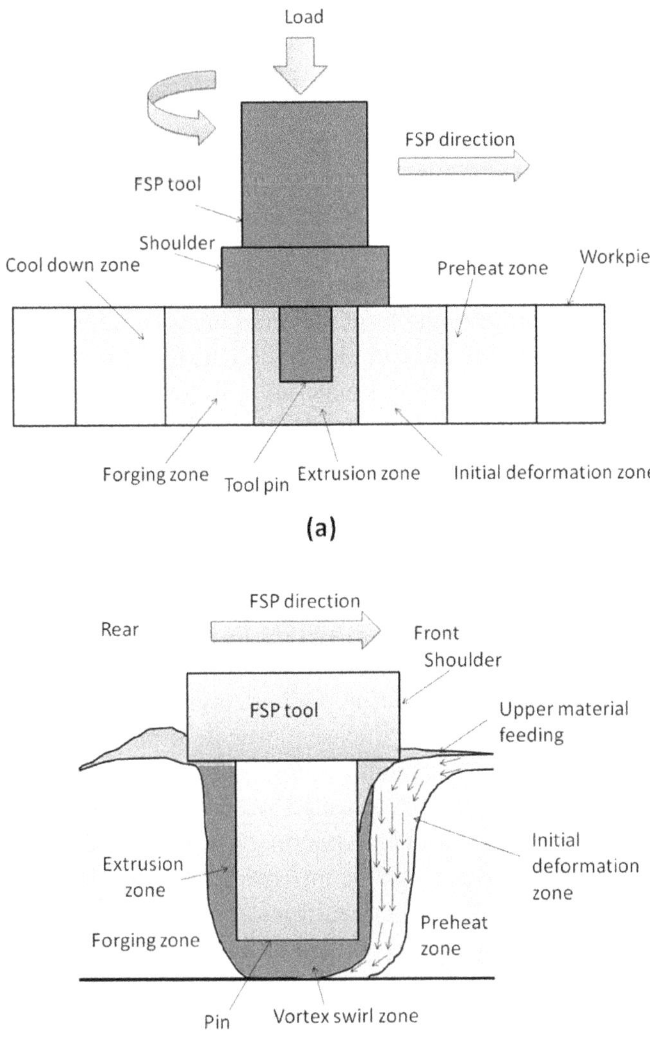

FIGURE 3.14 Schematic illustration of different zones resulted during FSP.

The work material in the pre-heat zone that is in front of the rotating pin experiences a rise in temperature due to the frictional heat and the adiabatic heating due to the plastic deformation of the material at the stir zone. The thermal characteristics of the work material and the process parameters such as tool rotational and traverse speed influence the level of temperature rise in this zone. In the initial deformation zone, the material experience stresses beyond the flow stress as the rotating FSP tool advances. Therefore, the material started deformation and moved towards the shoulder and the extrusion zone. In the extrusion zone, the material at the vicinity of the pin surface experience extrusion from the front to the rear of the pin. A swirl motion of the material can also be seen at the bottom of the pin in a small region. Next, to the extrusion zone, plastically deformed material is forged due to the load applied by the rotating shoulder. The material from the front region of the rotating pin is extruded to fill the cavity created by the tool pin travel. Due to the hydrostatic pressure imposed by the shoulder, the material is forged and restricted from any escape from the processed zone. After this step, the processed zone is cooled down in the post heat/cool zone. Overall, the material flow behavior in FSP is complex and is greatly influenced by the process parameters and the tool design.

3.4 GRAIN REFINEMENT MECHANISM IN FSP

Microstructural modification during FSP involves principles of severe plastic deformation. Introducing intense plastic deformation coupled with high temperature lead to recrystallization and microstructural modification in FSP [33–37]. If a material is heated to a certain temperature to rearrange the grain boundaries, the corresponding temperature can be called as recrystallization temperature, and the microstructure-modified material is said to be recrystallized. If the recrystallization happens during the plastic deformation of a material, then it is said to be dynamic recrystallization that is the prime mechanism behind the microstructure modification during FSP. Based on the modified microstructure in the processed zone, three distinct regions are denoted as a nugget or stir zone, TMAZ and HAZ. Dynamic recrystallization is seen in the nugget zone, and hence this zone can also be called a dynamically recrystallized zone. Nugget zone contains a high density of sub-grains evolved within the grains and dislocations [38–40]. Usually, dynamic recrystallization results in finer and equiaxed grains. From the works carried out to understand the mechanism dynamic recrystallization in FSP, it was observed that three kinds are known as discontinuous dynamic recrystallization (DDRX), continuous dynamic recrystallization (CDRX), and geometric-dynamic recrystallization (GDRX) [41–43] play as the important reasons behind the grain refinement during FSP. If a new grain is evolved from a high-angle grain boundary, then it can be called as DDRX [41]. DDRX is lower in materials which show higher recovery such as aluminum alloys. However, in the presence of larger secondary phase particles, aluminum may show DDRX [42, 43]. In the case of CDRX, it is believed that subgrains rotate and attain a high orientation change with a slight boundary migration [44–46]. It is also an important observation that the variation in the grain size within the nugget zone that can be attributed to the variation in the temperature in the processed zone [47]. Particularly, in the thickness direction of an FSP region, a variation in the average grain size can be observed. The important reasons are the difference in the amount of heat dissipation in the thickness direction which directly influences the material flow mechanisms.

KEYWORDS

- **continuous dynamic recrystallization**
- **discontinuous dynamic recrystallization**
- **geometric-dynamic recrystallization**

REFERENCES

1. Mishra, R. S., Mahoney, M. W., McFadden, S. X., Mara, N. A., & Mukherjee, A. K., (2000). High strain rate superplasticity in a friction stir processed 7075 Al alloy. *Scripta Mater., 42,* 163–168.
2. Xinggang, J., Jianzhong, C., & Longxiang, M., (1993). *Acta Metall. Mater., 41*, 2721.
3. G. V. V. Surya Kiran, K. Hari Krishna, Sk. Sameer, Bhargavi, M., B. Santosh Kumar, G. Mohana Rao, Naidubabu, Y., Ravikumar Dumpala, B. Ratna Sunil, Machining characteristics of fine-grained AZ91 Mg alloy processed by friction stir processing, Trans. Nonferrous Met. Soc. China 27(2017) 804−811.
4. Ratna, S. B., Sampath, K. T. S., & Uday, C., (2012). Bioactive grain refined magnesium by friction stir processing. *Materials Science Forum, 710,* pp. 264–269.
5. Thomas, W. M., Nicholas, E. D., & Smith, S. D., (2001). Friction stir welding-tool developments. *Proc. Aluminum Automotive and Joining Sessions* (pp. 213–224). Warrendale, PA, TMS.
6. Zhang, Y. N., Cao, X., Larose, S., & Wanjara, P., (2012). Review of tools for friction stir welding and processing. *Canadian Metallurgical Quarterly, 51*(3), 250–261.

7. Rai, R., De, A., Bhadeshia, H. K. D. H., & DebRoy, T., (2011). Review: Friction stir welding tools. *Science and Technology of Welding and Joining, 16*(4), 325–342.
8. Dawes, C. J., Threadgill, P. L., Spurgin, E. J. R., & Staines, D. G., (1995). '*Development of the New Friction-Stir Technique for Welding Aluminum Phase II.*' TWI member report, Cambridge, UK.
9. Thomas, W. M., Staines, D. G., Norris, I. M., & De Frias, R., (2002). 'Friction stir welding – tools and developments.' *Weld. World, 47*(11/12), 10–17.
10. Prado, R. A., Murr, L. E., Soto, K. F., & McClure, J. C., (2003). 'Selfoptimization in tool wear for friction stir welding of Al 6061z20% Al2O3 MMC.' *Mater. Sci. Eng. A., A349*, 156–165.
11. Reynolds, A. P., Seidel, T. U., & Simonsen, M., (1999). In: *Proceedings of the First International Symposium on Friction Stir Welding.* Thousand Oaks, CA, USA.
12. Reynolds, A. P., (2000). *Sci. Technol. Weld. Joining,5,* 120.
13. Seidel, T. U., & Reynolds, A. P., (2001). *Metall. Mater. Trans. A., 32,* 2879.
14. Guerra, M., McClure, J. C., Murr, L. E., & Nunes, A. C., (2001). In: Jata, K. V., Mahoney, M. W., Mishra, R. S., Semiatin, S. L., & Filed, D. P., (eds.), *Friction Stir Welding and Processing* (p. 25). TMS, Warrendale, PA, USA.
15. Colligan, K., (1999). In: *Proceedings of the First International Symposium on Friction Stir Welding* (pp. 14–16). Thousand Oaks, CA, USA.
16. Colligan, K., Material Flow Behavior during Friction Stir Welding of Aluminum, Weld, J., (1999). 78, 229S–237S.
17. London, B., Mahoney, M., Bingel, W., Calabrese, M., Bossi, R. H., & Waldron, D., (2003). In: Jata, K. V., Mahoney, M. W., Mishra, R. S., Semiatin, S. L., & Lienert, T., (eds.), *Friction Stir Welding and Processing II* (p. 3). TMS.
18. Midling, O. T., (1994). Material flow behaviour and microstrucrural integrity of friction stir butt weldments In: Sanders, T. H., Jr., & Strake, E. A., Jr., (eds.), *Proceedings of the Fourth International Conference on Aluminum Alloys* (Vol. 1, pp. 451–458).
19. Ouyang, J. H., & Kovacevic, R., (2002). *J. Mater. Eng. Perform., 11,* 51.
20. Krishnan, K. N., (2002). *Mater. Sci. Eng. A., 327*, 246.
21. Biallas, G., Braun, R., Donne, C. D., Staniek, G., & Kaysser, W. A., (1999). In: *Proceedings of the First International Symposium on Friction Stir Welding* (pp. 14–16), Thousand Oaks, CA, USA.
22. Mahoney, M. W., Rhodes, C. G., Flintoff, J. G., Spurling, R. A., & Bingel, W. H., (1998). *Metall. Mater. Trans. A., 29,* 1955.
23. Sutton, M. A., Yang, B., Renolds, A. P., & Taylor, R., (2002). *Mater. Sci. Eng. A., 323,* 160.
24. Ma, Z. Y., Tjong, S. C., & Geng, L., (2000). *Scripta Mater., 42,* 367.
25. Ma, Z. Y., Tjong, S. C., Geng, L., & Wang, Z. G., (2000). *J. Mater. Res., 15*, 2714.
26. Xu, S., Deng, X., Reynolds, A. P., & Seidel, T. U., (2001). *Sci. Technol. Weld. Joining, 6*, 191.
27. Dong, P., Lu, F., Hong, J. K., & Cao, Z., (2001). *Sci. Technol. Weld. Joining, 6*, 281.
28. Goetz, R. L., & Jata, K. V., (2001). In: Jata, K. V., Mahoney, M. W., Mishra, R. S., Semiatin, S. L., & Filed, D. P., (eds.), *Friction Stir Welding and Processing* (p. 35). TMS, Warrendale, PA, USA.

29. Stewart, M. B., Adamas, G. P., Nunes, A. C. Jr., & Romine, P., (1998). *Developments in Theoretical and Applied Mechanics* (pp. 472–484). Florida Atlantic University, USA.
30. Nunes, A. C. Jr., (2001). In: Das, S. K., Kaufman, J. G., & Lienert, T. J., (eds.), *Aluminum 2001* (p. 235). TMS, Warrendale, PA, USA.
31. Ke, L., Xing, L., & Indacochea, J. E., (2002). *Joining of Advanced and Specialty Materials IV* (pp. 125–134). ASM International, Materials Park, USA.
32. Arbegast, W. J., (2003). In: Jin, Z., Beaudoin, A., Bieler, T. A., & Radhakrishnan, B., (eds.), *Hot Deformation of Aluminum Alloys III* (p. 313). TMS, Warrendale, PA, USA.
33. Benavides, S., Li, Y., Murr, L. E., Brown, D., & McClure, J. C., (1999). *Scripta Mater., 41*, 809.
34. Li, Y., Murr, L. E., & McClure, J. C., (1999). *Mater. Sci. Eng. A., 271*, 213.
35. Ma, Z. Y., Mishra, R. S., & Mahoney, M. W., (2002). *Acta Mater., 50*, 4419.
36. Mahoney, M. W., Rhodes, C. G., Flintoff, J. G., Spurling, R. A., & Bingel, W. H., (1998). *Metall. Mater. Trans. A.,29*, 1955.
37. Kwon, Y. J., Shigematsu, I., & Saito, N., (2003). *Scripta Mater., 49*, 785.
38. Sato, Y. S., Kokawa, H., Enmoto, M., & Jogan, S., (1999). *Metall. Mater. Trans. A., 30*, 2429.
39. Heinz, B., & Skrotzki, B., (2002). *Metall. Mater. Trans. B., 33*(6), 489.
40. Charit, I., & Mishra, R. S., (2004). In: Zhu, Y. T., Langdon, T. G., Valiev, R. Z., Semiatin, S. L., Shin, D. H., & Lowe, T. C., (eds.), *Ultrafine Grained Materials III*. TMS.
41. Humphreys, F. J., & Hotherly, M., (1995). *Recrystallization, and Related Annealing Phenomena*. Pergamon Press, New York.
42. Doherty, R. D., Hughes, D. A., Humphreys, F. J., Jonas, J. J., Jensen, D. J., Kassner, M. E., et al., (1997). *Mater. Sci. Eng. A., 238*, 219.
43. Gourder, S., Konopleva, E. V., McQueen, H. J., & Montheillet, F., (1996). *Mater. Sci. Forum, 217–222*, 441.
44. Bricknell, R. H., & Edington, J. W., (1991). *Acta Metall. A., 22,* 2809.
45. Hales, S. J., & McNelley, T. R., (1988). *Acta Metall., 36,* 1229.
46. Liu, Q., Huang, X., Yao, M., & Yang, J., (1992). *Acta Metall. Mater., 40,* 1753.
47. Pao, P. S., Lee, E., Feng, C. R., Jones, H. N., & Moon, D. W., (2003). In: Jata, K. V., Mahoney, M. W., Mishra, R. S., Semiatin, S. L., & Lienert, T., (eds.), *Friction Stir Welding and Processing II* (p. 113). TMS, Warrendale, PA, USA.

CHAPTER 4

Influencing Factors

The success of any process is governed by certain influencing factors. For FSP, there are several factors including tool design (pin, shoulder, ratio of pin and shoulder diameters), processing parameters (tool rotational speed, travel speed, penetration depth, and tilt angle) material type, and workpiece temperature which control the quality of the process. Figure 4.1 schematically shows different factors which contribute to the success of FSP.

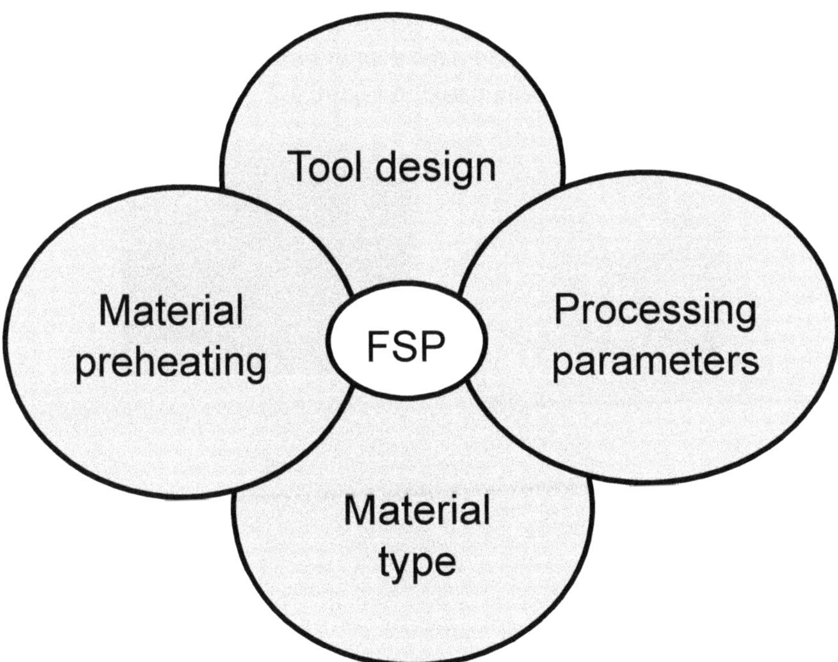

FIGURE 4.1 Schematic diagram of different influencing factors in FSP.

4.1 TOOL DESIGN

As discussed in Chapter 3, the FSP tool plays a crucial role in the success of FSP. Pin and shoulder are the two important elements of the tool which need careful design. If the shoulder diameter is higher, the area that is in contact with the work material is increased, and hence, more amount of heat is generated. If the shoulder diameter is lower, the amount of generated heat is insufficient to plastically deform the material in the stir zone and lead to defects. If the diameter of the pin is higher, the size of the cavity produced by the pin penetration and travel is higher that may not be completely filled with the stirring material and leaves a defect. If the pin diameter is too small, the amount of the material that is stirred is limited, and a very thin and narrow nugget is resulted. Therefore, designing the FSP tool shoulder and pin with appropriate dimensions is crucial to avoid defects and to develop a modified surface by FSP successfully. The relative dimensions of the shoulder and pin also influence the amount of heat generation and the level of grain refinement. From the works of Saikrishna et al. [1] and Ratna Sunil et al. [2], it was observed that the ratio of shoulder diameter to the pin diameter had played an important role on grain refinement of AZ31 Mg alloy as compared in Figure 4.2.

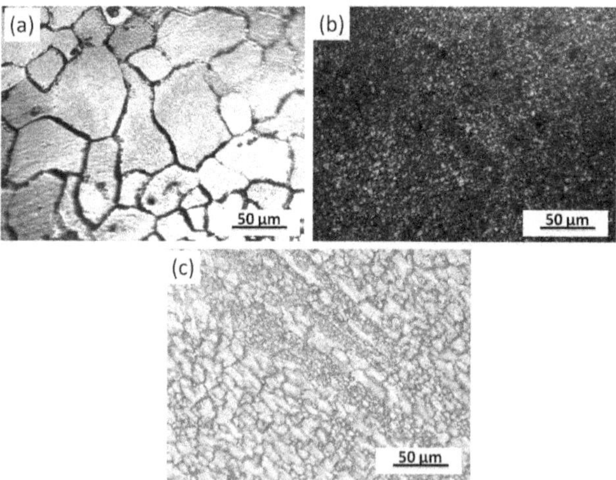

FIGURE 4.2 Comparison of grain refinement in AZ31 Mg alloy processed by two different FSP tools with different shoulder diameter to pin diameter ratios: (a) unprocessed AZ31 Mg alloy, (b) FSPed, AZ31 Mg alloy, processed with FSP tool shoulder to pin diameter ratio 3:1 and (c) FSPed, AZ31 Mg alloy, processed with FSP tool shoulder to pin diameter ratio 6:1.

When the ratio of the shoulder diameter to pin diameter was 3:1, grain refinement was observed as higher. Whereas for the same alloy at the similar processing conditions by using an FSP tool with shoulder diameter to pin diameter ratio as 6:1, grain growth was observed in the stir zone. Variation in the heat generation due to the variation in the contact area of the FSP tool with workpiece was observed as the significant reason behind the different level of grain refinement. Additionally, the level of grain modification was also observed as bimodal in AZ31 Mg alloy processed with FSP tool with a shoulder to pin ratio of 6:1. This kind of behavior is common in FSPed metals. The appearance of modified grains after FSP in two different scales which contain small and coarse grains called as onion ring pattern is usually observed in FSPed materials. This kind of difference in grain refinement is occurred due to the difference in the level of material flow during the stirring of the FSP tool.

4.2 PROCESSING PARAMETERS

Processing parameters including tool rotational speed, travel speed, penetration, and tool tilt angle govern the heat generation and distribution, thermal profile and cooling during FSP. Understanding the effect of these parameters is essential in FSP as they influence the material flow and microstructure evolution.

4.2.1 Tool Rotation Speed

The amount of heat generation during FSP is proportional to the tool rotational speed. The higher amount of heat is generated if the tool rotational speed is higher due to the higher levels of friction. Several studies were conducted to understand the effect of tool rotational speed on obtaining defects in free processed zones. The rotating pin results in stirring of the material around the pin and lead to proper mixing to fill the cavity produced by the pin traverse motion. However, increasing the tool rotational speeds may not always increase the rate of material stirring. Here, the generated heat also plays a role on material flow mechanism and further the level of grain refinement. Given heat input is increased with higher tool rotational speeds. On the other hand, if the tool rotational speed is decreased, the generated heat is insufficient to initiate the required material flow. Tunneling defect is resulted if the material flow during FSP is inappropriate. Figure 4.3 shows the photograph of FSPed AZ31 Mg alloy. The samples were processed with two different tool rotational speeds. The surface of the workpiece was observed as sound, but at the cross section, tunneling defect was observed for the workpiece processed with lower tool rotational speed (720 rpm). As the tool rotational speed was increased to 1100 rpm, the cross sections of the workpiece demonstrated sound nugget zone without tunneling defect. However, by increasing tool rotational speeds, tunneling defect may be eliminated but other adverse effects may results as given below:

i) grain growth;
ii) narrowing the width of nugget zone;
iii) deposition of base material to the tool shoulder.

Grain growth during FSP can be possible as reported by Saikrishna et al., [1] due to recrystallization. Not always, dynamic recrystallization results in fine grains. It depends on the amount of heat generation in the stir zone and the type of material. Figure 4.4 shows the grain growth at the bottom of the Nugget zone compared with the grain refinement within the stir zone. Due to the difference in the heat dissipation with the nugget zone different levels of grain sizes were observed.

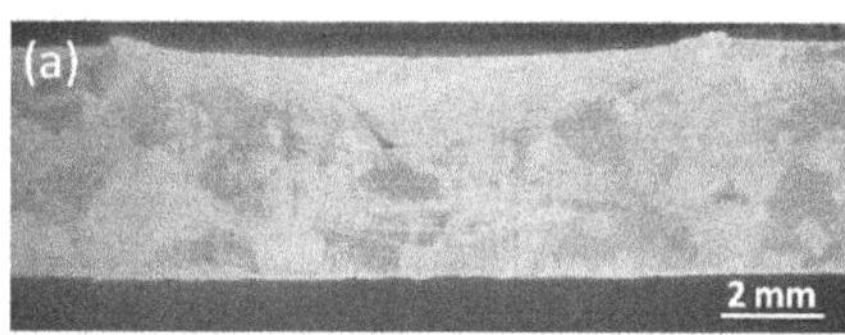

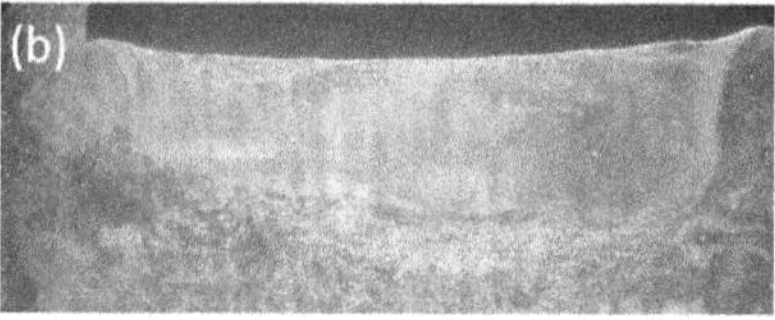

FIGURE 4.3 Photograph showing a cross section after FSPed of AZ31 Mg alloy: (a) defect and (b) defect free stir zones.

The narrow width of the nugget zone results with increased tool rotational speeds as a lower amount of material is stirred around the rotating pin. For example, as shown in Figure 4.5 the width of the stir zone was decreased as the tool rotational speed was increased from 1100 to 1800 processed at a constant tool travel speed (16 mm/min). Additionally, the width of the region containing fine microstructure within the nugget zone was also observed as decreased as the tool rotational speed was increased to 1800 rpm.

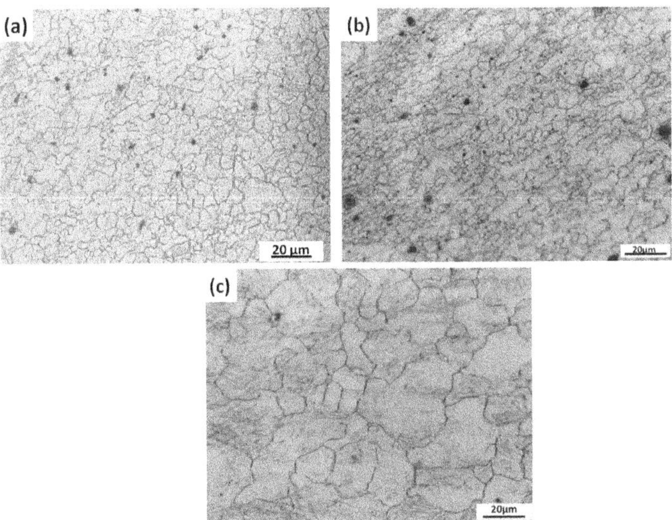

FIGURE 4.4 Optical microscope images: (a) unprocessed AZ31 Mg alloy, (b) fine grains within the nugget zone and (c) at the bottom of FSPed AZ31 Mg alloy.

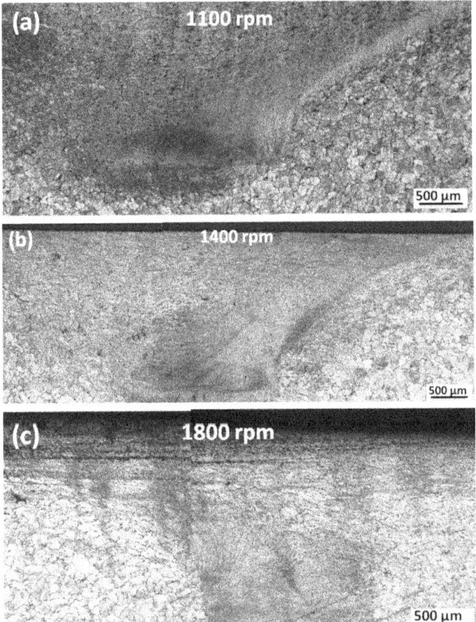

FIGURE 4.5 Optical microscope images of FSPed ZE41 alloy showing different stir zone widths processed at: (a) 1100 rpm, (b) 1400 rpm and (c) 1800 rpm.

Deposition of material to the tool shoulder is the third important observation that can be made with increased tool rotational speed. As the generated heat is higher at the increased tool rotational speeds, the material in the stir zone undergoes plastically soft state and easily transferred to the surface of the tool shoulder. Figure 4.6 shows typical photographs of FSPed AZ91 Mg alloy. Initially, the material from the stir zone was deposited on the shoulder of the FSP tool as shown in Figure 4.6 (b). Later, the material has back deposited to the workpiece as shown in Figure 4.6 (c). Due to the deposited material from the workpiece to the tool shoulder, the processed zone was also observed as defective and at the end of the process; a large mass of the workpiece material was observed as deposited.

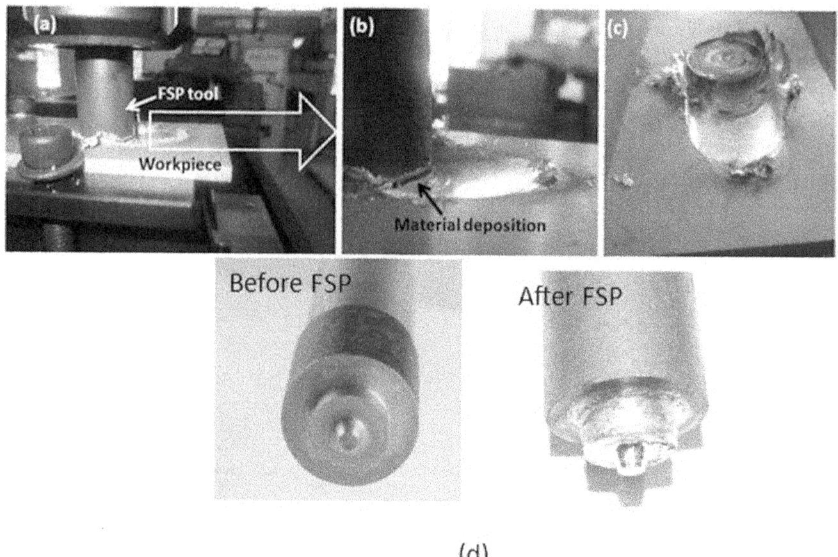

FIGURE 4.6 Photographs showing the material deposition on the tool shoulder: (a) FSP of AZ91 Mg alloy, (b) material deposition to the tool shoulder and (c) Back deposition of the coated material from the tool shoulder to the workpiece, (d) photographs showing the FSP tool before and after FSP of Mg. (*Source:* (a), (b), (c): Reprinted with permission from Kondaiah et al., [3]. © Elsevier 2017.)

Most of the cases, deposition of a thin layer of workpiece material is observed on the surface of the tool shoulder which is inevitable in FSP. This thin layer may eliminate the further diffusion of elements from the FSP tool to the workpiece material. As shown in Figure 4.6 (d), the FSP tool was coated with pure Mg after FSP which further reduced the dissolution of iron into the base material and helped to prevent the galvanic corrosion [5]. Afrin et al., [6] studied the effect of tool rotational and travel speeds on microstructure and mechanical properties of friction stir welded AZ31 Mg alloy. Due to the lower heat input at the lower welding speeds, fine microstructure and increased tensile properties were observed. The influence of tool rotation speed is also significant in other material systems. While processing pure copper, S. Cartigueyen and K. Mahadevan were also observed the significant role of tool rotational speed on obtaining defect free stir zone [7]. Experiments were conducted at three tool rotational speeds (250, 350 and 500 rpm) with a constant feed (50 mm/min). Tunneling defect was noticed in the nugget zone processed with 250 rpm speed and defect free stir zones were observed at 350 and 500 rpm. The poor material flow is the results of lower heat input which caused the tunneling defect in stir zone processed at 250 rpm. However, lower speeds resulted in fine grains and higher hardness compared with the higher speeds. Further, good mechanical properties were recorded for the FSPed copper done at 500 rpm tool rotational speed. Similarly, Leal et al., [8] also reported decreases material discontinuities in FSPed copper when the tool rotational speed was increased from 400 to 750 rpm with the same travel speed (25 mm/min). On the contrary, Zhang et al., [9] processed 2219-T6 alloy by FSP at different tool rotational speeds and formation of a void defect in the processed zones above 1400 rpm was observed. Furthermore, the grain size was measured as increased at higher speeds. Grain growth is a common observation associated with the increased tool rotational speeds, but the formation of the void with increased tool rotation speed is a significant observation. Hence, optimizing tool rotational speed is crucial in developing defect free grain refined processed zones by FSP.

4.2.2 Tool Travel Speed

Tool travel speed is another crucial parameter that dictates the success of FSP in modifying the microstructure. Initially, rotating FSP tool is plunged into the workpiece and allowed to raise the temperature in the stir zone due to friction to facilitate a sufficient rate of material flow around the tool pin. The tool is then traveled across the designed path by providing an appropriate feed rate. During the initial dwell time, a certain amount of heat is generated in the stir zone which further distributed through the nugget zone, TMAZ and HAZ. The amount and rate of heat dissipation depend on feed rates. A higher amount of heat is concentrated in the stir zone with lower feed rates and influences the material flow rates. Whereas the given heat input is decreased with lower tool travel speeds. If the heat input is lower, fine grains are evolved during FSP [10].

On the other hand, higher feed rates result in inappropriate material filling in the cavity produced by the tool pin. Therefore, adopting optimized tool travel speed is crucial to develop defect-free stir zones. Figure 4.7 shows the optical microscope images of ZE41 Mg alloy processed at a constant tool rotational speed with four different feed rates. As the feed rate was increased from 16 to 25 and 50 mm/min, the flow rate of the material was observed as improved and a defect-free nugget zone was obtained. However, due to insufficient material flow at 100 mm/min feed rate, similar to the stir zone produced at 16 mm/min, a discontinuity in the material was observed as shown in Figure 4.7 (d). The right combination of tool rotational speeds and travel speeds plays an important role in FSP. Therefore, optimizing these two parameters by preliminary experiments or by following statistical analysis with a set of data of preliminary experiments is a common practice in finalizing the tool rotational and travel speeds for a given material system with a specific FSP tool.

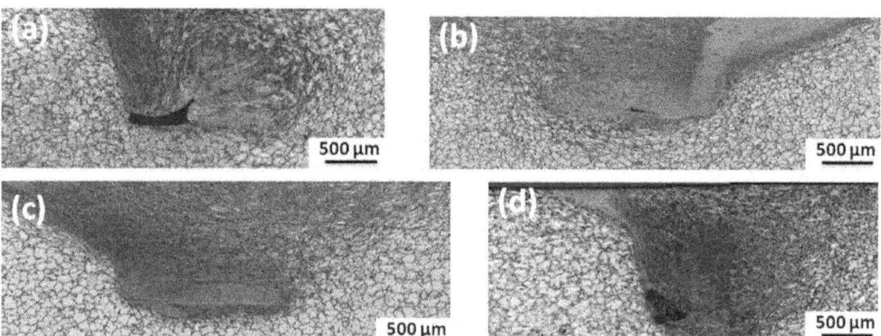

FIGURE 4.7 Optical microscope images showing the effect of tool travel speed on material flow in the stir zone of ZE41 Mg alloy processed at 1100 rpm tool rotational speed: (a) 16 mm/min, (b) 25 mm/min, (c) 50 mm/min and (d) 100 mm/min.

FIGURE 4.8 Photographs of different FSPed regions on pure Mg sheets: (a) poor penetration of FSP tool, (b) sufficient penetration and (c) excessive penetration of FSP tool.

4.2.3 Tool Penetration

During FSP, in the initial step, the rotating FSP tool is sufficiently penetrated into the workpiece and allowed to stir up to certain dwell time before the tool is plunged along the traverse direction. In order to generate a sufficient amount of heat, establishing appropriate contact between the tool shoulder and the workpiece surface is crucial. Otherwise, the rotating pin simply produces a groove instead of stirring the workpiece material around it. Therefore, appropriate penetration of the FSP tools into the workpiece play an important role in modifying the surface without defects. If the tool penetrates excessively into the workpiece, the material in the form of flash is resulted. Figure 4.8 shows both the effects resulting from poor and excessive penetration of the FSP tool. Along with optimizing the tool rotational and traverse speeds, adopting sufficient tool penetration is important in FSP.

4.2.4 Tilt Angle

The tool tilt angle is the angle measured between the axis of the FSP tool and the axis perpendicular to the surface of the workpiece as schematically shown in Figure 4.9. Tool tilt is crucial particularly when the shoulder is concave type in achieving appropriate material flow beneath the shoulder [11]. Tool tilt facilitates the stirring material to be confined under the shoulder as the tool travels during FSP [12, 13]. However, by using other kinds of shoulder designs, without tool tilt FSP can be successfully carried out. Usually, the tool tilt angle is maintained from 0 to 3°.

4.3 MATERIAL TYPE

Usually, metals respond to external forces differently based on their crystal structure. Due to the presence of more number of close-packed slip systems, materials of FCC crystal structure exhibit a higher level of plastic deformation compared with the metals of other crystal structure. In FSP, the material flow depends on the level of plastic deformation that a base material can undergo. This is the reason why optimizing the process parameters is required for every individual metal or alloy. By using the same tool design and process parameters, FSP may not give defect-free processed regions for two different metals or alloys. Even within the same kind of alloy systems, for example, within the aluminum alloys or magnesium alloys; optimizing the process parameters is required for every individual alloy. If the material is ductile in nature, the heat generation and conduction mechanisms are different compared with the materials of brittle type. Similarly, the tool design also must be modified based on the material type. Ductile materials can be easily processed by using FSP tools in which, the ratio of the shoulder diameter to the pin diameter is higher. But this ratio must be in lower end when FSP is carried out on brittle materials. Otherwise, inappropriate material flow can be resulted and further results the similar kind of surface defect in the processed zone as shown in Figure 4.8.

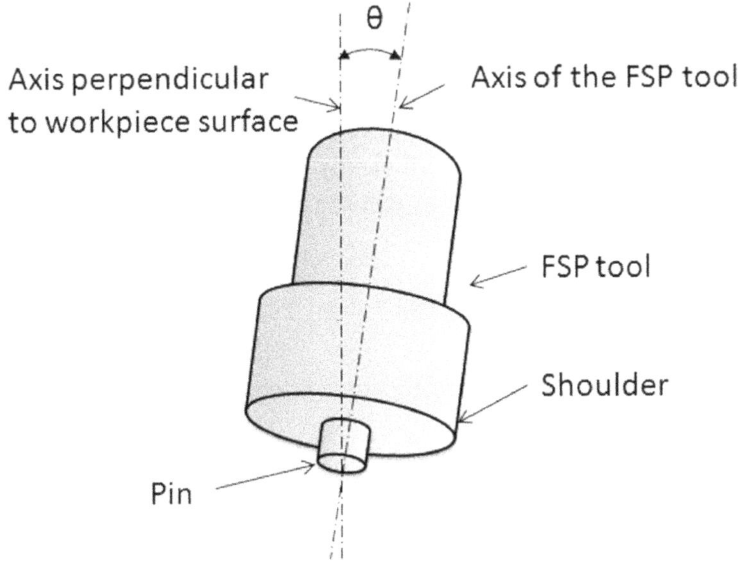

FIGURE 4.9 Schematic representation of tool tilt.

4.4 MATERIAL PREHEATING

Preheating of workpiece material reduces the load required to initiate the plastic deformation during FSP. Particularly, while processing materials of high melting temperature such as steels and titanium alloys, preheating the workpiece offers several advantages. Furthermore, materials of higher heat conductivity such as copper-based materials also give better processing results by preheating. Heating decreases the yield strength of a material and therefore at lower loads material can be easily deformed to modify the microstructure. In addition, the preheating of workpiece increases the tool travel speed and decreases the tool wear [14]. Usually, tool wear is another crucial factor needed to be considered while processing high strength alloys. Tool wear is higher in the initial stage of the process as the workpiece is at the lower temperature. As the rotating tool is penetrated into the workpiece, gradually heat is generated, and the material is stirred around the pin [15, 16]. By providing preheating to the workpiece, tool wear in the initial stage of the process can be reduced.

Even though there are several available reports on the influence of process parameters on FSP, every time selecting appropriate process parameters for a given material is really a difficult task as the same process parameters may not yield the same results as the variations in the material composition and processing equipment working efficiency. Hence, understanding the role of each process parameter is important before designing FSP experiments for a given material system. Still, there are several alloys, processing of them has not yet sufficiently understood. In-depth studies in the future may open valuable insights to make FSP as a viable tool in surface engineering.

KEYWORDS

- **material preheating**
- **material type**
- **processing parameters**
- **tool tilt**

REFERENCES

1. Saikrishna, N., Pradeep, K. R. G., Balakrishnan, M., & Ratna, S. B., (2016). Influence of bimodal grain size distribution on the corrosion behavior of friction stir processed biodegradable AZ31 magnesium alloy. *Journal of Magnesium and Alloys, 4*, 68–76.
2. Ratna, S. B., Sampath, K. T. S., Uday, C., Nandakumar, V., & Mukesh, D., (2014). Nano-hydroxyapatite reinforced AZ31 magnesium alloy by friction stir processing: A solid-state processing for biodegradable metal matrix composites. *J. Mater Sci. Mater. Med., 25*,975–988.
3. Kondaiah, V. V., Pavanteja, P., Afzal, K. P., Anand, K. S., Ravikumar, D., & Ratna, S. B., (2017). Microstructure, hardness and wear behavior of AZ31 Mg alloy – fly ash composites produced by friction stir processing. *Materials Today: Proceedings, 4*, 6671–6677.
4. Ratna, S. B., (2014). *Fine-Grained Magnesium-Based Materials Processed by Severe Plastic Deformation Techniques for Degradable Implant Applications*. Doctoral thesis, IIT Madras, Chennai, India.
5. Ratna, S. B., Sampath, K. T. S., Uday, C., Nandakumar, V., & Mukesh, D., (2014). Friction stir processing of magnesium–nanohydroxyapatite composites with controlled in vitro degradation behavior. *Materials Science and Engineering C., 39*, 315–324.
6. Afrin, N., Chen, D. L., Cao, X., & Jahazi, M., (2008). Microstructure and tensile properties of friction stir welded AZ31B magnesium alloy. *Mater. Sci. and Engg. A., 472*, 179–86.

7. Cartigueyen, S., & Mahadevan, K., (2015). Influence of rotational speed on the formation of friction stir processed zone in pure copper at low-heat input conditions. *Journal ofManufacturing Processes, 18,* 124–130.
8. Leal, R. M., Galvão, L. A., & Rodrigues, D. M., (2015). Effect of friction stir processing parameters on the microstructural and electrical properties of copper.*Int. J. Adv. Manuf.Technol., 80*(9–12), 1655–1663.
9. Zhang, H. J., Liu, H. J., & Yu, L., (2011). Microstructure and mechanical properties as a function of rotation speed in underwater friction stir welded aluminum alloy joints. *Material Design, 32,* pp. 4402–4407.
10. Vivek, P., Vishvesh, B., & Abhishek, K., (2016). Influence of friction stir processed parameters on superplasticity of Al-Zn-Mg-Cu alloy. *Materials and Manufacturing Processes, 31*(12), 1573–1582.
11. Fuller, C. B., (2007). Friction stir tooling: Tool materials and design, Ch. 2. In: Mishra, R. S., & Mahoney, M. W., (eds.), *Friction Stir Welding and Processing* (pp. 7–35). ASM International, Materials Park, OH.
12. Sutton, M. A., Reynolds, A. P., Yang, B., & Taylor, R., (2003). Mode I fracture and microstructure for 2024-T3 friction stir welds.' *Mater. Sci. Eng. A., A354*(6–16), 22.
13. Lumsden, J., Pollock, G., & Mahoney, M., (2005). 'Effect of tool design on stress corrosion resistance of FSW AA7050-T7451.' In: '*Friction Stir Welding and Processing III*' (pp. 19–25). San Francisco, CA, TMS.
14. Posada, M., Deloach, J., Reynolds, A. P., & Halpin, J. P., (2003). In: David, S. A., DebRoy, T., Lippold, J. C., Smartt, H. B., & Vitek, J. M., (eds.), *Proceedings of the Sixth International Conference on Trends in Welding Research* (pp. 307–312). Pine Mountain, GA, ASM International.
15. Lienert, T. J., & Gould, J. E., (1999). In: *Proceedings of the First International Symposium on Friction Stir Welding*. Thousand Oaks, CA, USA.
16. Lienert, T. J., Stellwag, Jr. W. L., Grimmett, B. B., Warke, R. M., Friction Stir Welding Studies on Mild Steel, Weld, J., (2003). 82(1), 1s.

CHAPTER 5

Material Systems Processed by FSP

Wide varieties of materials have been processed by FSP to develop surface modified structures. Aluminum, magnesium, titanium, copper, and steels are the most important group of materials processed by friction stir processing. This chapter presents the work done in FSP of different materials.

5.1 ALUMINUM

Aluminum alloys are the most widely used material group processed by FSP. Based on the alloying elements, aluminum alloys are categorized as 1xxx, 2xxx, 3xxx, 4xxx, 5xxx, 6xxx, 7xxx, and 8xxx. Deferent alloys were developed within the same group of aluminum alloys based on the follow-up heat treatment processes. For the first time, Mishra et al., [1] demonstrated developing grain refined Al7075 alloy by FSP. Increased high strain rate superplasticity in FSPed Al7075 was reported processed at 1×10^{-2} S^{-1} and 490°C. Later on, several aluminum alloys were processed by FSP.

Patel et al., [2] processed Al7075 by FSP and the effect of process parameters on the formation of defects, material flow and microstructure was investigated. At the low heat input due to optimizing the process parameters, lower grain growth was observed. Formation of the cavitations was observed at the interface of the thermomechanical affected zone (TMAZ) and stir zone due to inadequate material flow. Similarly, Navaser and Atapour [3] also processed AA7075 alloy by FSP and grain size was observed as increased with decreased intermetallic particles with increased tool rotational speeds. Further, the homogeneous distribution of intermetallic particles was observed with increased tool rotational speed. From the corrosion studies, increased pitting corrosion was noticed with increased tool rotational speed. Intergranular corrosion was not observed at the heat affected zone and TMAZ.

Moghaddam et al., [4] done a comparative study to investigate the microstructure evolution during the early stage of the severe plastic deformation by conducting multi-axial forging (MAF), accumulative back extrusion (ABE), and friction stir processing (FSP) under "predetermined thermomechanical condition based on the Zener Hollomon value." After grain refinement, increased yield strength and ultimate tensile strength was noticed for all the processed samples. Interestingly, FSP developed more isotropic properties compared with the samples processed with MAF and ABE routes. FSP lead to fade out the initial rolling texture at a higher level compared with the other two methods. Wenjing Yang et al. [5] demonstrated double-sided multi-pass FSP to eliminate HAZ and TMAZ. Figure 5.1 shows a schematic representation of double-sided multi-pass FSP in which, FSP is carried on two sides with an overlap of 30–35%. By adopting this variation in processing, equiaxed fine grains of around 2.5 μm were successfully produced by eliminating HAZ and TMAZ.

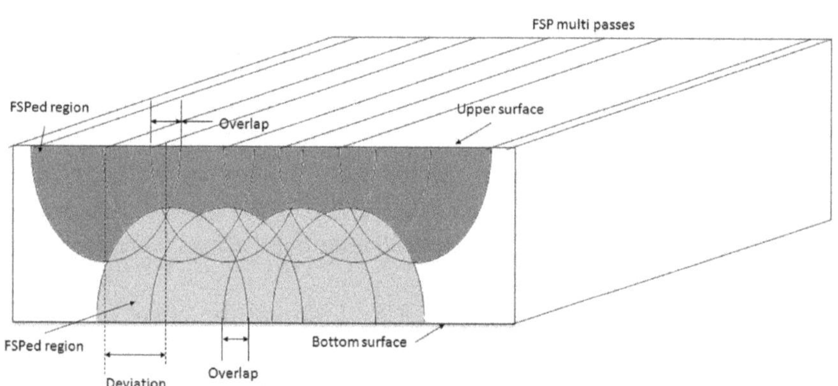

FIGURE 5.1 Schematic representation of double-sided multi passes FSP.

Al5083 is another group of Al alloys processed by FSP. Magnesium is the main constituting element in 5083 alloy and widely used in aerospace, marine and automobile industries due to high strength to weight ratio and excellent corrosion resistance. Al5083 is a solution-strengthening alloy and therefore, requires work hardening to alter the mechanical properties. Sadegh Rasouli et al. [6] developed fine-grained 5083 alloys and corrosion performance was investigated. After FSP, recrystallized fine and equiaxed grains were successfully developed. The size distribution of the new grains was observed as uniform compared with the starting condition. Due to the modified microstructure, better corrosion behavior was observed for the FSPed Al5083. Chen et al., [7] produced fine-grained Al5083 alloy by FSP and studied the role of number of FSP passes on microstructure evolution and mechanical properties. Additionally, abnormal grain growth was also studied during annealing. After a single FSP pass, fine grains were produced. The new grains were observed as equiaxed, and the grain boundaries were categorized as high angle grain boundaries. Abnormal grain growth was observed in FSPed alloy after annealing. However, the stir zone produced at high rotational speeds has shown more resistance to the grain growth during the annealing. Interestingly, as the number of FSP passes were increased to 2 and 3, the level of grain refinement was not significant as shown in Figure 5.2. It was also a significant finding that the grain growth was also inhibited in the FSPed samples as the number of passes was increased.

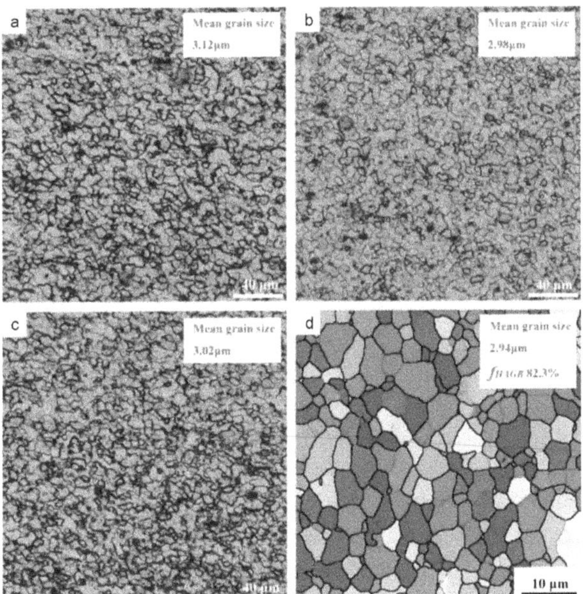

FIGURE 5.2 (See color insert.) Optical microscope images of FSPed Al5083 alloy obtained at stir zone at 1200 tool rotational speed and 360 mm/min tool travel speed: (a) after 1st pass, (b) after 2nd pass (c) after 3rd pass and (d) EBSD map of sample after 3 FSP passes. (*Source:* Reprinted with permission from Chen et al., [7]. © 2016 Elsevier.)

Reza Abdi Behnagh et al.[8] also produced ultra fine grain structure at the surface of Al5083 and increased hardness up to 50% was achieved after FSP. Similar to the other observations, increased tool rotational speed resulted in grain growth and decreased tool rotational speeds resulted in the fine and homogeneous microstructure. Miori et al., [9], processed Al5083 alloy and developed fine-grained structure. From the experiments and simulation carried out using Abaqus software clearly demonstrated increased formability after FSP. However, the FSPed sheets containing defects have failed before the targeted pressure.

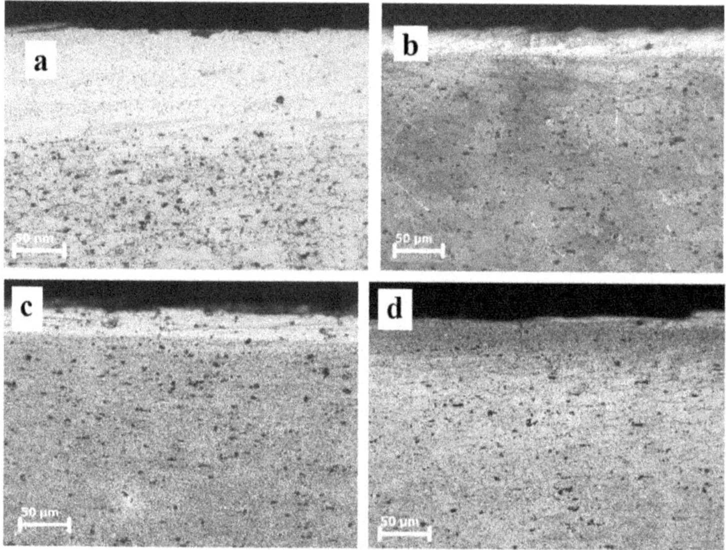

FIGURE 5.3 Optical microscope images of 1050 alloy processed by FSP: (a) 500 rpm with 15 mm/min feed, (b) 500 rpm with 30 mm/min feed (c) 20 mm/min feed with 500 rpm and (d) 20 mm/min with 1000 rpm. (*Source:* Reprinted with permission from Kurt et al., [11]. © 2011 Elsevier.)

Sachin Kumar [10] introduced using ultrasonic vibrations during FSP of Al6063 alloy. Experiments were conducted on Al6063 by applying ultrasonic and without ultrasonic vibrations at different sets of process parameters. Mechanical properties, surface roughness, temperature profiles, and forces resulted during FSP were studied. Mechanical properties, surface properties, surface properties were observed as better for the ultrasonic-assisted FSP samples compared with FSP. Kurt et al., [11] processed 1050 alloy by FSP and due to the microstructure modification, improved mechanical properties particularly, increased formability was observed.

Further, the effect of process parameters on the depth of the surface layer was investigated. It was observed that as the tool travel speed is increased from 15 to 30 mm/min, the depth of the affected surface was observed as decreased s shown in Figure 5.3. Several other aluminum alloys were also processed by FSP, and most of the studies summarized the influence of the modified microstructure on mechanical, surface and corrosion properties.

5.2 MAGNESIUM

Magnesium is the other group of nonferrous metals which has been widely investigated by processing through FSP. Magnesium alloys are categorized as difficult to process materials. However, using FSP, it has been clearly demonstrated that several Mg alloys can be successfully processed by FSP to develop grain refined surfaces. Among all Mg alloys, AZ series (Aluminum and Zinc) is the most common group of alloys widely processed by FSP. A few studies also demonstrated the promising grain size effect to enhance mechanical, surface, corrosion and bio-properties of pure Mg processed by FSP [12–15].

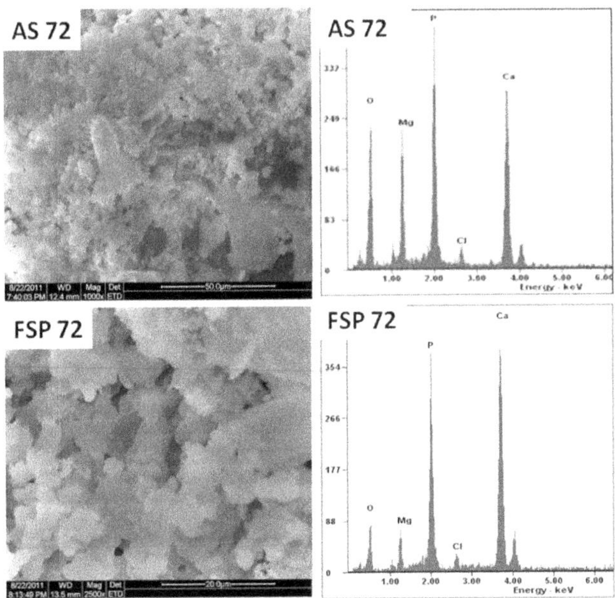

FIGURE 5.4 SEM images and corresponding EDS analysis indicates higher biomineralization on FSPed Mg compared with unprocessed Mg (as Mg) after 72 h of immersion.

In our previous study, grain refinement was achieved from 1500 μm to 6.2 μm after FSP. From the contact angle measurements, the hydrophilicity of grain refined Mg surface was observed. Due to increased surface energy and smaller grain size, a higher level of biomineralization was achieved on the FSPed Mg compared with unprocessed Mg after immersed in simulated body fluids [12]. Further, the amount of minerals which were deposited on the FSPed Mg were identified as hydroxyapatite and magnesium phosphate from the XRD studies which is an indication of enhanced bioactivity [13]. Figure 5.4 shows the SEM images of biomineralization on FSPed Mg compared with unprocessed Mg. This enhanced biomineralization achieved by grain refinement through FSP can control the degradation and promotes a higher level of healing rate in biomedical applications.

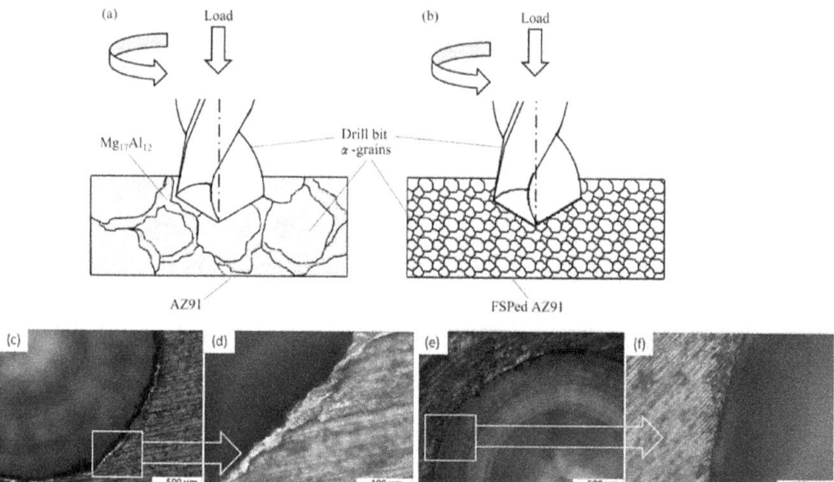

FIGURE 5.5 Machining observations of FSPEd AZ91 Mg alloy: (a) schematic representation of drilling coarse AZ91 Mg alloy, (b) schematic representation of FSPed AZ91 Mg alloy, (c) and (d) edge damage after drilling of coarse-grained AZ91 Mg alloy, (e) and (f) edge damage of FSped AZ91 Mg alloy. (*Source:* Reprinted with permission from Surya Kiran et al., [16]. © 2017 Elsevier.

Ahmadkhaniha et al., [15] processed as cast pure Mg by FSP to investigate the effect of grain refinement on mechanical and biocorrosion properties. With successive passes the level of grain refinement was observed as increased. From the microhardness measurements, increased hardness was observed up to two times after FSP. After three FSP passes, yield strength and ultimate tensile strength was observed as increased as the number of passes was increased to 3. However, percentage of elongation was observed as increased after the first pass but further decreased after two and three passes. From the corrosion studies, better degradation performance was observed for the FSPed samples.

In AZ series Mg alloys, AZ31 Mg alloy is another common alloy used to develop surface modified structures by FSP. AZ31 Mg alloy is a wrought product commercially available in the form of extruded rods and rolled sheets. Arab et al., [17] produced fine-grained AZ31 Mg alloy by a stepped FSP tool, and defect-free FSP region was produced. By adopting a modified tool, the material escape was prevented, and the uniform shear strain was introduced in the Mg alloy. Due to this novel modification in the tool design, a higher level of grain refinement was achieved. Darras et al., [18] developed fine-grained AZ31 Mg alloy by FSP and increased hardness was measured by decreasing the tool rotational speed and increasing the tool travel speed. The reason behind the increased grain refinement level was explained as shorter thermal contact due to higher tool travel speed. Similarly, at higher tool rotating speed, more heat is generated, and grain growth is resulted. This is similar to what reported by Pradeep Kumar Reddy et al., [19] while welding AZ31 similar alloy sheets by friction stir welding (FSW). Welding was performed at 1200, 1400, 1600 and 1800 rpm tool rotational speeds and optimum grain refinement was observed at 1400 rpm. As the tool rotational speed is increased to 1600 and 1800 rpm, grain growth was observed due to the concentration of the heat in the stir zone. It was also observed that the grain refined AZ31 Mg alloy by FSP demonstrated a higher level of biomineralization when immersed in simulated body fluids [20]. Additionally, enhanced corrosion resistance and higher bioactivity were also observed. From the cell culture studies, the authors demonstrated a higher level of cell activities which is an indication for enhanced bioactivity due to grain refinement. The results strongly suggested adopting FSP to develop grain refined Mg alloys for biomedical applications.

Loke et al., [21] processed AZ91 Mg alloy, and grain refinement (<10 μm) was successfully achieved. The intermetallic ($Mg_{17}Al_{12}$) phase present at the grain boundaries was dissolved after FSP in the stir zone. The hardness measurements done at the stir zone indicated increased hardness (84.8 Hv) from a starting size of 69.9 Hv. Further, the ultimate shear strength was also measured as increased from 123. 9 MPa to142.4 MPa. Similarly, Feng and Ma [22] processed cast AZ91 Mg alloy and primarily they observed breakup of the secondary phase and also the dissolution of the $Mg_{17}Al_{12}$ phase network into the grains. A significant grain refinement (15 μm) was observed in FSPed AZ91. The authors then subjected FSPed AZ91 to aging, and increased mechanical properties were observed due to precipitation. In one of the recent studies, the role of FSP on altering the phase amount and distribution was studied and clearly observed that secondary phase was decreased. The formation of supersaturated grains with preferred orientation due to FSP has shown the prominent effect on corrosion behavior. Furthermore, the effect of grain refinement on machining was studied, and the machinability of fine-grained AZ91 Mg alloy was observed as improved [16]. It was a clear demonstration that the grain refinement significantly influences the material removal mechanisms as schematically represented in Figure 5.5 (a) and (b). When the grain size was large and containing hard, and brittle phases, the formation of chips and the resulting cutting forces were observed as completely different compared with the formation of chips and cutting forces resulted in FSPed AZ91 Mg alloy. Due to improved machinability after grain refinement as also observed in fine-grained titanium reported by Lapovok et al., [23], the edge damage was also observed as decreased after grain refinement.

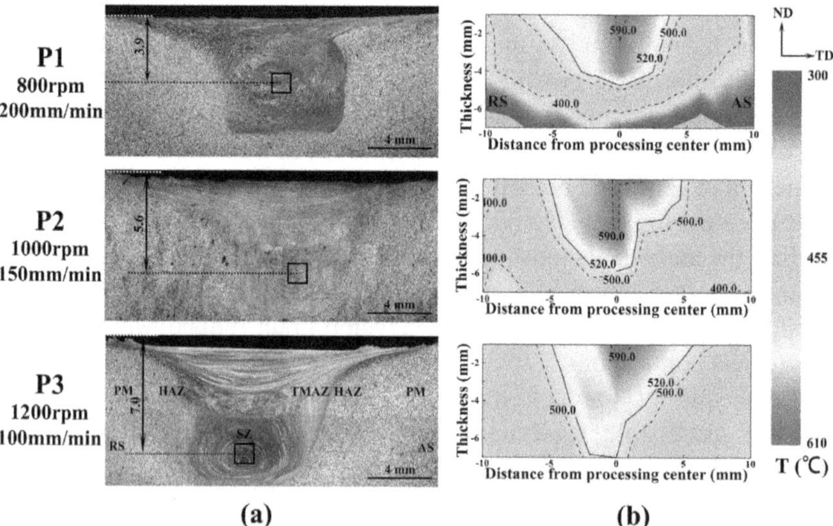

FIGURE 5.6 (See color insert.) (a) Cross-sectional macrostructures of the FSPed Mg alloy, and (b) thermographs showing the temperature distribution. (*Source*: Reprinted with permission from Han et al., [24]. © 2016 Elsevier.)

Mg alloys with rare earth (RE) as important constituting alloying elements are another significant Mg alloys exhibit promising properties and hence, attracted as suitable candidates for several automobile and aerospace applications. In Mg-RE alloys, several alloying elements such as Gd, Ce, Nd, Y are added in specific fractions. Han et al., [24] FSPed as-cast Mg–Nd–Zn–Zr alloy at different process parameters and with the help of thermocouples, the temperature distribution was measured during FSP. Figure 5.6 shows the cross section macrostructure of FSPed Mg alloy and corresponding thermographs showing the distribution of the temperature. Peak temperatures were measured as around 600°C in the nugget zone in FSPed alloy. Xin et al., [25] processed Mg-Gd-Y-Nd-Zr alloy by FSP and achieved a grain refinement up to 3–4 μm. The role of REs on microstructure evolution was observed as significant. The texture studies also revealed the formation of a week texture in the stir zone as also reported by several other authors [26–28].

Cao et al., [29] studied the precipitation behavior of Mg-Y-Nd alloy while carrying FSP. Later, FSPed alloy was aged to alter the microstructure and to investigate the mechanical properties. Grain refinement was achieved up to 2.7 μm. Due to the temperature which was generated during FSP resulted in secondary phases in the process of cooling. Improved ultimate tensile strength (303 MPa), yield strength (290 MPa), and % elongation (11) were measured after FSP compared with the as-cast condition. Sabbaghian and Mahmudi [30] studied the influence of microstructure modification of Mg-3Gd-1Zn by FSP on mechanical properties. As reported by several other authors, the coarse dendrite structure usually appears in cast structure was disappeared after FSP with fine grain sizes of 4 μm and 2.5 μm after one and two passes of FSP. Hardness and shear strength of the FSPed alloy was measured as increased compared with the starting condition which was attributed to the smaller grain size. Cao Genghua and Zhang Datong [31] processed Mg-Y-Nd alloy by FSP, and significant grain refinement was achieved. The intermetallic $Mg_{12}Nd$ was observed as a network like structure at the grain boundaries which was broken as discontinuous particles after FSP. After FSP, due to microstructure modification, the tensile properties were remarkably increased to 305 MPa with 22% of elongation. The research work carried out in processing different Mg alloys by FSP demonstrated the promising role of FSP in enhancing mechanical, corrosion, degradation and bioproperties targeted for several applications.

5.3 COPPER

Copper is the other group of materials processed with FSP. Xue et al., [32] processed pure Cu T3 (99.9%) and high purity oxygen free Cu TU1 (99.99%). In order to achieve better heat monitoring, cooling effect was added by providing flowing water. Ultrafine-grained (UFG) pure Cu of equiaxed grains with weak texture was achieved with good strength and ductility after FSP. The grain size was also observed as uniform. Additionally, isotropy and tension-compression symmetry were achieved in processed Cu. The same research group also demonstrated developing UFG, and nanostructured Cu-Al alloy by FSP [33] and better mechanical properties were observed compared with unprocessed Cu alloy.

5.4 TITANIUM

Titanium is another nonferrous metal can be found in many industrial applications. Titanium and its alloys are also belonging to difficult to process materials group. Fattah-al Hosseini et al., [34], processed pure titanium of grade-2 by FSP and microstructural observations were done and studied the influence of modified microstructure on the corrosion behavior of Ti in borate buffer solution. Similar to the other metals, with an increase in the number of FSP passes, grain refinement was observed. From the electrochemical studies by conducting potentiodynamic polarization tests, corrosion and passive current densities were observed as decreased due to grain refinement. From their work, it was demonstrated that with multi-pass FSP corrosion resistance of pure Ti can be improved by forming more resistive, thicker, and yet less defective passive films.

Li et al., [35] processed Ti-6Al-4V in the presence of nitrogen and surface nitriding was successfully achieved. Starting material was Ti-6Al-4V alloy containing duplex phase. The produced nitriding surface layer was observed as strongly bonded to the substrate with a thickness of 300 μm. Dendrite formed grains were not observed in the surface layer after FSP due to the nature of the process. FSP is a solid-state method which completely eliminated the formation of the dendrites. Additionally, the higher rate of chemical reaction and the lower amount of heat availability at the processed zone is limited, and the rapid dissipation of heat during FSP also helped to eliminate the formation of the dendrites.

The formation of TiN phase with smaller grains (2–3 μm) was observed on the top of the surface layer up to 150 μm depth. After removing the crowns in the top surface layer, a composite like structure was observed as shown in Figure 5.7. The microhardness of the top coated surface layer was observed as 1105 HV compared with the nitriding layer 982 HV. The mechanism behind the formation of the nitriding layer was explained by the authors as the formation of a dramatic thermal chemical reaction between the nitrogen and the substrate.

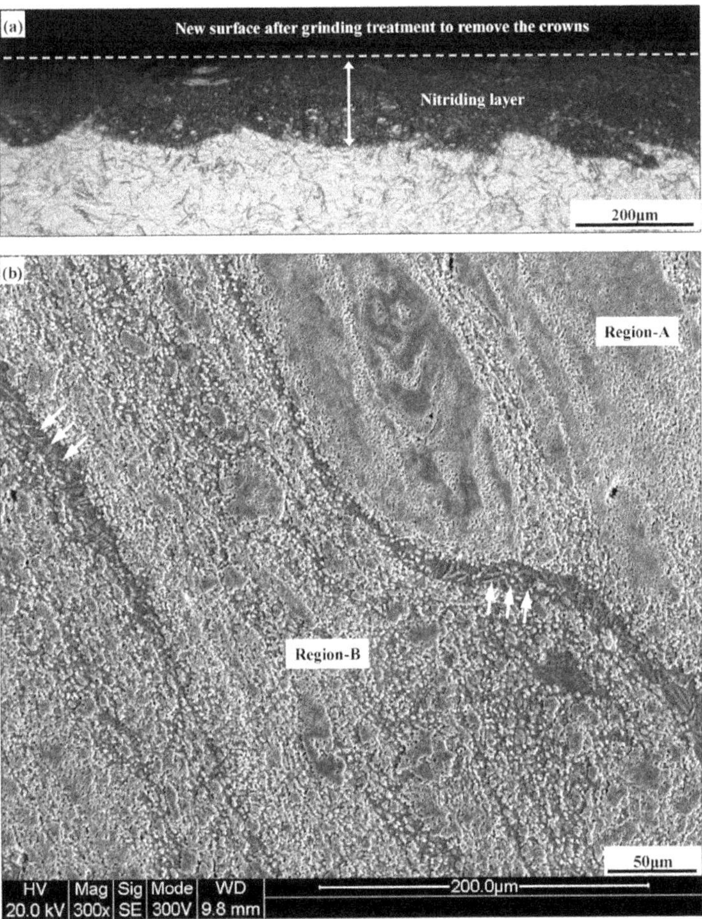

FIGURE 5.7 (a) Optical microscope image showing the cross-section of the surface nitriding layer and (b) SEM image showing the composite like structure within the nitriding layer. (*Source:* Reprinted with permission from Li et al., [35]. © 2013 Elsevier.)

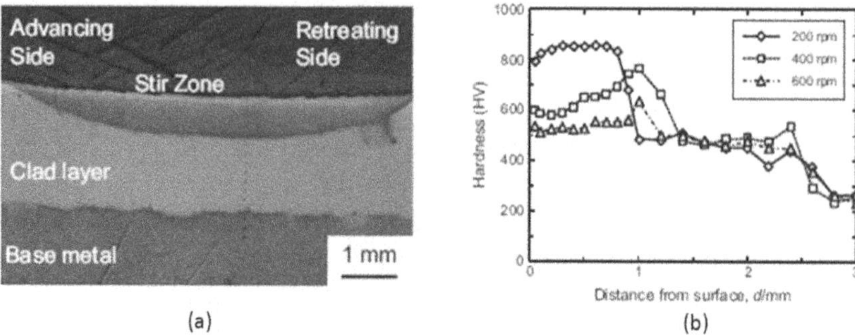

FIGURE 5.8 (a) Cross section microscope image showing FSPed region and the cladding layer, and (b) hardness profile of FSPed surface down at different tool rotational speeds. (*Source:* Reprinted with permission from Nagaoka et al., [40]. © 2015 Elsevier.)

5.5 IRON

Iron is the most widely used material system to alter the properties by using FSP. Several kinds of steels have been used as base materials, and structures with improved material properties were produced through FSP route. Razmpoosh et al., [36] studied the evolution of microstructure in twinning-induced plasticity steel during FSP by conducting the experiments at different speeds of 800, 1600, and 2000 rpm at a constant feed of 50 mm/min. 1600 rpm was identified as optimum tool rotational speed to develop ultrafine grains. Ferrite to austenite transformation was accelerated due to FSP and resulted to decrease in the amount of ferrite after FSP. Similar to other metallic systems, grain refinement lead to increase the hardness in the nugget zone.

Xue et al. [37] have done FSP on mild steel, and ultrafine ferrite/martensite dual phase steel was developed. The refined ferrite grains were distributed around the martensite phase in the size of 200 nm. Yasavol and Jafari [38] processed AISI D2 tool steel and microstructure, mechanical and corrosion properties were investigated. After FSP, homogeneous distributed fine carbides were noticed throughout the matrix that is composed of ferrite and martensite. Due to the more number of low angle grain boundaries resulted from FSP, increased hardness and corrosion resistance was measured. Tarasov et al., [39] observed coarse grains in the stir zone and HAZ after FSP in high carbon steel. Additionally, globular pearlite grains similar to base material were observed in the HAZ. Due to the presence of cementite and bainite in the stir zone, increased hardness was observed after FSP.

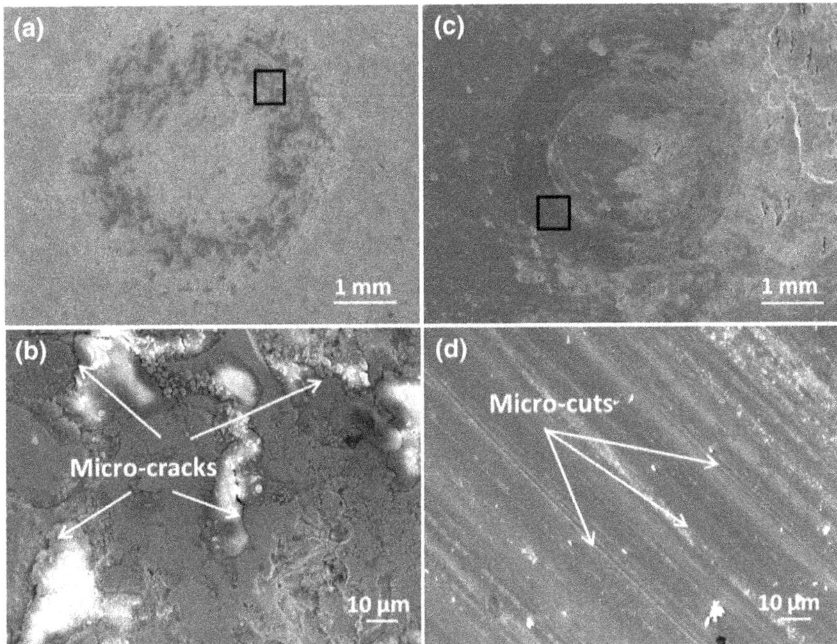

FIGURE 5.9 SEM images showing surface wear tracks on the samples: (a) as-sprayed, (b) as-sprayed higher magnification, (c) FSPed sample, and (d) FSPed sample higher magnification. (*Source:* Reprinted with permission from Rahbar-Kelishami [41]. © 2015 Elsevier.)

Nagaoka et al., [40] done FSP on D2 tool steel layer developed by cladding. Figure 5.8 shows FSPed region on the cladding layer and the hardness profile in the thickness direction. Carbide particles were formed in the interdendritic regions after cladding. The distribution of M_7C_3 carbide particles was observed as homogenously distributed after FSP. The cladding layer was subjected to austenitizing heat treatment, and the microstructure and hardness were compared with FSPed surface. FSPed surface exhibited higher hardness compared with the homogenized cladding layer due to the decreased fraction of austenite after FSP. Due to the low heat input, a high volume fraction of M_7C_3 and martensite was observed. Similarly, Rahbar-Kelishami et al., [41] developed a surface layer of WC-12%Co on 52100 steel and subjected to FSP. Wear properties before and after FSP were investigated. Due to FSP, the microstructure of the surface has been transformed into martensite with retained austenite from the starting microstructure containing ferrite and pearlite. This significant change in the microstructure increased the hardness and toughness after FSP compared with the starting condition. As shown in Figure 5.9, increased wear resistance was noticed for the FSPed surface. Uniform wear tracks can be seen on FSPed surface compared with a coated surface. The depth of the wear tracks was observed as lower on the FSPed surface. Breaking down of WC particle after FSP formed a hard surface layer and contributed to increase the wear resistance up to 1.5 times compared with the as coated surface. It was also observed that the formation of a new secondary phase and carbides due to FSP also contributed to increase the wear resistance. Langlade et al., [42] also demonstrated increased tribological behavior for FSPed AISI 1050 steel. Tinubu et al., [43] also developed a high wear-resisting surface on A-286 stainless steel by FSP. The increased wear properties were attributed to increased hardness due to the microstructure modification after FSP. Other reports [44, 45] also demonstrate the promising role of FSP in enhancing mechanical properties due to microstructure modification by FSP. Compared with other processing routes, developing surfaces by FSP is in its infant stage. For the past decade, several interesting and novel process variations were tried and demonstrated by several authors. Significant information in connection with FSP is available in a few material systems. But, Future research contributions to provide information related to several other commercial alloys which are not yet processed by FSP may explore more possibilities to adopt FSP in the manufacturing industry.

KEYWORDS

- **aluminum**
- **copper**
- **friction stir processing**
- **iron**
- **magnesium**
- **titanium**

REFERENCES

1. Mishra, R. S., Mahoney, M. W., McFadden, S. X., Mara, N. A., & Mukherjee, A. K., (2000). High strain rate superplasticity in a friction stir processed 7075 al alloy. *Scripta Mater., 42,* 163–168.
2. Patel, V., Badheka, V., & Kumar, A., (2015). Influence of friction stir processed parameters on superplasticity of Al-Zn-Mg-Cu alloy. *Mater. Manuf. Process., 31*(12), 1573–1582
3. Navaser, M., & Atapour, M., (2016). Effect of friction stir processing on pitting corrosion and intergranular attack of 7075 aluminum alloy. *Journal of Materials Science & Technology*. http://dx.doi.org/doi: 10.1016/j.jmst.2016.07.008.
4. Moghaddam, M., Zarei-Hanzaki, A., Pishbin, M. H., Shafieizad, A. H., & Oliveira, V. B., (2016). Characterization of the microstructure, texture and mechanical properties of 7075 aluminum alloy in the early stage of severe plastic deformation. *Materials Characterization, 119,* 137–147.
5. Wenjing, Y., Jizhong, L., Xue, W., & Hua, D., (2015). Improvement of microstructure and mechanical properties of 7050-T7451 aluminum by a novel double-sided friction stir processing. *Materials Science Forum, 838–839*, 385–391.
6. Sadegh, R., Reza, B. A., Abdolrahman, D., & Noureyeh, S. H., (2014). Improvement in corrosion resistance of 5083 aluminum alloy via friction stir processing. *Proceedings of the Institution of Mechanical Engineers, Part L: Journal of Materials Design and Applications*. doi: 10.1177/1464420714552539.
7. Yu, C., Hua, D., Jizhong, L., Zhihui, C., Jingwei, Z., & Wenjing, Y., (2016). Influence of multi-pass friction stir processing on the microstructure and mechanicalpropertiesofAl-5083alloy. *Materials Science and Engineering A., 650,* 281–289.
8. Reza, A. B., Ninggang, S., Masoud, A., Hongtao, D., Reza, A. B., et al., (2016). Ultrafine-grained surface layer formation of aluminum alloy 5083 by friction stir processing. *Procedia CIRP, 45*, 243–246.
9. Miori, G. F., Bordinassi, E. C., Delijaicov, S., & Batalha, G. F., (2015). The sheet metal formability of AA-5083-O sheets processed by friction stir processing. *Advances in Materials Science and Engineering*, 1–21. http://dx.doi.org/10.1155/2015/716165.
10. Sachin, K., (2016). Ultrasonic assisted friction stir processing of 6063 aluminum alloy. *Archives of Civil and Mechanical Engineering, 16,* 473–484.
11. Adem, K., Ilyas, U., & Eren, C., (2011). Surface modification of aluminum by friction stir processing. *Journal of Materials Processing Technology, 211,* 313–317.
12. Ratna, S. B., Sampath, K. T. S., & Uday, C., (2012). 'Bioactive magnesium by friction stir processing.' *Mater. Sci. Forum, 710*, 264–269.
13. Ratna, S. B., Sampath, K. T. S., & Uday, C., Nandakumar, V., & Mukesh, D., (2014). 'Friction stir processing of magnesium – nanohydroxyapatite composites with controlled in vitro degradation behavior.' *Mater. Sci. Eng. C., 39*, 315–324.
14. Ma, C., Chen, L., Xu, J., Fehrenbacher, A., Li, Y., Pfefferkorn, F. E., et al., (2013). Effect of fabrication and processing technology on the biodegradability of magnesium nanocomposites. *J. Biomed. Mater. Res.*, 2–9. http://dx.doi.org/10.1002/jbm. b.32891.

15. Ahmadkhaniha, D., Jarvenpaa, A., Jaskari, M., & Heydarzadeh, S. M., (2016). Microstructural modification of pure Mg for improving mechanical and biocorrosion properties. *Journal of the Mechanical Behavior of Biomedical Materials, 61,* 360–370.
16. Surya, K. G. V. V., Hari, K. K., Sameer, B. S. K., Santosh, K. M. B., Mohana, R. G., Naidubabu, Y., Ravikumar, D., & Ratna, S. B., (2017). Machining characteristics of fine-grained AZ91 Mg alloy processed by friction stir processing. *Trans. Nonferrous Met.Soc. China, 27,* 804–811.
17. Seyed, M. A., Seyed, A., Jenabali, J., & Seyed, M. Z., (2016). The effect of friction stir processing by stepped tools on the microstructure, mechanical properties and wear behavior of an Mg-Al-Zn alloy. *Journal of Materials Engineering and Performance.* doi: 10.1007/s11665–016–2291–1.
18. Darras, B. M., Khraisheh, M. K., Abu-Farha, F. K., & Omar, M. A., (2007). Friction stir processing of commercial AZ31 magnesium alloy. *J. Mater. Process. Technol., 191*, pp. 77–81.
19. Pradeep, K. R. G., Ratna, S. B., & Balakrishna, B., (2017). Joining of AZ31 Mg alloy sheets by friction stir welding and investigating corrosion initiated failure. *Materials Today: Proceedings, 4,* 6712–6717.
20. Ratna, S. B., Sampath, K. T. S., Chakkingal, U., & Mukesh, D. N. V., (2014). Nano-hydroxyapatite reinforced AZ31 magnesium alloy by friction stir processing: A solid-state processing for biodegradable metal matrix composites. *J. Mater. Sci. Mater. Med., 25*, 975–988.
21. Wai, H. L., Raafat, I., & Sri, L., (2015). Improving the microstructure and mechanical properties of a cast Mg- 9Al-1Zn alloy using friction stir processing. *Materials Science Forum,838–839*, pp. 214–219.
22. Feng, A. H., & Ma, Z. Y., (2007). Enhanced mechanical properties of Mg-Al–Zn cast alloy via friction stir processing. *Scripta Materialia, 56,* 397–400.
23. Lapovok, R., Molotnikov, A., Levin, Y., Bandaranayake, A., & Estrin, Y., (2012). Machining of coarse-grained and ultrafine-grained titanium. *Journal of Materials Science*, *47*, 4589–4594.
24. Jingyu, H., Juan, C., Liming, P., Feiyan, Z., Wei, R., Yujuan, W., & Wenjiang, D., (2016). Influence of processing parameters on the thermal field in Mg–Nd–Zn–Zr alloy during friction stir processing. *Materials and Design, 94,* 186–194.
25. Renlong, X., Xuan, Z., Zhe, L., Dejia, L., Risheng, Q., Zeyao, L., & Qing, L., (2016). Microstructure and texture evolution of a MgeGdeYeNdeZr alloy during friction stir processing. *Journal of Alloys and Compounds, 659,* 51–59.
26. Liu, D. J., Xin, R. L., Zheng, X., Zhou, Z., & Liu, Q., (2013). *Mater. Sci. Eng. A., 561,* 419.
27. Xin, R. L., Liu, D. J., Li, B., Sun, L. Y., Zhou, Z., & Liu, Q., (2013). *Mater. Sci. Eng. A., 565,* 333.
28. Xin, R. L., Li, B., Liao, A. L., Zhou, Z., & Liu, Q., (2012). *Metall. Mater. Trans. A., 43,* 2500.
29. Genghua, C., Datong, Z., Xicai, L., Weiwen, Z., & Wen, Z., (2016). Effect of aging treatment on mechanical properties and fracture behavior of friction stir processed Mg–Y–Nd alloy. *J. Mater. Sci., 51*, 7571–7584.

30. Sabbaghian, M., & Mahmudi, R., (2016). Microstructural evolution and local mechanical properties of friction stir processed Mg-3Gd-1Zn cast alloy. *Journal of Materials Engineering and Performance, 25*, 1856–1863.
31. Cao, G., & Zhang, D., (2015). Microstructure and mechanical properties of submerged friction stir processing Mg-Y-Nd alloy. *Materials Science Forum, 816,* pp. 404–410.
32. Xue, P., Wang, B. B., Chen, F. F., Wang, W. G., Xiao, B. L., & Ma, Z. Y., (2016). Microstructure and mechanical properties of friction stir processed Cu with an ideal ultrafine-grained structure. *Materials Characterization, 121,* 187–194.
33. Xue, P., Xiao, B. L., & Ma, Z. Y., (2014). Microstructure and mechanical properties of friction stir processed ultrafine-grained and nanostructured Cu-Al alloys. *Acta Metall.Sin., 50,* 245.
34. Arash, F. H., Mojtaba, V. A., & Meysam, H. On the passive and electrochemical behavior of severely deformed pure Ti through friction stir processing. *Int. J. Adv. Manuf. Technol.*, doi: 10.1007/s00170–016–9420–8.
35. Bo, L., Yifu, S., & Weiye, H., (2013). Surface nitriding on Ti–6Al–4V alloy via friction stir processing method under a nitrogen atmosphere. *Applied Surface Science, 274,* 356–364.
36. Razmpoosh, M. H., Zarei-Hanzaki, A., Heshmati-Manesh, S., Fatemi-Varzaneh, S. M., & Marandi, A., (2015). The grain structure and phase transformations of TWIP steel during friction stir processing. *Journal of Materials Engineering and Performance*, *JMEPEG 24*, 2826–2835.
37. Xue, P., Xiao, B. L., Wang, W. G., Zhang, Q., Wang, D., Wang, Q. Z., & Ma, Z. Y., (2013). Achieving ultrafine dual-phase structure with superior mechanical property in friction stir processed plain low carbon steel. *Mater. Sci. Eng. A., 575*, pp. 30–34.
38. Noushin, Y., & Hassan, J., (2015). Microstructure, mechanical and corrosion properties of friction stir-processed AISI D2 tool steel. *Journal of Materials Engineering and Performance*. doi: 10.1007/s11665–015–1484–3.
39. Yu, T. S., Melnikov, A. G., & Rubtsov, V. E. (2015) Friction stir processing on high carbon steel U12. *Advanced Materials with Hierarchical Structure for New Technologies and Reliable Structures AIP Conf. Proc.,* 1683, 020229–1–020229–4, doi: 10.1063/1.4932919.
40. Toru, N., Yoshihisa, K., Hiroyuki, W., Masao, F., Yoshiaki, M., & Hidetoshi, F., (2015). Friction stir processing of a D2 tool steel layer fabricated by laser cladding. *Materials & Design, 83,* 224–229.
41. Rahbar-Kelishami, A., Abdullah-Zadeh, A., Hadavi, M. M., Banerji, A., Alpas, A., & Gerlich, A. P., (2015). Effects of friction stir processing on wear properties of WC–12%Co sprayed on 52100 steel. *Materials and Design, 86,* 98–104.
42. Langlade, C., Roman, A., Schlegel, D., Gete, E., & Folea, M., (2015). Formation of a tribologically transformed surface (TTS) on AISI 1045 steel by friction stir processing. *Materials and Manufacturing Processes*, doi: 10.1080/10426914.2015.1090584.
43. Tinubu, O. O., Das, S., Dutt, A., Mogonye, J. E., Ageh, V., Xu, R., Forsdike, J., Mishra, R. S., & Scharf, T. W., (2016). Friction stir processing of A-286 stainless steel: Microstructural evolution during wear, *Wear, 356–357,* 94–100.
44. Xue, P., Li, W. D., Wang, D., Wang, W. G., Xiao, B. L., & Ma, Z. Y., (2016). Enhanced mechanical properties of medium carbon steel casting via friction stir processing and subsequent annealing. *Materials Science & Engineering A., 670,* 153–158.

45. Cui, H. B., Xie, G. M., Luo, Z. A., Ma, J., Wang, G. D., & Misra, R. D. K., (2016). *Microstructural Evolution and Mechanical Properties of the Stir Zone in Friction Stir Processed AISI201 Stainless Steel.* doi: 10.1016/j.matdes.2016.05.106.

CHAPTER 6

Applications, Challenges, and Future Scope

Friction stir processing (FSP) is a new method which does not melt the substrate and therefore can become a promising tool in several applications in the manufacturing industry. However, there are challenges involved in adopting FSP in the manufacturing field. This chapter provides an overview of the potential applications of FSP and challenges involved in developing structures using FSP. Further, the future scope and the possible process variations which can be developed based on the basic FSP principle are also brought out.

6.1 FSP TO INTRODUCE SUPERPLASTICITY

Superplasticity can be defined as the ability of a material which can undergo large amounts of deformation, for example, more than 200% without failure [1–3]. A higher level of formability is possible with superplastic forming by which complex shaped structures can be produced. For the past two decades, superplastic forming has attracted great attention in the metal forming field to produce near net shape structures at low cost. Automobile and aerospace industries are the two important areas where superplastic forming helped to produce low weight high strength superplastic structures at a relatively lower cost. Grain size is one of the crucial factors dictates the level of superplasticity. In addition to grain size, high angle grain boundary, the presence of very fine secondary phase particles at the grain boundaries to arrest the grain growth are the other favoring factors which promote superplasticity. Grain boundary sliding is the important mechanism in superplastic forming.

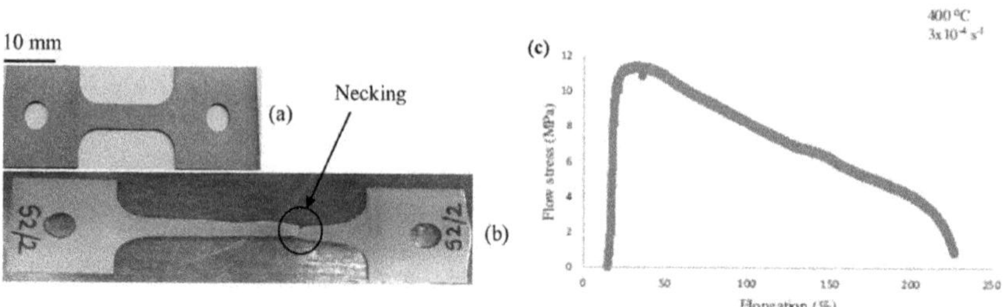

FIGURE 6.1 (a) Photograph of the tensile specimen produced from FSP and (b) Photograph showing elongated specimen after tensile test and (c) stress-strain curve of FSPed samples. (*Source:* Reprinted with permission from Patel et al., [4]. © 2017 Elsevier.)

Severe plastic deformations (SPD) techniques in which a large amount of strains are introduced to develop nano/ultra fine-grained structures are the processing routes widely adopted to develop a structure with superplasticity behavior. It was a clear demonstration from the literature that FSP also can introduce superplasticity by refining the microstructure. It was Mishra et al., [3, 5], for the first time, demonstrated superplastic behavior in FSPed 7075 Al alloy. Later on different material systems such as Al-Zn-Mg-Cu alloy [6–8], 7075 Al alloy [9–11, 27–32], 7075-T651 [12, 13], 5083 Al alloy [14–17], A356 Al alloy [18], 2219 alloy [19], 2024 alloy [20], Al-Mg-Sc alloy [21–26], AZ31 Mg alloy [33], AM60 alloy [34], and AZ91 Mg alloy [2] were successfully processed by FSP and superplastic behavior was clearly demonstrated. A typical photograph and corresponding stress-strain curve of AA7075 alloy subjected to a hot tensile test as shown in Figure 6.1 demonstrate the superplastic behavior of FSPed AA7075 alloy due to the grain refinement [35].

Along with developing bulk superplastic structures, FSP can be used in applications where selective area modification to introduce site-specific superplasticity. Wherever the structure must exhibit superplastic behavior, such localized regions can be processed, and fine grain structure can be achieved. This kind of selective area superplasticity is difficult to achieve with other processing routes. Figure 6.2 shows the schematic diagram of selective area superplasticity by FSP. Enhancing room-temperature formability of metals by FSP is another area of application where the modified microstructure helps to increase the formability not only at a higher temperature but also at the room temperature. Thick as well s thin sheets or plates can be successfully processed to improve the formability.

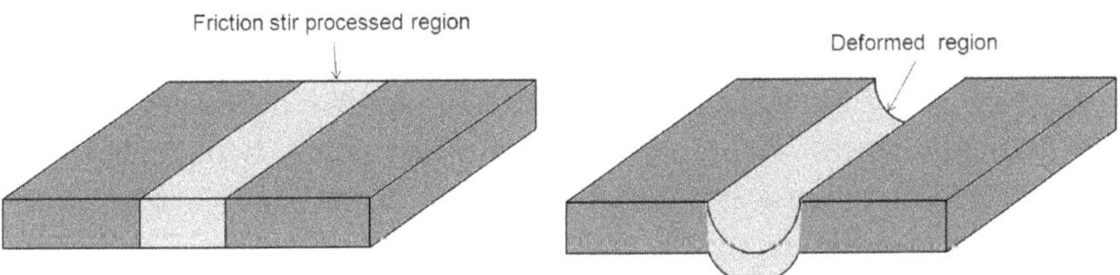

FIGURE 6.2 Schematic representation of selective area superplastic deformation of FSPed region.

6.2 TREATING CAST COMPONENTS

Casting is one of the basic manufacturing processes through which cost-effective components and structures with complex geometries can be successfully produced. However, compared with wrought products; components produced by casting exhibit poor mechanical properties such as strength, toughness and fatigue resistance. This is due to the type microstructure (coarse), dendrite structure and presence of porosity in the cast components compared with the wrought products. Heat treatment is one of the methods used to alter the microstructure and to reduce the level of porosity. Here, FSP can be a potential tool used to modify the microstructure and address the issue of porosity. Several reports on FSP to alter the microstructure and the porosity of cast structures can be seen in the literature [35–39]. By modifying the microstructure of metals through FSP, several mechanical properties such as yield strength, ultimate tensile strength, % elongation, fatigue life, toughness, and bending strength were clearly observed as increased.

6.3 MODIFICATION OF FUSION WELD ZONES

Fusion welding results cast microstructure that contains coarse grains with dendrites. Additionally, adjacent heat affected zones also contain a different kind of microstructure compared with the base microstructure. FSP can be used to modify the microstructure of fusion weld zones to achieve higher performance of the weld joint. In some cases, large structures and complicated joints where friction stir welding is not possible, fusion welding is adopted, and later the microstructure is altered by FSP.

Due to FSP, the cast structure in the fusion zone is completely changed into fine grain structure and eliminates defects or pores present in the weld zone. Processing the weld zones can be done by two different modes as schematically shown in Figure 6.3. In the first mode, two smaller FSP tools are used to process the interface region of fusion zone and base material as shown in Figure 6.3 (a) and in the second mode; the fusion zone itself is processed by FSP as shown in Figure 6.3 (b). By modifying the microstructure in the fusion zone and heat-affected zone, fatigue life and corrosion resistance of the weld joint can be enhanced [40, 41].

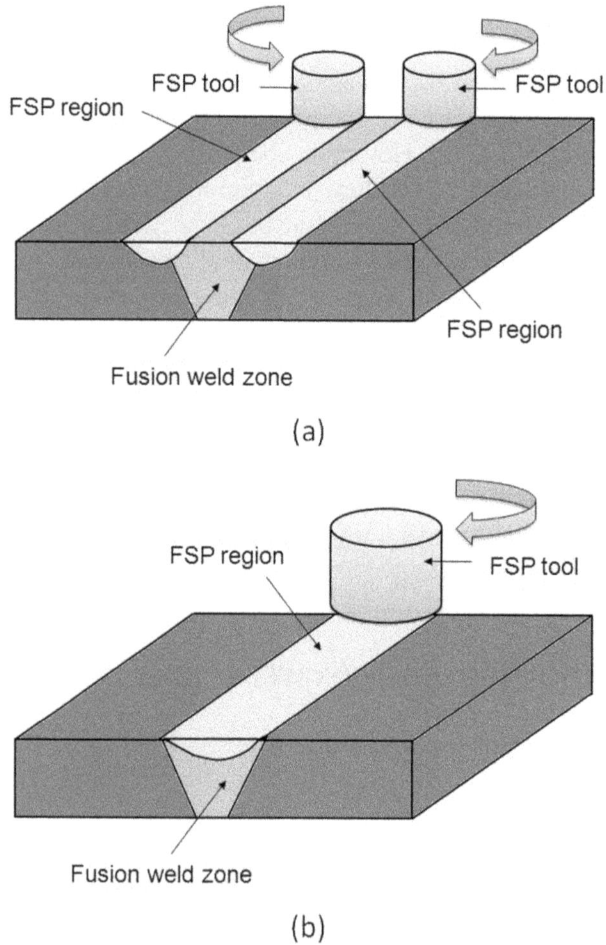

FIGURE 6.3 Schematic representation of processing fusion weld zones by FSP.

6.4 POTENTIAL BENEFITS, CHALLENGES, AND FUTURE PERSPECTIVES

By modifying the microstructure through FSP; increased strength, lower level of distortion and residual stresses, increased fatigue life, wear resistance, increased surface energy, superplasticity, etc. can be achieved in the metal without modifying the chemical composition. As the process also eliminates the melting state, several issues which are associated with molten metal can be completely eliminated by adopting FSP. The existing industries can easily adopt the technology as the operations, which are essential in completing the FSP process, are similar to other machine tool operations. Use of protective gas may not be required except while processing high reactive metals at high temperatures which reduces the complications, associated with shielding gas. Use of hazardous gases or fumes is not seen in FSP and therefore, qualifies to be called as green technology. In some case based on the type of processing base material, inert gases such as Argon or Nitrogen are sufficient to provide shielding atmosphere while carrying FSP.

On the other hand, several issues in connection with the materials, machines and industrial standards and regulations which must be addressed to make FSP as a promising tool in the manufacturing industry. Individual tool design for every new alloy is a crucial limitation to make FSP as a generalized process for various alloys. For every single alloy, process parameters must be optimized for specific tool design. The same process parameters may not be sufficient to adopt for an alloy of the same category with slight chemical composition modification. Thermal aspects and material flow during FSP and material consolidation after FSP are important areas which have been still under investigation. Unavailability of standard FSP tool designs, level of tool development and understanding of different tool specifications are in the preliminary stage which needs more investigations yet to be completed. Furthermore, lack of standards, specifications, designs, methods and trained workmen in friction-assisted processes is another important limitation. The geometry of the workpiece material and the amount of surface area that needs modification are the valid inputs from the materials perspective that requires additional research to explore the possibilities to adopt FSP in more number of applications. In order to make the process more automated, adopting control systems may further increase the capital cost which is another limitation may restrict the small and medium level industries to adopt FSP. It has been more than 17 years from the first report published on FSP. Several research groups significantly contributed in understanding the process and material flow mechanisms. Initially, FSP has been adopted to develop grain-refined surfaces to introduce improved mechanical properties. Later, several other properties were also improved by FSP such as better wear resistance, increased surface energy, and better corrosion properties. As the area of the processed region after a single pass of FSP is less, parallel multi passes are required to develop more FSPed surface area. Special tool holding and controls must be designed in order to make FSP as a promising tool to develop more surface area. Making the process automated with the help of control systems and application of robotics will take the FSP to the next level as an advanced manufacturing technique.

KEYWORDS

- **casting**
- **FSP**
- **severe plastic deformations**
- **superplasticity**

REFERENCES

1. Figueiredo, R. B., & Langdon, T. G., (2009). *Scr. Mater., 61*, 84–87.
2. Fang, C., Datong, Z., Yuanyuan, L., & Weiwen, Z. (2013) *High Strain Rate Superplasticity of a Fine-Grained AZ91 Magnesium Alloy Prepared by Submerged Friction Stir Processing,* Mater. Sci. Eng. A., 568, 40–48.
3. Mishra, R. S., Mahoney, M., McFadden, S., Mara, N., & Mukherjee, A., (1999). High strain rate superplasticity in a friction stir processed 7075 Al alloy. *Scripta Materialia., 42*, 163–168.
4. Vivek, P. V., Vishvesh, B., & Abhishek, K., (2017). Effect of polygonal pin profiles on friction stir processed superplasticity of AA7075 alloy. *Journal of Materials Processing Technology, 240*, 68–76.
5. Ma, Z., Mishra, R. S., & Mahoney, M. W., (2002). Superplastic deformation behavior of friction stir processed 7075Al alloy. *Acta Materialia., 50*, 4419–4430.
6. Liu, F., & Ma, Z., (2008). Low-temperature superplasticity of friction stir processed Al-Zn-Mg-Cu alloy. *Scripta Materialia., 58*, 667–670.

7. Wang, K., Liu, F., Ma, Z., & Zhang, F., (2011). Realization of exceptionally high elongation at high strain rate in a friction stir processed Al-Zn-Mg-Cu alloy with the presence of liquid phase. *Scripta Materialia., 64*, 572–575.
8. Orozco-Caballero, A., Cepeda-Jimńnez, C., Hidalgo-Manrique, P., Rey, P., Gesto, D., Verdera, D., Ruano, O., & Carreño, F., (2013). Lowering the temperature for high strain rate superplasticity in an Al-Mg-Zn-Cu alloy via cooled friction stir processing. *Materials Chemistry and Physics, 142*, 182–185.
9. Ma, Z., Mishra, R. S., & Liu, F., (2009). Superplastic behavior of micro-regions in two-pass friction stir processed 7075 Al alloy. *Materials Science and Engineering: A., 505*, 70–78.
10. Johannes, L., & Mishra, R., (2007). Multiple passes of friction stir processing for the creation of superplastic 7075 aluminum. *Materials Science and Engineering: A., 464*, 255–260.
11. Liu, F., & Ma, Z., (2009). Achieving high strain rate superplasticity in cast 7075Al alloy via friction stir processing. *Journal of Materials Science, 44*, 2647–2655.
12. Burgueño, A., Dieguez, T., & Svoboda, H., (2012). Effect of processing parameters on superplastic and corrosion behavior of aluminum alloy friction stir processed. In: *Materials Science Forum, Trans Tech Publ., 706*, pp. 965–970.
13. Dieguez, T., Burgueño, A., & Svoboda, H., (2012). Superplasticity of a Friction Stir Processed 7075-T651 aluminum alloy. *Procedia Materials Science, 1*, 110–117.
14. Charit, I., & Mishra, R. S., (2004). Evaluation of microstructure and superplasticity in friction stir processed 5083 Al alloy. *J. Mater. Res., 19*, 3329–3342.
15. El-Danaf, E. A., El-Rayes, M. M., & Soliman, M. S., (2011). Low-temperature enhanced ductility of friction stir processed 5083 aluminum alloy. *Bull. Mater. Sci., 34*, 1447–1453.
16. Johannes, L., Charit, I., Mishra, R. S., & Verma, R., (2007). Enhanced superplasticity through friction stir processing in continuous cast AA5083 aluminum. *Mater. Sci. Eng. A., 464*, 351–357.
17. Pradeep, S., & Pancholi, V., (2013). Effect of microstructural inhomogeneity on superplastic behavior of multipass friction stir processed aluminum alloy. *Mater. Sci. Eng. A., 561*, 78–87.
18. Ma, Z., Mishra, R. S., & Mahoney, M. W., (2004). Superplasticity in cast A356 induced via friction stir processing. *Scripta Mater., 50*, 931–935.
19. Liu, F., Xiao, B., Wang, K., & Ma, Z., (2010). Investigation of superplasticity in friction stir processed 2219Al alloy. *Mater. Sci. Eng. A., 527*, 4191–4196.
20. Charit, I., & Mishra, R. S., (2003). High strain rate superplasticity in a commercial 2024 Al alloy via friction stir processing. *Mater. Sci. Eng. A., 359*, 290–296.
21. Liu, F., & Ma, Z., (2008). Achieving exceptionally high superplasticity at high strain rates in a micrograined Al–Mg–Sc alloy produced by friction stir processing. *Scripta Mater., 59*, 882–885.
22. Liu, F., & Ma, Z., (2011). Superplasticity governed by effective grain size and its distribution in fine-grained aluminum alloys. *Mater. Sci. Eng. A., 530*, 548–558.
23. Liu, F., Ma, Z., & Zhang, F., (2012). High strain rate superplasticity in a micro-grained Al–Mg–Sc alloy with predominant high angle grain boundaries. *J. Mater. Sci. Technol., 28*, 1025–1030.

24. Liu, F., Ma, Z., & Chen, L., (2009). Low-temperature superplasticity of Al–Mg–Sc alloy produced by friction stir processing. *Scripta Mater., 60*, 968–971.
25. Liu, F., & Ma, Z., (2010). Contribution of grain boundary sliding in low-temperature superplasticity of ultrafine-grained aluminum alloys. *Scripta Mater., 62*, 125–128.
26. Smolej, A., et al., (2014). Superplasticity of the rolled and friction stir processed Al–4.5Mg–0.35 Sc–0.15 Zr alloy. *Mater. Sci. Eng. A., 590*, 239–245.
27. Ma, Z., Mishra, R. S., & Liu, F., (2009). Superplastic behavior of microregions in two-pass friction stir processed 7075Al alloy. *Mater. Sci. Eng. A., 505*, 70–78.
28. Liu, F., & Ma, Z., (2009). Achieving high strain rate superplasticity in cast 7075Al alloy via friction stir processing. *J. Mater. Sci., 44*, 2647–2655.
29. Wang, K., Liu, F., Ma, Z., & Zhang, F., (2011). Realization of exceptionally high elongation at high strain rate in a friction stir processed Al– Zn–Mg–Cu alloy with the presence of liquid phase. *Scripta Mater., 64*, 572–575.
30. Ku, M. H., Hung, F. Y., Lui, T. S., & Li-Hui, C., (2011). Embrittlement mechanism on tensile fracture of 7075 Al alloy with friction stir process (FSP). *Mater. Trans., 52*, 112–117.
31. Burgueno, A., Dieguez, T., & Svoboda, H., (2012). Effect of processing parameters on superplastic and corrosion behavior of aluminum alloy friction stir processed. *Materials Science Forum, 706*, pp. 965–970.
32. Dieguez, T., Burgueno, A., & Svoboda, H., (2012). Superplasticity of a friction stir processed 7075-T651 aluminum alloy. *Procedia Mater. Sci., 1*, 110–117.
33. Da-Tong, Z., Feng, X., Wei, W. Z., Cheng, Q., & Wen, Z., (2011). Superplasticity of AZ31 magnesium alloy prepared by friction stir processing. *Transactions of Nonferrous Metals Society of China, 21*(9), 1911–1916.
34. Cavaliere, P., & De Marco, P. P., (2007). Friction stir processing of AM60B magnesium alloy sheets. *Materials Science and Engineering A., 462*, 393–397.
35. Oh-Ishi, K., Cuevas, A. M., Swisher, D. L., & McNelley, T. R., (2003). The influence of friction stir processing on microstructure and properties of a cast nickel aluminum bronze material. *Mater. Sci. Forum, 426–432*(Part 4), pp. 2885–2890.
36. Sharma, S. R., Ma, Z. Y., & Mishra, R. S., (2004). Effect of friction stir processing on fatigue behavior of A356 alloy. *Scr. Mater., 51*(3), pp. 237–241.
37. Santella, M. L., Engstrom, T., Storjohann, D., & Pan, T. Y., (2005). Effects of friction stir processing on mechanical properties of the cast aluminum alloys A319 and A356. *Scr. Mater., 53*(2), pp. 201–206.
38. Ma, Z. Y., Sharma, S. R., & Mishra, R. S., (2006). Effect of multiple-pass friction stir processing on microstructure and tensile properties of a cast aluminum-silicon alloy. *Scr. Mater., 54*(9), pp. 1623–1626.
39. Oh-Ishi, K., Zhilyaev, A. P., & McNelley, T. R., (2006). A microtexture investigation of recrystallization during friction stir processing of As-cast NiAl bronze. *Metall. Mater. Trans. A., 37*(7), pp. 2239–2251.
40. Fuller, C., Mahoney, M., & Bingel, W., (2003). Friction stir processing of aluminum fusion welds. *Proceedings of the Fourth International Symposium on Friction Stir Welding,* (Park City, UT), TWI.
41. Fuller, C., & Mahoney, M., (2006). The effect of friction stir processing on 5083-H321/5356 Al arc welds. *Microstructural and Mechanical Analysis, Metall. Trans* (3712)3605-3615.

PART II

Surface Composites by Friction Stir Processing

CHAPTER 7

Introduction to Surface MMCs

Metals, ceramics, and polymers are the three basic groups of materials. The combination of two or more than two materials of different phases which may be metals, ceramics or polymers which are chemically and physically distinct gives a composite. Composites offer a wide variety of advantages compared with their constituting phases. The composite exhibits different behavior compared with the constituting phases. The performance of a single (primary) phase material can be improved by introducing a different phase (secondary) of the same material or different material, and such material can be called a reinforced composite. Usually, the secondary phase is dispersed in the primary phase in the form of long fibers, short and discontinuous fibers or fine particles. If we call the primary phase as the matrix, this newly produced matrix composite exhibit a different kind of behavior with respect to the initial primary phase. The dispersing phase serves as a strengthening element and improves the overall strength of the matrix material to withstand at the higher loads and extreme operational conditions.

Carbides, oxides, nitrides, and a combination of some ceramic powders are a few examples of secondary phase materials. The matrix material can be a metal, ceramic or polymer material. Based on the matrix materials composites are called as metal matrix composites (MMCs), ceramic matrix composites (CMCs) and polymer matrix composites (PMCs). These composites are different from laminated composites in which different phases are arranged in the lamellar structure.

7.1 METAL MATRIX COMPOSITES

Metal matrix composites (MMCs) are an important category of composite materials in the modern engineered materials which have clearly demonstrated their potential as promising candidates for various structural applications. In MMCs, the matrix material is a metal which can be in a pure phase or an alloy. The matrix material exhibits soft nature compared with dispersing secondary phase material. Aluminum, copper, magnesium, and titanium are a few examples of matrix materials and SiC, Al_2O_3, TiB_2, WC are a few examples for reinforcing secondary phase materials [1]. By adding an appropriate amount of secondary phase, the specific strength of the metals or alloys can be improved. In addition, higher fatigue and wear resistance can be achieved.

Figure 7.1 shows a schematic illustration of MMCs. The function of the matrix in MMCs is to hold the fibers and transfer the loads to the fibers. Based on the fiber arrangement in the composite, the appropriate matrix material is selected. For example in a long fiber reinforced matrix, the majority of the loads are transferred through the continuous fibers. In such conditions, the prime function of the matrix is to hold the fibers together and therefore metals of lower stiffness can be used as the matrix materials. Whereas in the composites of short and discontinuous fibers, matrix material must have sufficient amount of stiffness in order to hold the fibers and withstand the loads as the loads are transferred through the matrix. The thermal properties of the matrix and fiber material also influence the successful function of the composite during its function. Particularly at elevated temperatures, if the difference in the coefficient of thermal expansion is more, the mechanical integrity between the fibers and the matrix is affected, and the composite may experience a sudden failure during the working. Therefore, the selection of materials for matrix and fibers is

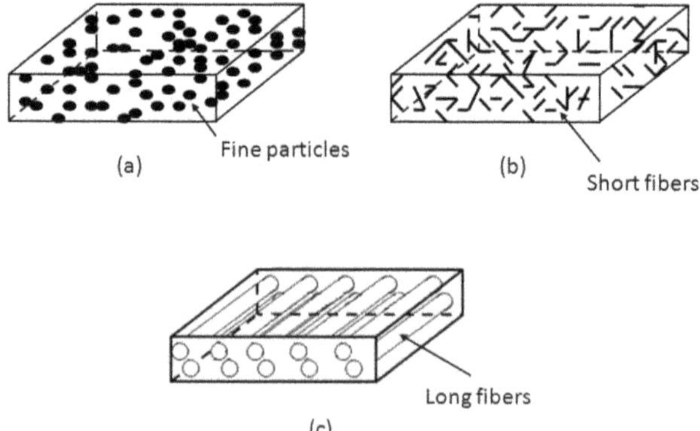

FIGURE 7.1 Different modes of secondary phase distribution in metal matrix composites (MMCs): (a) composites of fine particles, (b) composites of discontinuous short fibers and (c) composites of continuous and aligned fibers.

important that dictates the successful functioning of the structure made of such MMCs. The potential structural applications of MMCs can be found in automobile, aerospace and marine industries [12].

7.2 ADVANTAGES AND LIMITATIONS WITH MMCS

When compared with other composites, MMCs offer advantages such as high strength and stiffness, high electrical and thermal conductivity. MMCs are high in dimensional stability with high strength and high elastic modulus. Also, exhibit improved wear resistance, excellent creep properties, and fatigue resistance compared with that of pure matrix phase [3]. Since the matrix material in MMCs is a metal and which has an electron cloud due to the metallic bond, all MMCs shows high electrical and thermal conductivity. Whereas in PMCs and CMCs, the matrix material is an either polymer or ceramic-based material in which interatomic or intermolecular bonds exist, and no free electrons are available. Therefore, PMCs and CMCs exhibit poor electrical and thermal properties compared with MMCs. MMCs show higher ductility and toughness compared with CMCs and exhibit high thermal stability compared with PMCs.

The ductility and toughness of MMCs are influenced by the amount and the way of distribution of the secondary phase present in the matrix which is relatively brittle and hard. Up to the moderate level of reinforcement of dispersed phase, metal-forming methods can be applied to MMCs, but the level of the formability, ductility, and toughness are reduced if the quantity of dispersed phase is increased in the matrix. The manufacturing cost of MMCs is also higher which makes them not feasible for many applications where economical perspectives also play a major role along with the performance of the material. The machining of MMCs is another limitation. It is difficult to machine MMCs with ordinary tools and special cutting tools such as diamond tools are required [4]. Excess machining of MMCs results in higher amounts of scrap which is not desirable as the materials used to fabricate MMCs are expensive.

However, the recent developments in the manufacturing industry with high productivity at low cost enhanced the perimeter of the possible applications of these MMCs. Particularly, in a few areas such as aerospace, automobile, and power generation fields, MMCs have been already proven as the best moderns materials [5]. Engine components, piston, and brake disks are a few examples for automobile applications and components in aircraft, satellites, jet engines and missiles are a few examples for aerospace applications.

7.3 MANUFACTURING PROCESSES OF MMCS

Basically, the manufacturing processes of MMCs can be divided into two groups as primary and secondary. Primary methods produce the MMCs as a bulk material, and secondary methods involve processing of these bulk MMCs into a final product that is close to the desired dimensions. There are several primary manufacturing processes developed to fabricate MMCs. These processes can be grouped as liquid phase or solid-state processing techniques. Stir casting, squeeze casting, spray deposition, and *in situ* fabrication are a few examples for liquid phase processing, and powder metallurgy, diffusion bonding, and vapor deposition methods are a few examples for solid-state processing techniques commonly used to fabricate bulk MMCs [3, 6, 7]. Each method has its own merits and demerits when compared with other methods in fabricating MMCs. Secondary methods are finishing processes including the basic material processing techniques such as machining, bending, extrusion, milling, forging, etc. which bring the bulk MMCs to final desired shape and size. Figure 7.2 shows the schematic representation of developing MMCs by different processes.

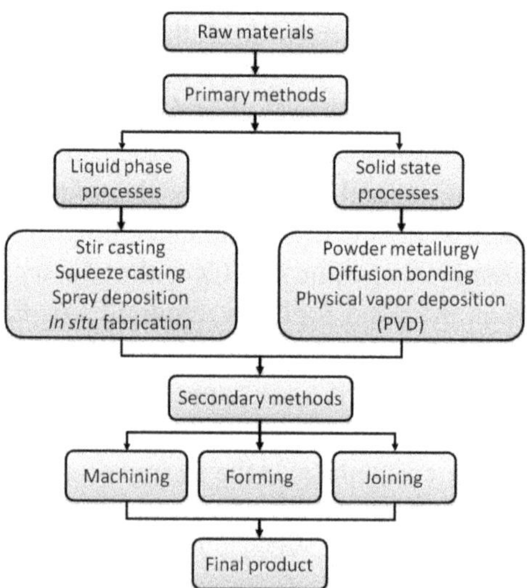

FIGURE 7.2 Schematic representation of the manufacturing processes of MMCs.

7.3.1 Casting

Casting is a well known ancient method, widely used in the manufacturing industry to fabricate different types of composites. *Stir casting* [8] and *squeeze casting* [9] are the most common examples for casting methods to fabricate MMCs. In *stir casting*, secondary phase particles are introduced into the molten metal, and the discontinuous fibers are dispersed throughout the molten metal with the help of a stirrer as shown in Figure 7.3(a). During the supply of these secondary phase particles into the molten metal or during the stirring operation, care is to be taken to avoid any gas entrapment to eliminate the formation of pores and voids. In *squeeze casting*, molten metal is injected into a pre-form which contains secondary phase particles or fibers as shown in Figure 7.3 (b) [10, 11]. Initially, the molten metal is pressed with low pressure into the preform, and then at the end of infiltration, the pressure is increased and allow the molten metal to get solidify to form the desired composite. Among all the casting methods, squeeze casting is the most common manufacturing method used to fabricate MMCs. Uniform dispersion of secondary phase is a limitation in melting and solidification (casting) process. If the secondary phase particles are in nanometer level, agglomeration of these particles is another issue which is associated with this method. Stability of the dispersing secondary phase is another limitation with casting practice.

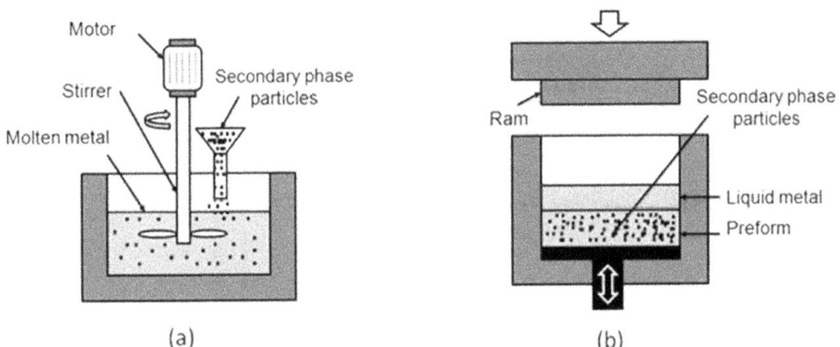

FIGURE 7.3 Schematic representation of MMCs fabrication by casting method: (a) stir casting and (b) squeeze casting.

7.3.2 *Spray Deposition*

In spray deposition method, matrix metal is sprayed in molten condition through a nozzle, and the secondary phase particles are introduced into the molten metal during the spraying and MMCs are fabricated as schematically shown in Figure 7.4 [12]. This method yields a high rate of production. During the spray deposition, the liquid droplet along with the secondary phase particles traveled with a high speed due to the flow of the pressurized gas and deposited on a substrate to form the composite. The rate of solidification is high in spray deposition method which reduces the level of reaction of the secondary phase with the matrix material. But the distribution of the secondary phase depends on various factors such as the size and shape of the particles and the time of introducing them into the spraying molten metal. Followed by spray atomization a secondary process is generally adapted to make the material denser and to achieve the homogeneous distribution of the secondary phase. This method is best suited for fabricating functionally graded composites, but the process is quite expensive.

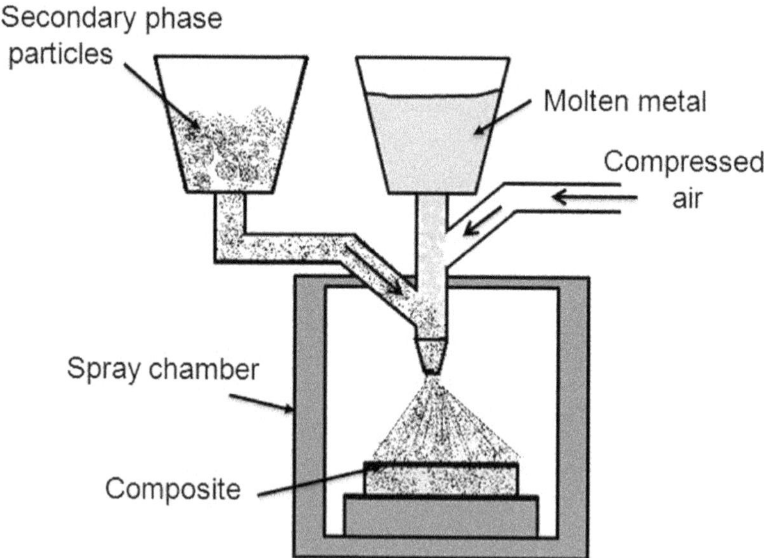

FIGURE 7.4 Schematic representation of MMCs fabrication by a spray deposition method.

7.3.3 In Situ Fabrication

Materials which can actively participate in the chemical reaction with the matrix or materials which exhibit phase transformation during solidification due to some isothermal reactions such as eutectic, monotectic and peritectic and develop secondary phase particles within the matrix and result in MMCs. Figure 7.5 shows the schematic representation of MMCs fabrication by *in situ* method. Composite fabrication by *in situ* method is not possible for all material systems. The problems with bonding strength at the interface of the secondary phase particles and matrix can be almost eliminated with *in situ* fabrication. The phases, which are formed due to reaction, show a strong interfacial bond with the surrounding matrix [13].

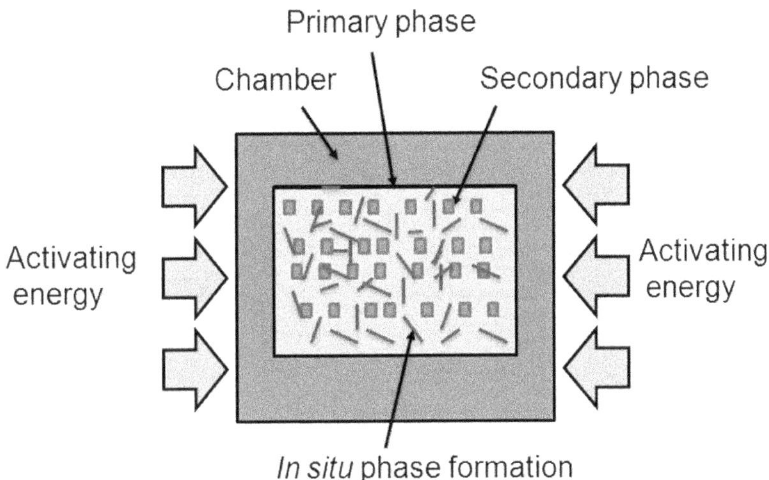

FIGURE 7.5 Schematic representation of MMCs fabrication by *in situ*(reaction) method.

7.3.4 Powder Metallurgy

Powder metallurgy is the best example for a solid-state processing technique to fabricate composite materials [14]. It is the most common and ancient method used to fabricate composites. Powder preparation, blending, compacting and hot pressing or sintering are the common basic steps involved in developing MMCs by powder metallurgy route as schematically shown in Figure 7.6 [15, 16]. After blending and compacting the powders, the product can be called as green compact or pellet which is then heated up to sufficient temperature below the melting temperature to initiate solid-state diffusion. Sometimes, secondary processing methods such as forging or extrusion are adapted after sintering or hot pressing to impart specific properties to the final product. Mixing the powders during the blending process dictates the level of uniformity of the secondary phase distribution and agglomeration in the matrix. If proper care is taken in powder preparation, uniform distribution of secondary phase can be achieved in powder metallurgy route, but porosity, grain growth and the size of the end product are the limitations.

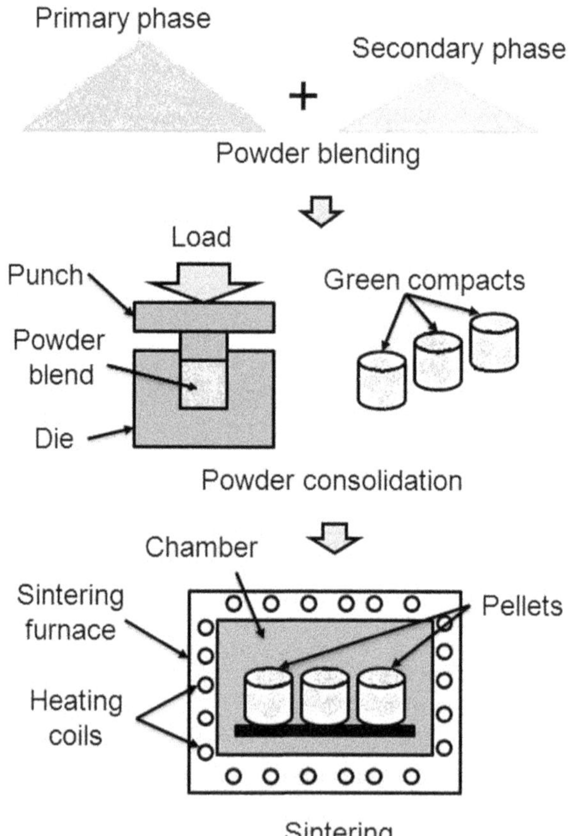

FIGURE 7.6 Schematic representation of MMCs fabrication by powder metallurgy route.

7.3.5 *Diffusion Bonding*

Diffusion bonding is another solid-state processing method used to develop composites without melting the material. In diffusion bonding, layers of thin metal sheets are sandwiched with fibers and pressed at elevated temperature by applying high pressure to fabricate the composite as schematically shown in Figure 7.7 [15]. Instead of fibers, different metal foils can be used to fabricate multilayered lamellar composites. Diffusion bonding involves expensive operations like long holding at elevated temperature by applying high pressure in order to diffusion to takes place at the interface of metal sheets and fibers. The diffusion of atoms is a function of temperature. Therefore, holding the material at elevated temperatures is a must procedure in diffusion bonding. Control over the distribution, orientation and volume fraction of the fibers in the matrix is an advantage with diffusion bonding. However, if the component geometry is complex, the difficulty level is increased in this process.

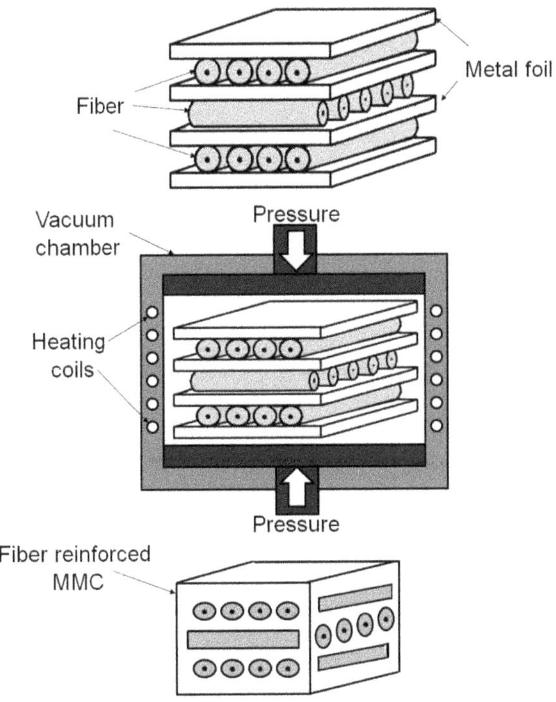

FIGURE 7.7 Schematic representation of MMCs fabrication by diffusion bonding.

7.3.6 Vapor Deposition Methods

In vapor deposition methods, fibers are coated with metallic vapor and then consolidated to develop MMCs. When the fibers are passed through a chamber consisting of metallic vapor, the fibers are coated with the metal due to the condensation of the metal on the fibers as shown in Figure 7.8. Then these metal coated fibers are consolidated by applying pressure and heat. Electron beam vapor deposition (EBVD) and physical vapor deposition (PVD) methods are the best examples of this category [3]. Laminated composites also can be fabricated by these techniques. The advantage with vapor deposition methods is a wide range of compositions can be prepared as matrix materials, but the process is expensive.

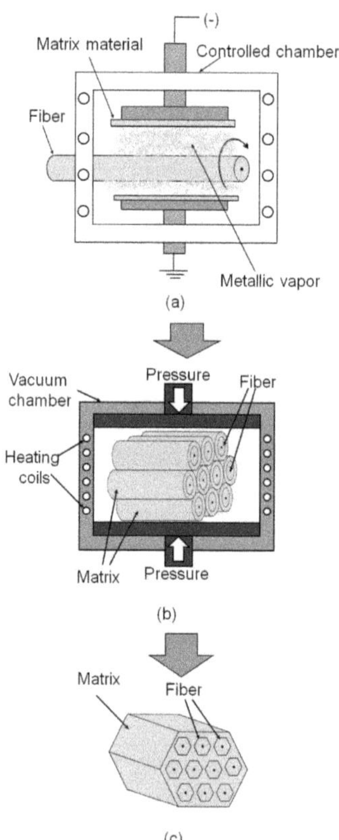

FIGURE 7.8 Schematic representation of vapor deposition method to fabricate MMCs: (a) fiber coated with a matrix material, (b) consolidating the coated fibers and (c) final product.

7.4 SURFACE MMCS

There are some important applications, where the surface properties play an important role on the functioning and the life of a component. Surface degradation due to wear and corrosion is an example of such properties which depend on the material's composition and structure at the surface. Surface MMCs which are having secondary phase particles dispersed at the surface level up to some depth improve the surface properties and at the same time core of the material is unchanged. In such situations, only the surface layers are modified in connection with its chemical composition by introducing a secondary phase, while the structure and composition at the core of the component remain the same. Therefore, the surface exhibits higher hardness and wear resistance and the material in bulk experience negligible loss to its toughness.

7.5 MANUFACTURING PROCESSES OF SURFACE MMCS

Altering the surface composition to produce surface composites requires methods of the special category which affect only the surface up to certain depth. Laser melt treatment, centrifugal casting, and plasma spraying are a few examples of such methods developed to fabricate surface MMCs. Among all of these techniques, laser melt treatment is well known and widely used method to modify the surfaces.

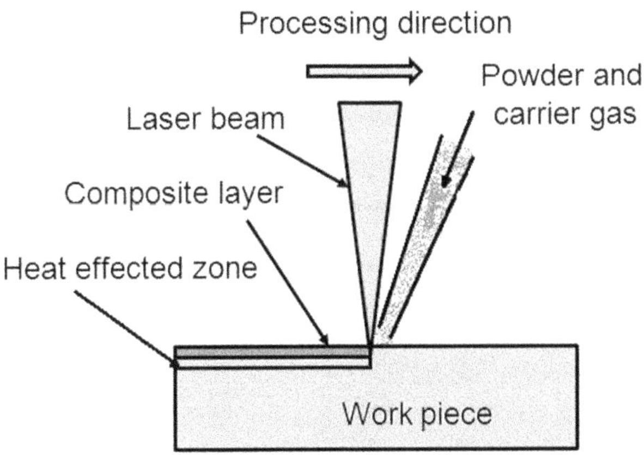

FIGURE 7.9 Schematic representation of surface MMCs fabricated by laser melt technique.

7.5.1 Laser Melt Treatment

The schematic representation of the laser melt treatment process is shown in Figure 7.9. During laser melt treatment, a laser beam melts the surface of the substrate as the secondary phase particles are introduced through a nozzle by a carrier inert gas. During the laser melt treatment, the surface undergoes a transformation from solid to liquid phase in which the secondary phase particle is dispersed [17]. A reaction may take place between the matrix material and the introduced secondary phase particles. The phase stability of the secondary phase particles at the processing temperature also needs to be considered in designing the surface composites using laser melt treatment. Since the solidification happens at the surface level, critical control over the processing parameters is required to achieve optimum microstructure at the surface layer.

7.5.2 Centrifugal Casting

In centrifugal casting, the secondary phase particles which are intended to be placed at the surface or circumference of an axe symmetric cast product added to the molten metal and poured into the rotating hot mold as schematically shown in Figure 7.10. These secondary phase particles are pushed in the radial direction to the circumference of the mold due to the centrifugal force that is applied on them as the density of the particles is more when compared with the density of the molten metal. But the geometry of the object is restricted in centrifugal casting. Only symmetric axe objects can be produced by centrifugal casting [18].

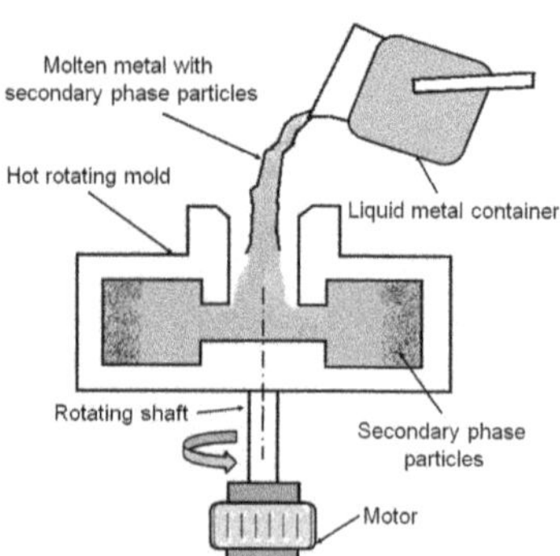

FIGURE 7.10 Schematic representation of surface MMCs fabricated by centrifugal casting.

7.5.3 Plasma Spraying

Plasma spraying is an advanced surface coating process which can be used to develop surface composites. In plasma spraying, the materials intended to be formed as a composite coating are introduced into the plasma jet at the entrance of the nozzle as shown in Figure 7.11. The material can be introduced in the form of powder or a continuous rod. The feed material is melted due to the high temperature in the plasma jet and forced towards the substrate as liquid droplets. These molten droplets rapidly solidify during the deposition on the substrate and form lamellar coatings. If the feed material is a combination of different phases, the substrate is coated with a composite layer.

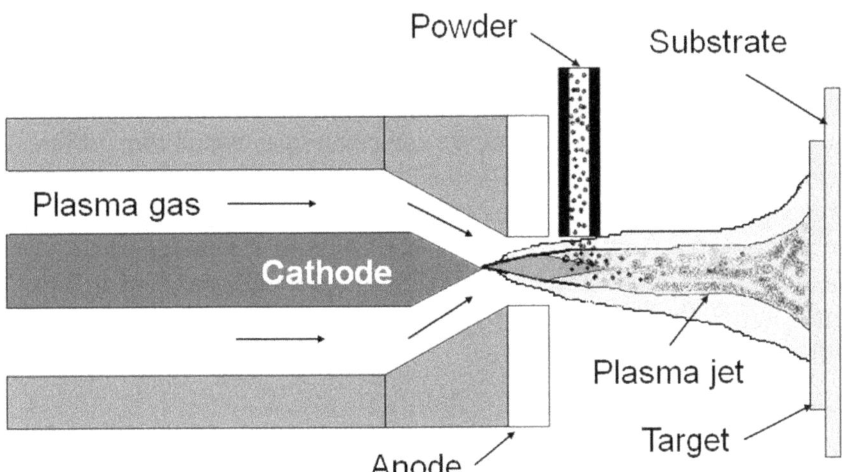

FIGURE 7.11 Schematic representation of plasma spraying process.

The other surface composite fabrication techniques also involve the material transformation from solid to liquid or vapor state. On the other hand, solid-state processing techniques which do not result in transformation of the material to liquid phase during the process offer relatively many advantages over conventional liquid phase material processing techniques. Friction stir processing is one of such best examples for solid-state processing techniques which can be used to develop surface composites.

Solid-state methods eliminate several issues associated with liquid state methods such as handling of liquid metal, dendrite segregation during solidification, coarse grain size after solidification, dispersed particles agglomeration and oxidation. At this juncture, friction stir processing (FSP), a solid-state processing technique that has been evolved from the basic principles of FSW offers a great advantage as a promising tool to develop surface MMCs without melting the matrix. FSP has emerged as a promising tool to alter the surface microstructure of metallic sheets and plates [19] as elaborately discussed in Section 7.1. Along with the grain refinement, particle dispersed surface MMCs can be fabricated using FSP. The stirring action of the rotating tool can be utilized to disperse secondary phase particles at the surface of metal during FSP. Development of FSP to fabricate composites can't replace the other composite manufacturing processes, but offers many advantages as one of the promising solid-state processing methods to fabricate the surface MMCs. The following chapters focus mainly on the fabrication of surface MMCs by FSP, and the effect of the fine grain size and presence of secondary phase at the surface on various properties of the composites developed form different material systems. Information on the hybrid composites produced by FSP using two or more than two dispersing phases is also provided. The effect of processing factors on successful fabrication of the composites is also discussed. The promising applications and challenges involved in developing surface MMCs by FSP are discussed at the end of Section 7.2 with a precise summary.

KEYWORDS

- **ceramic matrix composites**
- **electron beam vapor deposition**
- **metal matrix composites**
- **physical vapor deposition**
- **polymer matrix composites**
- **squeeze casting**
- **stir casting**

REFERENCES

1. Deborah D. L Chung (2010). *Composite Materials Science and Applications*. 2nd edn, Springer, New York.
2. Kaczmar, J. W., Pietrzak, K., & Wøosin Âski W (2000). The production and application of metal matrix composite materials. *Journal of Materials Processing Technology106,* 58–67.
3. Nikhilesh, C., & Krishan KC (2013). Metal *Matrix Composites.* 2nd edn. Springer, New York.
4. Paulo Davim, J., (2012). *Machining of Metal Matrix Composites*, Springer, New York.
5. Kainer, K. U. (2006). Metal *Matrix Composites: Custom-made Materials for Automotive and Aerospace Engineering.* Wiley-VCH Verlag GmbH & Co. Weinheim.
6. Guo, X., & Derby, B. (1995). Solid-state fabrication and interfaces of fiber reinforced metal matrix composites. *Progress in Materials Science 39,* 411–495.
7. Surappa, M. K. (1997). Microstructure evolution during solidification of DRMMCs (Discontinuously Reinforced Metal Matrix Composites): State of Art, *Journal of Materials Processing Technology 63,* 325–333.
8. Hashim, J., Looney, L., & Hashmi, M. S. J. (1999). Metal matrix composites: production by the stir casting method, *Journal of Materials Processing Technology 92–93,* 1–7.
9. Ghomashchi, M. R., & Vikhrov A (2000). Squeeze casting: an overview. *Journal of Materials Processing Technology 101,* 1–9.
10. Yue, T. M., & Chadwick, G. A. (1996). Squeeze casting of light alloys and their composites. *Journal of Materials Processing Technology 58,* 302–307.
11. Vijayaram, T. R., Sulaiman, S., Hamouda, A. M. S., & Ahmad, M. H. M. (2006). Fabrication of fiber reinforced metal matrix composites by squeeze casting technology. *Journal of Materials Processing Technology 178,* 34–38.
12. Lavernia, E. J., & Grant, N. J. (1988). Spray deposition of metals: A review. *Materials Science and Engineering,98*, 381–394.
13. Daniel, B. S. S., Murthy, V. S. R., & Murty, G. S. (1997). Metal-ceramic composites via in-situ methods. *Journal of Materials Processing Technology 68,* 132–155.
14. Ghosh, A. K. (1993). in *Fundamentals of Metal Matrix Composites*, Butterworth-Heinemann, Stoneham, MA.
15. Davis, E. A., & Ward, I. M. (1993). An introduction to metal matrix composites. Cambridge University Press, New York, USA.
16. Hunt, W. H. (1994). *Processing and Fabrication of Advanced Materials,* The Minerals, and Metal Materials Society, Warrendale, PA.
17. Ayers, J. D., Tucker, T. R. (1980). Particulate-TiC-hardened steel surfaces by laser melt injection. *Thin Solid Films*, 73(1), 201–207.
18. Weisheit, A., Galun, G., & Mordike, B. L. (2014). *Comprehensive Materials Processing. 5,* 39–67.
19. Mishra, R. S., Ma, Z. Y. (2005). Friction stir welding and processing. *Mat Sci Eng R. 50*, 1–78.
20. Dawes, C., & Thomas, W. (1995). *TWI Bull., 6*, 124.

21. Hokaed, E., & Laverinya, J. (1999). Particulate reinforced metal matrix composite—A review. *Journal of Materials Science, 92* (1), 1–7.
22. Mishra, R. S., Mahoney, M. W., McFadden, S. X., Mara, N. A., & Mukherjee, A. K. (2000). High strain rate superplasticity in a friction stir processed 7075 Al alloy, *Scripta Mater. 42,* 163–168.
23. Ratna Sunil, B., Sampath Kumar, T. S., Uday, C., Nandakumar, V., & Mukesh, D. (2014). Nano-hydroxyapatite reinforced AZ31 magnesium alloy by friction stir processing: a solid-state processing for biodegradable metal matrix composites. *J Mater Sci: Mater Med 25,* 975–988.
24. Riabkina-Fishman, M., Rabkin, E., Levin, P., Frage, N., Dariel, M. P. (2001). Laser produced functionally graded tungsten carbide coatings on M2 high-speed tool steel. *Materials Science and Engineering A302,* 106–114.
25. Thomas, W. M., Nicholas, E. D., Needham, J. C., Murch, M. G., Templesmith, P., & Dawes, C. J., (1991). GB Patent 9125978.8.

CHAPTER 8

Surface MMCs by Friction Stir Processing

This chapter describes the fabrication of surface composites by FSP in detail. Starting from a brief explanation about using FSP to produce a composite, the possible defects in the processed regions, which are generally found in the composites fabricated by FSP and different ways to eliminate those defects, are brought out in detail. Advantages and limitations with FSP to fabricate surface composites compared with the other manufacturing methods are summarized at the end.

8.1 COMPOSITE FABRICATION BY FSP

Fabrication of surface composites by FSP is another development which has wide scope in developing tailored surfaces targeted for applications where surface properties play an important role. The stirring action of the FSP tool can be used to incorporate and distribute secondary phase particles in a metal in order to fabricate surface MMCs. Along with incorporating the secondary phase particles, grain size reduction at the surface can be achieved which is another advantage with FSP. For the first time, Mishra et al. [1] demonstrated the fabrication of 5083Al-SiC surface composite using FSP in the literature. The schematic representation of the process is shown in Figure 8.1. Usually, a narrow groove is produced on the surface of the sheet, and the secondary phase particles are filled, and FSP is carried out in order to produce the surface composite. During the process, the secondary phase particles are distributed within the nugget zone due to the plastic flow of the material and dispersed throughout the stirred region which is equal to the shoulder diameter of the FSP tool. The level of distribution of these particles within the matrix depends on the material plastic flow during the FSP process that is a dependent of processing parameters. There are many factors which influence the particle dispersion during FSP. Tool geometry which includes a pin and shoulder design, processing parameters which include load, tool rotational speed, traverse speed, tool penetration depth and type of workpiece are important factors which dictate the success of the process in developing surface MMCs by FSP. The introduced secondary phase particles bring several additional properties to the surface of a metal. Surface wear mechanisms are completely altered by introducing hard and wear-resisting ceramic powder. Additionally, surfaces with higher strength can be produced by finely dispersing nanoparticles. Corrosion properties, bioproperties, and tribological properties can be enhanced for a specific surface for a determined area. Localized surface composites that are limited to certain regions where special properties are required can also be produced by adopting this technique.

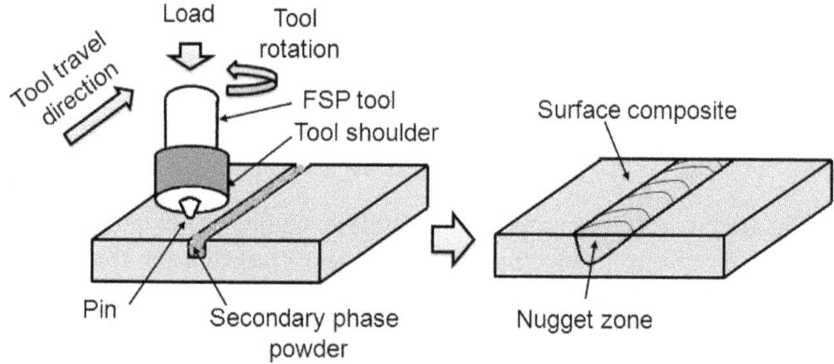

FIGURE 8.1 Schematic drawing showing surface composite fabrication by FSP.

8.2 MECHANISM BEHIND COMPOSITE FORMATION DURING FSP

The mechanism behind the composite formation during FSP can be explained by considering the material flow during FSP. It is almost the same in the context of plastic deformation how it happens in FSP except for the presence of an additional phase in the form of particles. It is obvious that the secondary phase particles incorporated in the matrix during the FSP influence the material flow rate, but the mechanisms are usually the same as observed in FSP. During the composite formation by FSP, in the context of particle addition, the following important observation can be made.

i. Particles that are filled in grooves or holes, dispersed along with the stirred material.
ii. Particles are entrapped within the stir zone during the dynamic recrystallization.
iii. Particles introduce pinning effect to arrest the grain growth and results in fine grains.

When the secondary phase powder is filled in the provision that is made on the surface of the sheet in the form of grooves or holes, it is not assured that the amount of powder actually filled and the fraction of the powder filled in the grooves that are introduced into the surface after FSP. In order to avoid the escape of the particles from the grooves and holes during FSP, another tool without a pin is employed on the surface before actual FSP. The rotating pin-less tool closes the powder containing groove or holes by modifying the surface of the substrate similar to friction surfacing. Here, the workpiece material is plastically stirred and covers the groove or hole opening to entrap the powder within the groove or hole. Now, the actual FSP is done on the surface processed region and the composite is produced. This kind of additional pre-processing avoids the spillage of the powder from the grooves and holes during FSP.

8.3 DEFECTS IN THE MMCS PRODUCED BY FSP

Usually, several defects are formed during the solidification process in liquid state methods such as porosity, solidification cracking, segregation of phases, microstructure modification and hot cracking. Additionally, the heat used to melt the surface in developing surface composites introduces distortion in the surrounding material. However, FSP is a solid-state method that eliminates these defects which result during solidification but results in other kinds of defects due to the following three factors.

i) Excessive heat generation;
ii) Insufficient heat generation; and
iii) Inappropriate processing conditions.

First two factors are directly influenced by the processing parameters (tool rotational speed, travel speed, penetration depth, tilt angle, etc.) and type of material to be processed. During FSP, inappropriate material flow results in a defect. Sufficient amount of heat is required which is generated due to the friction to deform the material to bring appropriate material flow plastically. At this temperature, microstructure also completely modified due to dynamic recrystallization.

8.3.1 Excessive Heat Generation

Excessive heat generation leads to an increase in the temperature at the processing zone close to that of melting point and leads to thermal softening of the work material. In such condition, several defects or flaws are observed in the processed zone which are mainly connected with the rate of material flow. Figure 8.2 shows the most common defects that resulted due to excessive heat generation. Surface galling or appearance of the processed surface as blisters is one of such defects commonly observed in FSPed surfaces. Figure 8.2 (a) shows FSPed AZ91 Mg alloy containing blisters on the processed surface. Further, the higher amount of heat generation in the stir zone may also affect the material of the adjacent regions to that of stir zone soften and expel in the form of flash as shown in Figure 8.2 (b). Due to the softening effect, the material beneath the FSP tool no longer resists the load applied by the tool and the tool penetrates into the workpiece, and the thickness of the workpiece is decreased (thinning effect) beneath the rotating tool. Another important microstructural feature that is seen after FSP is the distribution of the grain size at two different scales. The appearance of fine grains and coarse grains side by side within the stir zone as bands or in some literature termed as onion ring patterns. Of course, this may not be classified as a defect; however, it is undesirable particularly in the context of fatigue resistance. Figure 8.3 shows a typical microstructure of fine and coarse grains within the stir zone of FSPed AZ31 Mg alloy which appears as bands in lower magnification. From the recent studies, the increased heat generation, different rate of material flow in the workpiece thickness direction, happening of material flow in more than one shear patterns and the profile of the FSP tool pin is believed to be the influencing factors behind the formation of these onion ring patterns. Several studies demonstrated that the higher processing temperatures lead to develop these bands and decrease the resistance to crack path propagation in the processed region subjected to cyclic loads [2–6].

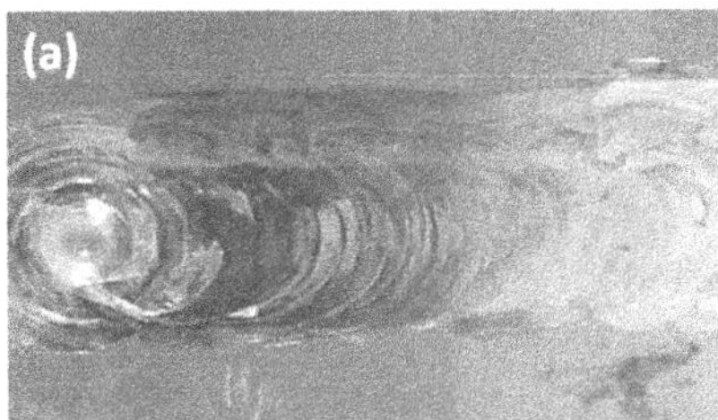

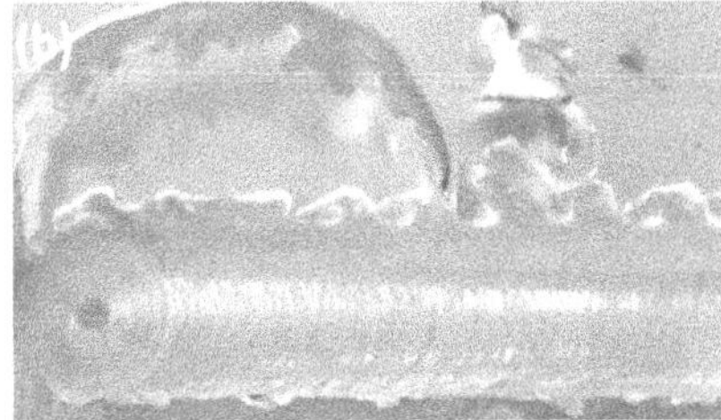

FIGURE 8.2 Defects in the composites produced by FSP: (a) surface galling and (b) excessive flash.

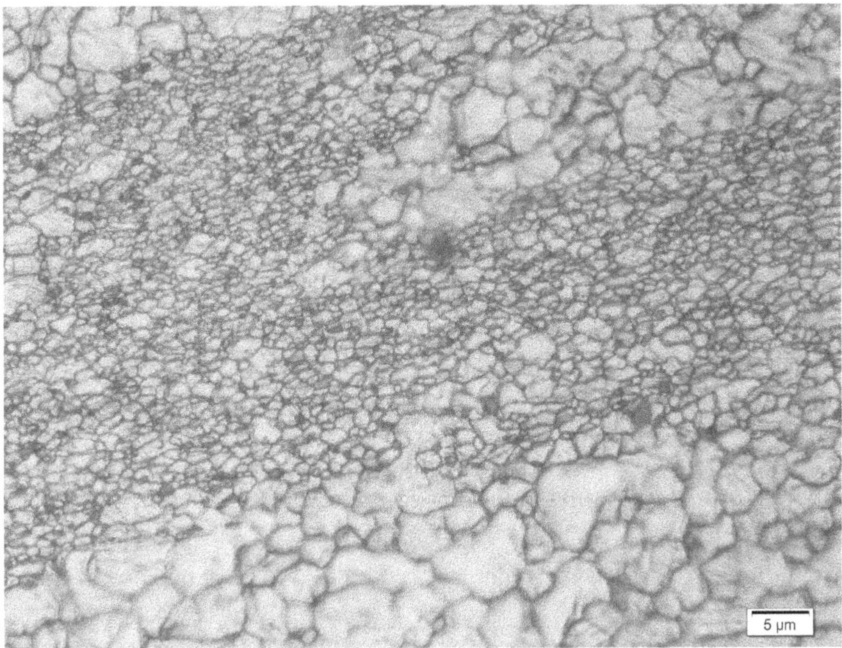

FIGURE 8.3 Microstructure of FSPed AZ31 Mg alloy.

The level of thermal conductivity of the work material governs the amount of heat concentration at the processing zone along with the other processing parameters which further directly influence the recrystallization rate and the size of the zone that undergoes microstructure modification. Colegrove et al. [7] demonstrated the decreased recrystallized region due to the higher amount of heat in the stir zone of an Al-Cu-Mg-Mn 2024 alloy with the increase of the tool rotation speed. Similarly, Zettler [8] also reported a decrease in the recrystallized region within the stir zone of Al-Si-Mg-Cu-Mn 6013 alloy after friction stir welding at a tool rotation speed of 3000 rpm. This behavior can be explained by considering the thermal softening effect brought by developing a higher amount of heat generation which causes a slip between the workpiece surface and the tool pin. This leads to decrease the strain rates in the stir zone. However, material flow can be seen due to the stirring action of the pin. This combination of material flow and heat may not be sufficient to activate recrystallization in the stir zone beneath the FSP tool. This may further lead to another undesirable feature called "Nugget collapse" where the nugget region appeared to be severely affected near the surface of the stir zone particularly at the retreating side of the stir zone.

8.3.2 *Insufficient Heat Generation*

If the combination of tool rotational speed and travel speed does not yield a sufficient amount of heat generation in the stir zone, defects and flaws are appeared due to insufficient material plastic flow. If the rate of heat loss from the processing zone is higher compared with the rate of heat generation, in such conditions, the material may not be soften and stirred properly. The load required to bring the material beneath the FSP tool shoulder is increased, and it may reach to a certain value where the tool pin may not withstand. Therefore, due to the excess demand for the load to plastically deform the workpiece, the tool pin may be damaged or worn out. If the stir zone is produced in cold conditions, e.g., at the temperatures lower than the desired, a flaw in the form of an interface due to insufficient metallurgical bonding is resulted. When this region is subjected to mechanical loading, if any failure is to be happened, is predominantly passes through this flaw. Lacks of fill and tunneling or wormhole are the two important defects observed due to insufficient heat generation (cold conditions).

Lack of fill can be visually seen on the surface of the FSPed region. Figure 8.4 (a) shows a typical photograph of FSPed pure Mg demonstrating lack of fill (as indicated with an arrow). These are the marks or indication left by the FSP tool travel on the surface after processing usually appeared close to the advancing side. Tunnel or warm hole defect is another category of defects usually observed due to insufficient flow of the material away from the surface within the stir zone as shown in Figure 8 (b). In the tunnel defect, the top surface material is completely metallurgically bonded, but in the thickness direction, the cavity generated by the tool travel is not sufficiently filled with the material, and a continuous hole is resulted. Addition of secondary phase particle influence the flow behavior of the matrix and the possibility to form a tunnel defect is increased.

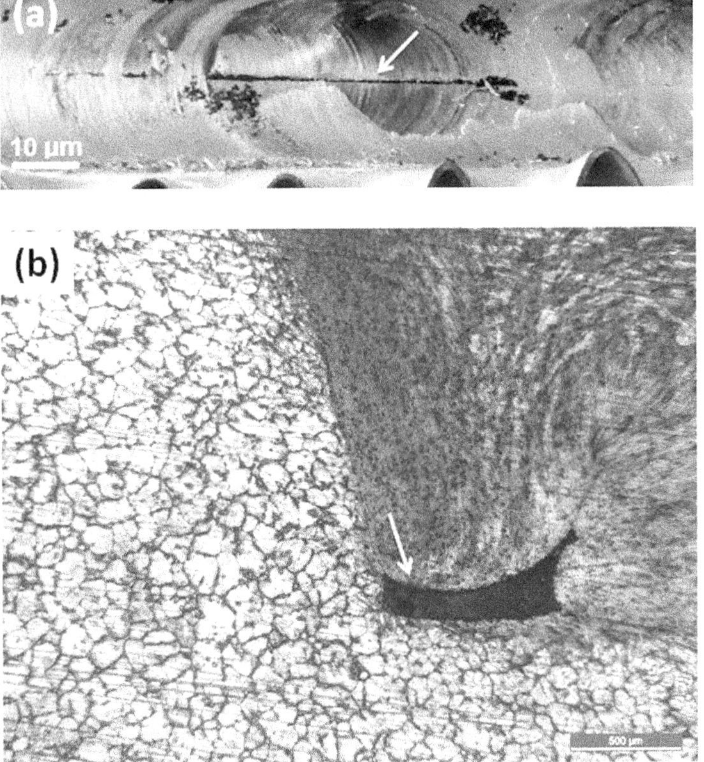

FIGURE 8.4 (a) Lack of fill defect in FSPed pure Mg, and (b) tunnel or wormhole defect observed in FSPed ZE41 Mg alloy (indicated with a white arrow).

To avoid these defects, flaws and undesirable microstructural features after FSP, understanding the thermo-mechanical behavior of the base material is crucial before doing FSP. Additionally, knowledge of the base material properties helps to design the experiments with appropriate processing inputs and increases the quality of the stir zone after FSP.

8.3.3 Inappropriate Processing Conditions

Defect due to poor design of FSP tool or the processing conditions is commonly observed in developing surface composites. Producing grooves or holes with appropriate dimensions, filling them with secondary phase powder, placing the region of the workpiece intended to be processed under the tool and adjusting the equipment are a few factors which influence the quality of the composites. Positioning the FSP tool is crucial otherwise; lack of penetration or cold processing conditions may results which decrease the rate of material flow. Designing the geometrical features of the FSP tool is also crucial which directly influence the rate of heat generation. While processing in the open air, reactive metals such as aluminum and magnesium which readily form an oxide layer when exposed to ambient condition at the elevated temperatures during FSP may leave a thin oxide layer at the surface. This oxide layer resulted due to FSP or naturally developed oxide layer before FSP if the surfaces are not properly prepared, may penetrate into the stir zone which is limited to a thickness of few micrometers. These flaws are appeared as thin snake-like lines which are difficult to find in the stir zone unless by a careful observation. This kind of microstructural features were reported by several authors in the friction stir weld joints of aluminum alloys [9–12]. From the observations, it is suggested to adopt proper surface preparation methods before doing FSP to avoid this kind of flaw.

8.4 SECONDARY PHASE DISTRIBUTION

In developing MMCs by liquid state methods, the added secondary phase particles are distributed in the matrix due to two reasons as explained by Zhao et al., [13]. During solidification, the particles are pushed by the solidification front and then segregated at the interdendritic regions.

On the other way, if the solidification front is engulfing the particles, the particles are uniformly dispersed. It was also brought by Youssef et al., [14] as observed during the formation of TiB_2reinforced aluminum composites that the critical velocity of the solidification front influences the type of distribution of the dispersed particles. If the velocity of the solidification front is below than a critical value, the particles are pushed to the interdendritic sites, and if the velocity is above the critical value, the particles are engulfed. In the liquid state methods, dispersion of the secondary phase is a challenging as different issues dictate the uniformity of the particle dispersion such as surface tension of the molten metal, the density of the dispersed phase, solid-liquid interfacial energies and the rate of solidification. However, in FSP, the dispersion is done within the solid-state and hence, the issues associated with the interfacial energies and solidification are completely eliminated. However, in FSP, the dispersion of the secondary phase particles is dependent of the material plastic flow within the solid-state.

It is true that the material flow in FSP is complex in nature as explained in several reports [15–21]. Also, the material flow behavior is different in both the advancing side and the retreating side [20]. Hence, the distribution of secondary phase powder that is added during FSP is asymmetrical and non-uniform in a single pass. Usually, after a single pass, more amount of dispersed phase is found in the advancing side compared with the retreating side in the FSPed zone. The incorporated particles become the initiation sites for recrystallization and introduce a pinning effect to arrest the grain growth [22, 23]. The dimensions of the FSP tool govern the size of the processed zone. The width of the stir zone depends on the diameter of the shoulder, and the depth of the composite formation depends on the length of the tool pin. Achieving uniform distribution of the secondary phase is a challenging task in FSP. Several approaches were proposed to uniformly distribute the secondary phase in the matrix during FSP. Tool off-set, multi-step FSP, multi-pass FSP and way of particle supply are the known strategies. Figure 8.5 (a) and (b) shows a typical scanning electron microscope (SEM) images of tungsten particles incorporated into 5083 aluminum alloy. The perfect interface of the distributed particle and the matrix is also shown in the magnified image in Figure 8.5 (c).

FIGURE 8.5 SEM images of tungsten distributed 5083 Al alloy in the composite produced by FSP: (a) low magnification, (b) high magnification, and (c) interface of tungsten particle and the matrix. (*Source:* Reprinted with permission from Shyam Kumar et al., [24]. © 2016 Elsevier.)

8.5 MULTI-PASS FSP

Multi-pass FSP was proposed by several authors to tailor the microstructure of the sheets. The defects are reduced with multi-passes, and the level of grain refinement also can be increased with the number of FSP passes. Nakata et al., [25] processed ADC12 aluminum die casting alloy by FSP up to 14 passes. The path of the processing was shifted to 4 mm towards the advancing side for every FSP pass as shown in Figure 8.6. The corresponding hardness profile and the macroscopic view at the cross-section are shown in Figure 8.6(c). Due to FSP, the cold flakes were eliminated, and uniform dispersion of finer Si particles was observed which significantly increased the tensile strength.

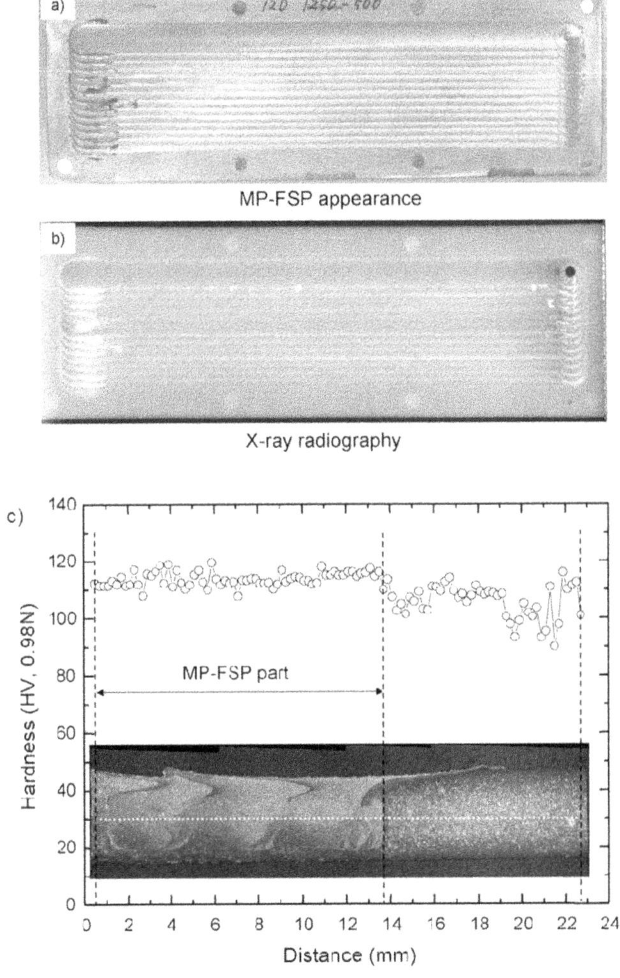

FIGURE 8.6 Surface morphology of multi-pass FSPed regions: (a) photograph showing the FSPed surface, (b) corresponding radiographs shows the defect-free surface, and (c) hardness profile and the cross-section of FSPed sheet. (*Source:* Reprinted with permission from Nakata et al., [25]. © 2006 Elsevier.)

Sing et al., [26] done three FSP passes and tribology and machining behavior was investigated for A356 alloy. Excellent wear resistance and low coefficient of friction were observed for 3 pass FSP sample. The authors have done machining by conducting drilling experiments. The cutting forces and the corresponding surface roughness were observed as a minimum for 2 pass FSP sample. Formation of built-up edge (BUE) was observed for the tools used for all the materials. The edge burr formation at the entry and exit of the hole was observed as lower for 3 pass sample compared with the other samples. From the studies, the increased FSP passes have shown a significant effect on enhancing the machinability and the wear behavior of FSPed materials. Alhosseini et al. [27] FSPed pure titanium up to 3 passes and corrosion studies were carried out using 0.5 M H_2SO_4 solution. The grain size was reached to sub-micrometer range due to continuous dynamic recrystallization after 3 passes. From the electrochemical studies, the corrosion resistance was observed as increased with a number of passes. Tripathi et al., [28] processed AZ31 magnesium alloy up to 4 FSP passes. Grain refinement was achieved up to a few hundred nanometers due to local shear and dynamic recovery/recrystallization.

John Baruch et al., [29] modified the microstructureofAS7U3G alloy by FSP up to 3 passes. Due to FSP, the Al-Si eutectic dendrite secondary phase particles were broken, and the network like distribution was disturbed as particles. Additionally, the size and aspect ratio of Al-Si particles were decreased with the increase of the FSP passes. The hardness in FSPed zone was measured as lower compared with the unprocessed alloy. Decreased dislocation density after FSP was explained as the reason behind the decreased hardness in spite of grain refinement. That indicates the significant effect of dislocations compared with grain boundary effect on hardness. Further, higher tensile and fracture properties were also observed for the multi-pass FSPed samples. Similarly increased corrosion resistance was reported in Al-Mg-Si alloys after 3 pass FSP as reported by Esmaily et al., [30]. Lu and Zhang [31] also studied the effect of multi-pass FSP in enhancing the mechanical properties of AZ91 Mg alloy. Processing was carried out for up to two passes. The average grain size of AZ91 after single pass and two passes was measured as 8.3 μm and 5.8 μm, respectively. After FSP, the network like distribution of secondary phase ($Mg_{17}Al_{12}$) was significantly decreased, and some amount of the secondary phase was found to be dissolved into the alpha Mg grains. In both cases, the microhardness in stir zone was observed as increased compared with the base material. Compared with the base material, tensile strength after single pass and two passes was almost doubled, and the % elongation was also increased to 2 and close to 3 times after FSP for single pass and two pass samples. Decreased grain size and the modified amount, size and distribution of secondary phase were the important factors that play a vital role in enhancing the mechanical properties.

TABLE 8.1 Effect of Number of FSP Passes on Different Properties of Different Materials

Material system	Number of passes	Observations	Reference
ADC12 aluminum die castingalloy	14	• Increased hardness and tensile strength	Nakata et al., [25]
A356 alloy	3	• Improved wear resistance • Decreased coefficient of friction • Improved machinability	Sing et al., [26]
Pure Ti	3	• Increased corrosion resistance due to the formation of more resistive, thick and less defective passive film	Alhosseini et al., [27]
AZ31 Mg alloy	4	• Significant texture effect as number of passes increased • Grain refinement up to a level of few hundreds of nanometer level	Tripathi et al., [28]
AS7U3G alloy	3	• Decreased Al-Si eutectic phase • Increased hardness, tensile and fracture strength	John Baruch et al., [29]
Al-Mg-Si alloy	3	• Improved corrosion resistance	Esmaily et al., [30]
AZ91 Mg alloy	2	• Higher level of grain refinement • Increased microhardness, tensile strength and % elongation.	Lu and Zhang [31]
Al 5083 alloy	3	• Improved resistance to grain growth during annealing after FSP done at higher rotational speeds	Chen et al., [32]

Chen et al., [32] have done 3 pass FSP of Al5083 alloy. After a single pass, interestingly, hardness and tensile strength of stir zone were decreased. Grain size was observed as increased when the rotational speed was increased from 600 to 1200 rpm. The mechanical properties were mostly closed even after 3 passes. In the stir zone produced at lower rotational speed, grain growth was noticed after annealing. Abnormal grain growth was suppressed after annealing the stir zone region produced at high rotational speed. The studies carried out in FSP with multiple passes clearly demonstrated the significant effect of multi-passes in enhancing the mechanical, corrosion and tribological properties. The important factors which play in the case of multi-passes are the modification of microstructure and the increase of the stir zone volume. It has been well understood that the multi-pass FSP has shown a significant effect on enhancing the material properties by modifying the microstructure. While preparing the surface composites by FSP, multi passes were done by several research groups, and promising results were also reported. Zhao et al., [33] produced a composite of 1060Al and AZ31B by using FSP. The two plates were stacked, and multi-pass FSP was carried out side by side to develop a composite surface. $Al_{12}Mg_{17}$ and Al_3Mg_2 intermetallic phases were formed in the composite zone which influences the failure mode and corrosion performance of the composite. $ZrSiO_4$ particles were introduced into AA5082 alloy by FSP up to four passes to develop composite plates as demonstrated by Rahsepar and Jarahimoghadam [34]. Due to the better distribution of the dispersed $ZrSiO_4$ particles as the number of passes was increased to four, improved tensile strength and corrosion performance were observed for the composite compared with the base material. Several other authors have also reported an increased level of uniformity in the distribution of the secondary phase by increasing the number of FSP passes. Table 8.1 briefly summarizes the work reported in multi-pass FSP.

8.6 NANO-COMPOSITES BY FSP

Nanomaterials are the important group of advanced materials which exhibit excellent properties compared with traditional materials due to the size effect. When the material size is decreased to nano-level (10–9 m), several material properties which are influenced by the quality of the lattice are severely affected. When the dispersing phase is supplied in the form of nanopowder, it offers several advantages in the materials perspective. Developing nanoparticles dispersed composites (nano-composites) through a liquid state method is complex due to the issue of high agglomeration of nanoparticles. As the surface energy for a given volume of material is increased to a great extent, nanomaterials poses very high surface energy and therefore tend to agglomerate more [35]. Presence of these agglomerates in the matrix lead to form voids and influences the interface bonding between the incorporated particles and the matrix. Hence, attaining uniform dispersion of nano-materials in the matrix through conventional methods is difficult. Here, FSP offers an advantage to develop nano-composites by uniformly dispersing the nanomaterial. Different authors demonstrated developing nanocomposites by FSP.

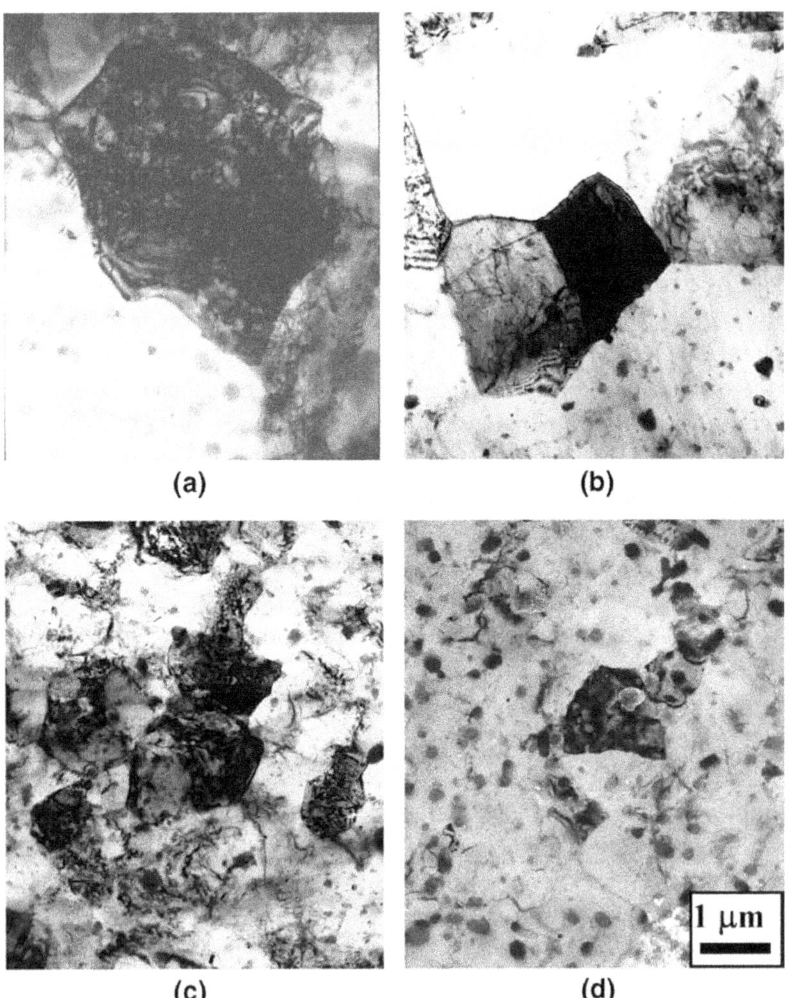

FIGURE 8.7 Transmission electron microscope (TEM) images of AZ61Mg-SiO_2 composites produced by FSP: (a) one groove one FSP pass, (b) one groove 4 FSP passes, (c) one grooved one FSP pass and (d) two grooves 4 FSP passes. (*Source:* Reprinted with permission from Lee et al., [36]. © 2006 Elsevier.)

Lee et al., [36] developed composites of AZ61 Mg alloy-nano-SiO_2 by FSP by introducing the nanoSiO_2 powder into the surface by adopting the groove filling method. One and two deep grooves were machined on the surface and filled with nano-SiO_2 amorphous powder. FSP was carried out for four passes. Excellent grain refinement and distribution of nano-SiO_2 was observed in the composite as shown in TEM images (Figure 8.7). With the addition of nano-SiO_2, grain refinement was observed as better compared with FSPed AZ61 Mg alloy without reinforcement. Excellent mechanical properties and high strain rate susceptibility (400%) was achieved in the composite. Shafiei-Zarghani et al. [37] produced nano-composite of AA6082-Al_2O_3 by FSP and processed up to four passes and the mechanical properties and wear properties were investigated. An increase of hardness up to 295Hv after fourth FSP pass was observed. Due to the presence of nano-Al_2O_3 particles, grain refinement was achieved up to 300 nm. From the wear studies, both the abrasion and adhesion wear were observed as the mechanisms behind surface wear and better wear resistance were noticed for the composite. Improved microhardness and better wear behavior for the composite were attributed to the Orowan mechanism and the grain size effect. Barmouz et al. [38] developed polymer nano-composite by FSP using high-density polyethylene (HDPE) as matrix and nano-clay as dispersing phase. From the hardness measurements, three times higher hardness was observed for FSPed composite compared with the one produced by mixing and melting. The important reason for increased hardness in FSPed composite was attributed to the better dispersion of the nano-clay particles in the polymer. Izadi and Gerlich [39] produced a composite of AA5059-multi-walled carbon nanotubes (MWCNT) and demonstrated the stability of CNT after two FSP passes. But the uniform distribution of CNT was achieved after three passes. The damage which was observed to CNT was explained due to the high shear stress at the elevated temperatures in the stir zone. The average microhardness of the composite was observed as two times compared with that of AA5059 alloy. This kind of behavior was also reported by other authors [40, 41]. Liu et al. [40] noticed breakage of MWCNTs during FSP which was attributed to intense stirring action due to FSP. Ultrafine grains (50–100 nm) were observed in some regions of stir zone. No interfacial reaction was observed between MWCNT and the matrix. By adding MWCNT, the hardness was increased up to 2.2 times compared with FSPed alloy without MWCNT. Similarly, slight damage was noticed in the composites of AZ31-MWCNT produced by FSP as reported by Saikrishna et al., [41].

The distribution of CNT was observed as better after FSP in the AA2009-CNT composite produced by powder forging route as reported by Liu et al. [42]. The cluster size of CNT was measured as decreased after four FSP passes. The damage to CNT after four FSP passes was observed as negligible. The CNTs in the composite were observed as shortened and formation of Al_4C_3 was noticed in the matrix.Liu et al. [43] have done rolling after developing composites by FSP. An amount of 1.5–4.5 vol. % CNT was reinforced in AA2009 alloy by conducting overlapping multi-pass FSP. Due to rolling, it was observed that the CNTs were individually well dispersed in AA2009 alloy and directionally aligned. The interfacial reaction was observed as negligible, and the CNTs were stable. Farnoush et al. [44] introduced nano-hydroxyapatite into Ti-6Al-4V alloy and developed surface composited by FSP. Later, a surface coating of HA on the FSPed Ti-HA surface composite was deposited by electrophoretic deposition. The bonding strength between the coating layer and the surface was observed as twice compared with the surface coating of HA provided on the as-received alloy. The reason behind increased bonding strength between the surface coating and the substrates was attributed to the presence of nano-HA particles in the composite which also reduced the thermal expansion mismatch between the matrix (Ti alloy) and the dispersing phase (HA). Ratna Sunil et al. [2, 45] produced Mg-nano-HA and AZ31-nano-HA composites by FSP targeted for degradable implant applications. Due to the grain refinement and the presence of nano-HA, enhanced corrosion resistance and increased bioactivity as reflected from the biomineralization and the cell response from the tissue culture studies was reported.

Compared with conventional composites, nano-composites offer several advantages due to the added effect of the dispersed nanoparticles. FSP can be a viable route to disperse nano-particles and relatively lower agglomeration, and better distribution can be achieved in FSP compared with liquid state methods. Additionally, incorporated nano-powder helps to retain the fine grains evolved due to dynamic recrystallization by introducing pining effect to arrest the grain growth.

8.7 IN-SITU COMPOSITES BY FSP

Another attractive route to develop surface composites is to develop in-situ composites by allowing reaction in the matrix to develop precipitates within the matrix material. In order to promote a suitable chemical reaction that produces desired secondary phase particles in a matrix, thermal energy is required from the external source or from the internal material by any means. The in-situ composite formation can be done through liquid state methods by melting and solidification. But, control over the amount of the secondary phase formation and the level of the distribution of the formed secondary phase is important limitations. Furthermore, agglomeration of the formed secondary phase particles within the matrix is more concerned. Availability of required thermal energy that is produced in the stirring action of the FSP tool and plastic deformation during the recrystallization process accelerates the formation of the secondary phases from the matrix in developing in-situ composites by FSP. Additionally, FSP offers several advantages such as smaller grain size and better distribution of the secondary phase during the formation itself compares with liquid state methods. The precipitated secondary phase particles during FSP are usually at the nano-level and hence, offers an additional advantage as the outcome is a nano-composite. Due to the stirring action of the FSP tool, material that flows around the pin and beneath the shoulder is also stirred, and the formed secondary phase particles are also stirred away from the location of formation and growth of the secondary phase particles is arrested, and nano-size is retained [46].

As explained by Sharma et al., [47] and illustrated in Figure 8.8, in the Aluminum system, in-situ intermetallic phase Al_xM is formed due to reaction at the interface of the aluminum matrix and the dispersed secondary phase particles during FSP. Later, Al_xM phase is dispersed as particles in the matrix due to plastic strain induced by FSP. Again, the freshly dispersed particle comes into contact with the aluminum matrix, and an intermetallic compound is formed. This kind of reaction is repeated, and in-situ phase formation leads to develop composite. Most of the studies were carried out by using an aluminum matrix. Hsu et al., [46] developed in-situ Al-Al_3Ti composite by dispersing Ti particles into the aluminum matrix. Ke et al., 2010 [48]introduced Ni particles into the Al matrix by FSP and successfully developed in-situ Al-Al_3Ni in-situ composite.Similarly, Lee et al., [49] and Hsu et al., [50], developed Al-Al3Fe, and Al-Al2Cu nanocomposites by using Al-Fe and Al-Cu respectively. Formation of several other phases such as Al_2O_3 or MgO was also reported by several authors in developing in-situ composites by FSP [51–53].

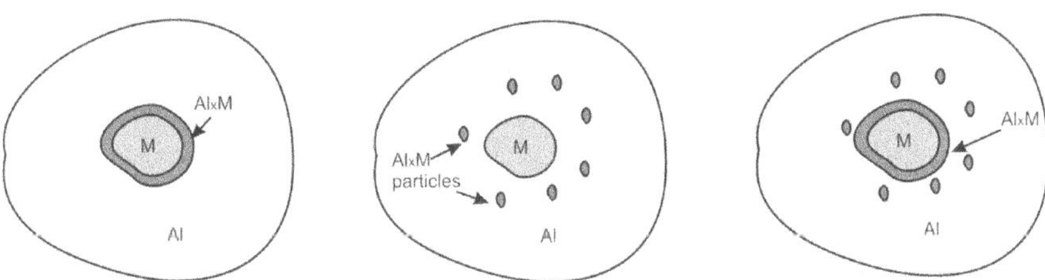

FIGURE 8.8 Schematic representation of aluminum-based in-situ composite formation during FSP. (*Source:* Reprinted with permission from : Sharma et al., [47]. © 2015 Elsevier.)

Chen et al. [51] mixed aluminum and cerium oxide powders and produced aluminum based in-situ composites. The distribution of the in-situ reinforcing phases characterized as $Al_{11}Ce_3$ of the average size of 1.4–3.5 μm and δ-Al_2O_3 with an average size of ~10 nm was observed as uniform after FSP. You et al. [54] used aluminum and copper oxide powder mixture and developed in situ composites followed by FSP. From the microstructural observation, the formation of nano-sized Al_2O_3 and Al_2Cu was noticed due to FSP. Lee et al. [55] used aluminum and molybdenum as starting materials and observed the formation of Al-Mo intermetallic phase particles after FSP. The authors demonstrated the exothermic reaction between aluminum and molybdenum as the important mechanism to produce fine Al-Mo intermetallic particles which were measured at a size of 200 nm.

Anvari et al. [56] applied the Cr_2O_3 powder on AA6061-T6 substrate by plasma spray method. Then FSP was carried out up to six passes on the surface. It was observed that due to FSP, Cr_2O_3 was reduced with aluminum and produced pure Cr and Al_2O_3. Furthermore, intermetallic compounds such as $Al_{13}Cr_2$ and $Al_{11}Cr_2$ were formed due to the reaction between Al and Cu. These in-situ composites produce by FSP exhibited excellent wear resistance. Rajan et al., [57] developed composites ofAA7075/TiB_2 by the in-situ reaction of inorganic salts K_2TiF_6 and KBF_4 with molten aluminum followed by FSP. Due to FSP, the secondary phases which were formed due to in-situ chemical reaction were uniformly distributed. The composites after FSP exhibited excellent mechanical and wear behavior. Figure 8.9 shows the fracture micrographs of unprocessed and composites with FSP and without FSP.

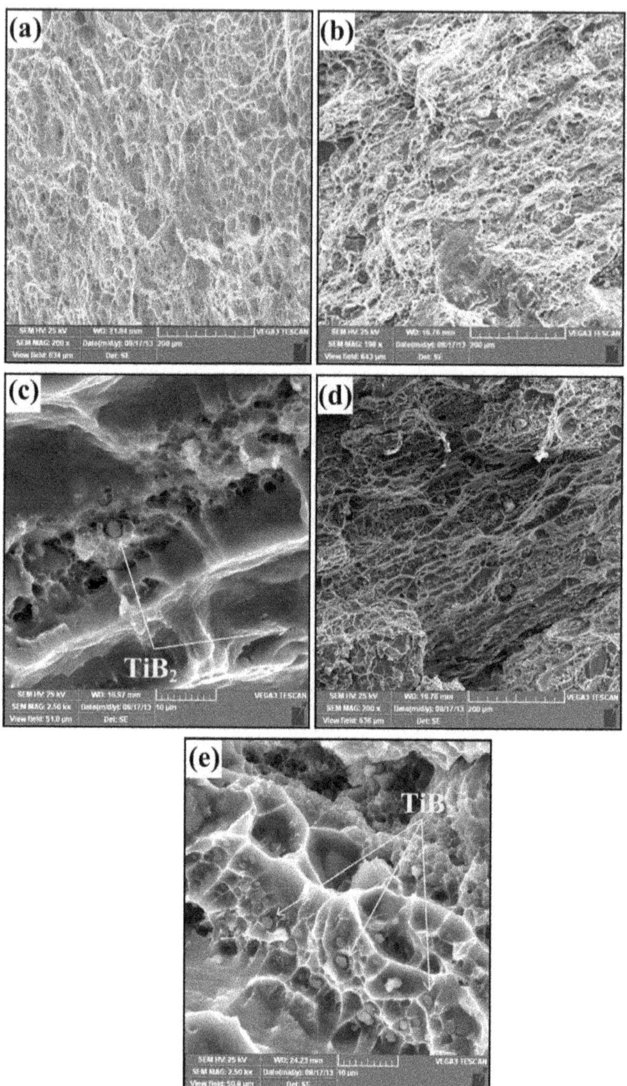

FIGURE 8.9 SEM images showing fracture micrographs of (a), unprocessed AA7075, (b) and (c) aa7075/TiB_2 composite, (d) and (e) FSPed AA7075/TiB_2 composite. Reprinted with permission from Rajan et al., [57]. © 2016 Elsevier.)

In order to form a secondary phase from the reaction between the matrix and the dispersed particles, sufficient time of exposure at elevated temperature is crucial. If the temperature and time to promote reaction during FSP are insufficient, in-situ composite do not develop. Khodabakhshi et al., [52] reinforced AA5052 alloy with TiO_2 by FSP and reaction to develop secondary phase was not seen in the composite significantly. The in-situ reaction was not completed, and around 5% of TiO_2 was observed as remained without any reaction in the composite after FSP. However, the authors adopted heat treatment (annealing at 400°C) and the amount of un-reacted TiO_2 particles were observed as decreased. After heat treatment, the composites exhibited excellent tensile properties and hardness without sacrificing the ductility. Similarly, Ke et al., [58], introduced Ni powder into Al matrix by FSP with an aim to develop in-situ composites and after three passes, un-reacted Ni particles were observed. The authors have done follow up heat treatment at 550°C for six hours, and the reaction between Ni and the matrix was observed as increased. Due to the formation of the in-situ nanocomposite by FSP and follow up heat treatment, the ultimate tensile strength was measured as increased up to 171% compared with FSPed aluminum.

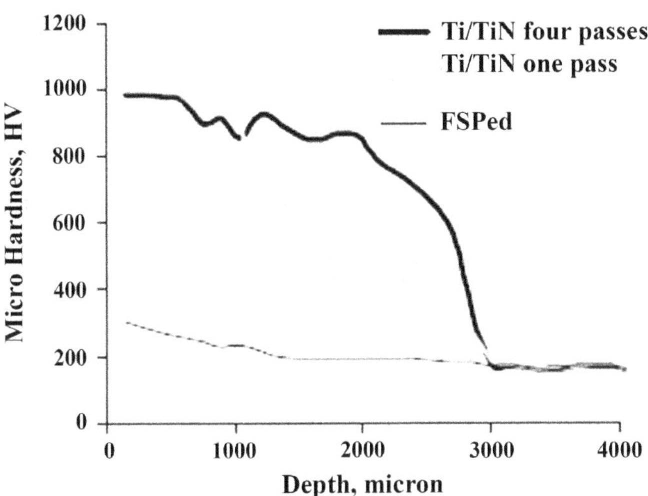

FIGURE 8.10 Microhardness profile of FSPed Ti and FSPed Ti/TiN composite. (*Source:* Reprinted with permission from Shamsipur et al., [62]. © 2013 Elsevier.)

Zhang et al., [59] also reported similar kind of observations in developing in-situ composites from Al and Ti by FSP. At high temperature, Ti reacted with Al and produced Al_3Ti particles of size varying from 0.5 to 3 μm. After four FSP passes, the size of the intermetallic particles was measured as decreased and the dispersion was observed as homogeneous. Other works of Zhang et al. [60, 61] in developing Al-TiO_2 and Al-Ti-Cu also demonstrated the importance of FSP on achieving uniform distribution. Shamsipur et al., [62] processed CP-Ti in the presence of Ni gas and formation of TiN precipitates of size 2 μm was observed with fine grains of size 4.5 μm. For comparison, FSP of Ti was also carried out in the presence of Argon. Due to the formation of hard TiN particles at the surface, six times of increment in microhardness was measured for the compared with unprocessed CP-Ti as shown in Figure 8.10. FSPed Ti has also shown two times increase in hardness compared with unprocessed Ti. The better mechanical properties for the composite were attributed to the combined effect of smaller grain size and the presence of TiN particles. Zhao et al., [63] produced in situ nano-ZrB_2/2024 composites by using 2024 Al-K_2ZrF_6–KBF_4 system at 1143 K and FSP was carried out to uniformly redistribute the ZrB_2 particles. The FSPed composite exhibited better mechanical properties, and in particular, superplasticity of the composite was excellent 292.5% at 750 K whereas the cast composite showed only 67.2% of elongation.

Developing in-situ composites though FSP or processing in-situ composites by FSP to achieve a better distribution of the formed secondary phases in the matrix, offer several advantages. The bonding strength between the secondary phase particle and the matrix is relatively higher compared with the composites produced by externally adding the secondary phase into the matrix by different routes. The overall work reported in this area is limited; however; the preliminary reports demonstrate the potential of FSP to develop in-situ composites.

8.8 HYBRID COMPOSITES BY FSP

Incorporating two or more than two secondary phase particles into the matrix produces hybrid composites. It is possible to develop hybrid composites through FSP. Several authors developed hybrid composites, and interesting findings were demonstrated. Hybrid composites give excellent material properties compared with single component reinforce composites as hybrid composites contain two or more than two reinforcing phases. Hybrid composites offer the combined advantages of constituting phases [64]. Among the phases which are added to the matrix, one phase is hard compared with the other. However, optimizing the relative % of each phase is crucial in developing hybrid composites to achieve better properties. Eskandari et al., [65] developed mono-composites by dispersing TiB_2 and Al_2O_3 particles into AA8026 matrix and also developed hybrid composites by dispersing both the phases in different proportions. Hybrid composites exhibited excellent hardness, tensile properties and wear properties compared with the nanocomposites and base material. Adding the two phases in appropriate ratios is crucial in developing hybrid composites. Mostafapour Asl and Khandani [66] produced a hybrid composite of AA5083-graphite/Al_2O_3 by FSP. The relative amounts of graphite and Al_2O_3 were altered and developedthe hybrid composites to study the effect of the ratio of the dispersing phases on mechanical and tribological properties. With a hybrid ratio of 3 (graphite/Al_2O_3), highest wear resistance and with a ratio of 1, optimum wear properties coupled with better tensile properties were observed.

Alidokht et al. [67] used molybdenum disulfide (MoS_2) and SiC to develop hybrid composites in A356 alloy matrix by FSP. Due to the presence of $MoS_{2,}$ decreased wear rate was observed as a stable mechanically mixed layer was formed due to FSP. The hardness of the hybrid composite was measured as lower compared with A356/SiC composite due to the presence of soft MoS_2 phase. Even though A356/SiC composite exhibited higher hardness, wear resistance for the hybrid composite was higher due to the presence of $MoS_{2.}$ Inthe absence of $MoS_{2,}$ the formation of stable mechanically mixed layer is not possible, and hence, lower wear resistance was noticed compared with the hybrid composite. Soleymani et al. [68] also introduced MoS_2 along with SiC particles into AA5083 alloy with a ratio of 1:2 by FSP process. Both the mono-composites (AA5083/SiC and AA5083/MoS_2) have shown poor wear resistance compared with the hybrid composite. Similar to the findings of Alidokht et al. [67], a solid lubricating layer was formed in the hybrid composite and promoted better wear resistance in the hybrid composites.

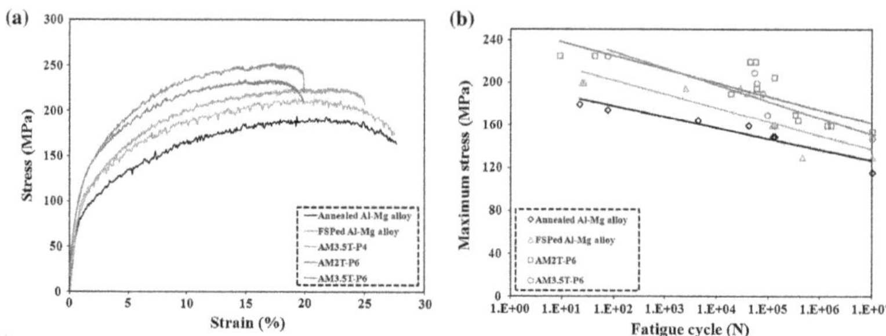

FIGURE 8.11 Mechanical behavior of base material, FSPed and hybrid composites: (a) stress-strain curves, and (b) S-N curves showing the fatigue response at R=0.1. (*Source:* Reprinted with permission from Zangabad et al., [70]. © 2016 Elsevier.)

Devaraju et al. [69] produced hybrid composites of AA6061/SiC + graphite with an aim to investigate the effect of dispersing two different phases on the mechanical and tribological properties. It was reported that with an increase in the fraction of graphite up to certain limit, hardness and wear resistance of the composite was observed as decreased. This behavior was explained by observing the formation of mechanically mixed layer which usually acts as a solid lubricant and decreases the direct metal to metal contact. Further by increasing the graphite fraction in the hybrid composite, wear resistance was observed as increased.

Zangabad et al., [70] developed hybrid composites of Al–Al_3Ti–MgO by FSP and fatigue behavior was investigated. FSP produced fine grains, and after 6 passes, improved dispersion was observed in the hybrid composites. Mechanical behavior of the composites was assessed by tensile tests and fatigue test, and the results are shown in Figure 8.11. The tensile strength of the composites was increased to a great extent, but ductility was decreased up to 30%. The yield strength was increased by 90%, and the ultimate tensile strength was increased by 31% in the hybrid composites. Higher fatigue strength was noticed for the hybrid composite. The two important properties which were observed as greatly influenced by hybrid composites are mechanical and wear properties. However, in most of the reported work, decreased ductility is a common observation which also needs attention. The effect of the presence of combined phases, which possesses different electrochemical behavior on the bulk corrosion behavior of the hybrid composites, is another issue which needs in-depth studies.

8.9 ADVANTAGES AND LIMITATIONS WITH FSP IN DEVELOPING SURFACE MMCS

Being a solid-state processing technique, FSP offers many advantages in the fabrication of composites compared with other manufacturing techniques. Phase stability is the prime advantage in FSP compared with melting and solidification practice. The FSP process needs simple equipment which is already established well in the manufacturing industry. Automated vertical type milling or drilling machines can be easily used as FSP equipment by providing an extra attachment for tool holding. FSP technique can be readily adopted by the industries in fabricating surface composites as it doesn't require much alteration to the existing machine tools which are already widely used in various manufacturing industries. Furthermore, FSP doesn't result in workpiece melting. Therefore, the oxidation problem during the solidification of the liquid metal can be eliminated. Precise control over the handling of the liquid phase of the metal also can be completely eliminated. The temperature that is generated during the process is easily dissipated to the base as the base metal which is relatively large compared with the FSP region act as a sink. Thermal stresses which arise during FSP are minimum compared with other conventional methods. Along with fabricating the composite or developing alloy phases at the surface, FSP leads grain refinement. At the end of the process, the final surface after FSP attains microstructural refinement with secondary phase particles embedded. The secondary phase particles are dispersed within the fine grain structured matrix. The combined advantage of grain refinement and particle dispersion can be achieved by FSP. The two prominent strengthening mechanisms: (i) grain boundary strengthening, and (ii) particle dispersion strengthening can be achieved at the surface level itself, and the core metal still exhibits ductile and toughness compared with the surface as shown in Figure 8.12.

In spite of the advantages, several limitations are also associated with FSP in developing surface MMCs. Similar to casting or any other liquid state methods, the composites cannot be produced in bulk by FSP. The geometry of the processed material is usually in the form of sheets or plates. The surfaces with irregular features cannot be processed. The effected thickness which is dispersed with the secondary phase is limited as it is a dependent of the tool pin dimensions. The width of the composite surface which is produced after each pass is also limited which is equal

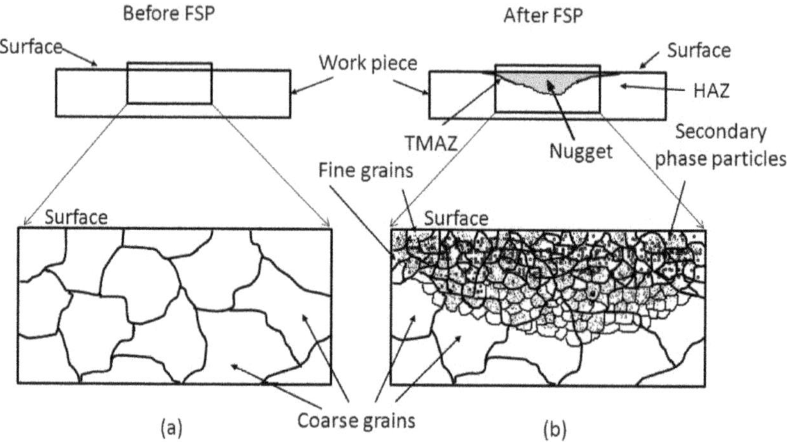

FIGURE 8.12 Schematic representation of the microstructure at the cross section of sheets: (a) before FSP and (b) after FSP.

to the diameter of the tool shoulder. In order to develop the more surface composite area, multi-pass FSP side by side is required. The distribution of the dispersed phase is again an important concern which may require overlap multi-pass FSP. Tool design is crucial, and the success of the composite formation by using a certain tool design may not guarantee as the optimum tool design for other alloys which specifically demands special designs for different materials.

KEYWORDS

- **FSP**
- **hybrid composites**
- **in situ composites**
- **MMC**
- **multi-pass FSP**
- **nano-composites**
- **scanning electron microscope**

REFERENCES

1. Mishra, R. S., Ma, Z. Y., & Charit, I., (2003). Friction stir processing: A novel technique for fabrication of surface composite. *Mater. Sci. Eng. A., 341*, 307–310.
2. Ratna, S. B., Sampath, K. T. S., Chakkingal, U., Nandakumar, V., & Doble, M., (2014). Nano-hydroxyapatite reinforced AZ31 magnesium alloy by friction stir processing: A solid-state processing for biodegradable metal matrix composites. *J. Mater. Sci. Mater. Med., 25*. 975–988.
3. Dickerson T. L., & Przydatek J., (2003). Fatigue of friction stir welds in aluminum alloys that contain root flaws. *International Journal of Fatigue, 25,* 1399–1409.
4. Zhou, C., Yang, X., & Luan, G., (2006). Effect of root flaws on the fatigue property of friction stir welds in 2024-T3 aluminum alloys. *Materials Science and Engineering A., 418*, 155–160.
5. Zettler, R., Lomolino, S., Dos Santos, J. F., Donath, T., Beckmann, F., Lipman, T., & Lohwasser, D., (2004). A study on material flow in FSW of an AA 2024-T351 and AA6056-T4 alloys. *5th International FSW Symposium, Metz, France, 5, 1–13.*
6. Lombard, H., Hattingh, D. G., Steuwer, A., & James, M. N., (2008). Optimizing FSW process parameters to minimize defects and maximize fatigue life in 5083-H321 aluminum alloy. *Engineering Fracture Mechanics, 75*, 341–354.
7. Colegrove, P. A., Shercliff, H. R., & Zettler, R., (2007). Model for predicting heat generation and temperature in friction stir welding from the material properties. *Science and Technology of Welding and Joining,12*(4), 284–297.
8. Zettler, R., (2008). *PhD Thesis*, GKSS Forschungszentrum, Geesthacht, Germany.
9. Sato, Y. S., Takauchi, H., Park, S. H. C., & Kokawa, H., (2005). 'Characteristics of the kissing-bond in friction stir welded Al alloy 1050.' *Materials Science and Engineering, A., A405*, 333–338.
10. Okamura, H., Aota, K., Sakamoto, M., Ezumi, M., & Ideuchi, K., (2002). 'Behavior of oxides during friction stir welding of aluminum alloy and their effect on its mechanical properties.' *Welding International, 16*(4), 266–275.
11. Palm, F., Steiger, H., & Henneböhle, U., (2003). 'The origin of particle (oxide) traces in friction stir welds.' *Proceedings of 4th International Symposium on Friction Stir Welding, Park City, Utah, USA*, ISBN 1–903761–01–8.
12. Threadgill, P. L., Leonard, P. L., Shercliff, H. R., & Withers, P. J., (2009) *Friction Stir Welding of Aluminium alloys*, Int. Mater. Review., *54* (2), 49–93.
13. Zhao, D., Liu, X., Liu, Y., & Bian, X., (2005). *In-situ* preparation of Al matrix composites reinforced by TiB2 particles and sub-micron ZrB2. *J. Mater. Sci., 40*, pp. 4365–4368.
14. Youssef, Y. M., Dashwood, R. J., & Lee, P. D., (2005). Effect of clustering on particle pushing and solidification behavior in TiB2 reinforced aluminum PMMCs. *Composites, A., 36*, pp. 747–763.
15. Dixit, M., Newkirk, J. W., & Mishra, R. S., (2007). *Scripta Mater., 56,* pp. 541–544.
16. Heydarian, A., & Dehghani, K., (2011). *3rd International Conference on Ultrafine Grained and Nanostructured Materials.*
17. Colligan, K., Material Flow Behavior during Friction Stir Welding of Aluminum, Weld, J., (1999). *78*, 229S–237S

18. Lorrain, O., Favier, V., Zahrouni, H., & Lawrjaniec, D., (2010). *J. Mater. Process. Technol., 210*, pp. 603–609.
19. Guerra, M., Schmidt, C., McClure, J. C., Murr, L. E., & Nunes, A. C., (2003). *Mater. Charact.,49*, pp. 95–101.
20. Schneider, J. A., & Nunes, A. C., (2004). *Metall. Mater. Trans. B., 35B*, pp. 777–783.
21. Liechty, B. C., & Webb, B. W., (2008). *J. Mater. Process. Technol., 208*, pp. 431–443.
22. Zhang, W. L., Wang, J. X., Yang, F., Sun, Z. Q., & Gu, M. Y., (2006). *J. Compos. Mater.,40*, pp. 1117–31.
23. Yazdipour, A., Shafiei, A. M., & Dehghani, K., (2009). *Mater. Sci. Eng. A., 527*, pp. 192–97.
24. Shyam, K. C. N., Ranjit, B., & Devinder, Y., (2016). Wear properties of 5083 Al – W surface composite fabricated by friction stir processing. *Tribology International,101*, 284–290.
25. Nakata, K., Kim, Y. G., Fujii, H., Tsumura, T., & Komazaki, T., (2006). Improvement of mechanical properties of aluminum die casting alloy by multi-pass friction stir processing. *Materials Science and Engineering A., 437*, 274–280.
26. Singh, S. K., Immanuel, R. J., Babu, S., Panigrahi, S. K., & Janaki, R. G. D., (2016). Influence of multi-pass friction stir processing on wear behavior and machinability of an Al-Si hypoeutectic A356 alloy. *Journal of Materials Processing Technology, 236*, 252–262.
27. Arash, F. A., Farid, R. A., & Mojtaba, V. A., (2016). Effect of multi-pass friction stir processing on the electrochemical and corrosion behavior of pure titanium in strongly acidic solutions. *Metallurgical and Materials Transactions, A.* doi: 10.1007/s11661–016–3854–3.
28. Tripathi, A., Tewari, A., Kanjarla, A. K., Srinivasan, N., Reddy, G. M., Zhu, S. M., Nie, J. F., Doherty, R. D., & Samajdar, I., (2016). Microstructural evolution during multi-pass friction stir processing of a magnesium alloy. *Metallurgical, and Materials Transactions, A.* doi: 10.1007/s11661–016–3403–0.
29. John, B. L., Raju, R., Balasubramanian, V., Rao, A. G., & Dinaharan, I., (2016). Influence of multi-pass friction stir processing on microstructure and mechanical properties of die cast Al–7Si–3Cu aluminum alloy. *Acta Metall. Sin. (Engl. Lett.), 29*(5), 431–440.
30. Esmaily, M., Mortazavi, N., Osikowicz, W., Hindsefelt, H., Svensson, J. E., Halvarsson, M., Thompson, G. E., & Johansson, L. G., (2016). Influence of multi-pass friction stir processing on the corrosion behavior of an Al-Mg-Si alloy. *Journal of The Electrochemical Society, 163*(3), C124–C130.
31. Zhilong, L., & Datong, Z., (2016). Microstructure and mechanical properties of a fine-grained '91 magnesium alloy prepared by multi-pass friction stir processing. *Materials Science Forum, 850*, 778–783.
32. Yu, C., Hua, D., Jizhong, L., Zhihui, C., Jingwei, Z., & Wenjing, Y., (2016). Influence of multi-pass friction stir processing on the microstructure and mechanical properties of Al-5083 alloy. *Materials Science & Engineering A., 650*, 281–289.
33. Yadong, Z., Zhimin, D., Changbin, S., & Ying, C., (2016). Interfacial microstructure and properties of aluminum-magnesium AZ31B multi-pass friction stir processed composite plate. *Materials and Design, 94*, 240–252.

34. Mansour, R., & Hamed, J., (2016). The influence of multi-pass friction stir processing on the corrosion behavior and mechanical properties of zircon-reinforced Al metal matrix composites. *Materials Science Engineering, A., 671*, 214–220.
35. Tjong, S. C., & Ma, Z. Y., (2000). Microstructural and mechanical characteristics of in situ metal matrix composites. *Materials Science and Engineering R: Reports, 29*, 49–113. doi:10.1016/S0927–796X(00)00024–3.
36. Lee, C. J., Huang, J. C., & Hsieh, P. J., (2006). Mg-based nano-composites fabricated by friction stir processing. *Scripta Materialia., 54*, 1415–1420.
37. Shafiei-Zarghani, A., Kashani-Bozorg, S. F., & Zarei-Hanzaki, A., (2009). Microstructures and mechanical properties of Al/Al2O3 surface nano-composite layer produced by friction stir processing. *Materials Science and Engineering: A., 500*, 84–91. doi: 10.1016/j.msea.2008.09.064.
38. Barmouz, M., Seyfi, J., Givi, M. K. B., Hejazi, I., & Davachi, S. M., (2011). A novel approach for producing polymer nanocomposites by in-situ dispersion of clay particles via friction stir processing. *Materials Science and Engineering: A., 528*, 3003–3006. doi: 10.1016/j.msea.2010.12.083.
39. Izadi, H., & Gerlich, A. P., (2012). Distribution and stability of carbon nanotubes during multi-pass friction stir processing of carbon nanotube/aluminum composites. *Carbon, 50,* 4744–4749. doi: 10.1016/j.carbon.2012.06.012.
40. Liu, Q., Ke, L., Liu, F., Huang, C., & Xing, L., (2013). Microstructure and mechanical property of multi-walled carbon nanotubes reinforced aluminum matrix composites fabricated by friction stir processing. *Materials and Design 45,* 343–348. doi: 10.1016/j.matdes.2012.08.036.
41. Saikrishna, A. N. E. G., Pradeep, K. R. B., Balakrishnan, M., Ravikumar, D., Jagannatham, C. M., & Ratna, S. B., (2018). An investigation on the hardness and corrosion behavior of MWCNT/Mg composites and grain refined Mg. *Journal of Magnesium and Alloys*, (in Press).
42. Liu, Z. Y., Xiao, B. L., Wang, W. G., & Ma, Z. Y., (2012). Singly dispersed carbon nanotube/aluminum composites fabricated by powder metallurgy combined with friction stir processing. *Carbon, 50*, 1843–1852. doi: 10.1016/j.carbon.2011.12.034.
43. Liu, Z. Y., Xiao, B. L., Wang, W. G., & Ma, Z. Y., (2013). Developing high-performance aluminum matrix composites with directionally aligned carbon nanotubes by combining friction stir processing and subsequent rolling. *Carbon, 62,* 35–42. doi: 10.1016/j.carbon.2013.05.049.
44. Farnoush, H., Sadeghi, A., Abdi, B. A., Moztarzadeh, F., & Aghazadeh, M. J., (2013). An innovative fabrication of nano-HA coatings on Ti-CaP nanocomposite layer using a combination of friction stir processing and electrophoretic deposition. *Ceramics International, 39*, 1477–1483. doi: 10.1016/j.ceramint.2012.07.092.
45. Ratna, S. B., Sampath, K. T. S., Uday, C., Nandakumar, V., & Mukesh, D., (2014). 'Friction stir processing of magnesium – nanohydroxyapatite composites with controlled in vitro degradation behavior.' *Mater. Sci. Eng., C.,39*, 315–324.
46. Hsu, C. J., Chang, C. Y., Kao, P. W., Ho, N. J., & Chang, C. P., (2006). Al-Al3Ti nanocomposites produced in situ by friction stir processing. *Acta Materialia., 54*, 5241–5249. doi: 10.1016/j.actamat.2006.06.054.
47. Vipin, S., Ujjwal, P., & Manoj, K. B. V., (2015). Surface composites by friction stir processing: A review. *Journal of Materials Processing Technology, 224*, 117–134.

48. Ke, L., Huang, C., Xing, L., & Huang, K., (2010). Al-Ni intermetallic composites produced in situ by friction stir processing. *Journal of Alloys and Compounds, 503*, 494–499. doi: 10.1016/j.jallcom.2010.05.040.
49. Lee, I. S., Kao, P. W., & Ho, N. J., (2008). Microstructure and mechanical properties of Al-Fe in situ nanocomposite produced by friction stir processing. *Intermetallics, 16*, 1104–1108. doi: 10.1016/j.intermet.2008.06.017.
50. Hsu, C. J., Kao, P. W., & Ho, N. J., (2005). Ultrafine-grained Al-Al2Cu composite produced in situ by friction stir processing. *Scripta Materialia., 53*, 341–345. doi: 10.1016/j.scriptamat.2005.04.006.
51. Chen, C. F., Kao, P. W., Chang, L. W., & Ho, N. J., (2009). Effect of processing parameters on microstructure and mechanical properties of an Al-Al11Ce3-Al2O3 in-situ composite produced by friction stir processing. *Metallurgical and Materials Transactions A., 41*, 513–522. doi: 10.1007/s11661–009–0115–8.
52. Khodabakhshi, F., Simchi, A., Kokabi, A. H., Gerlich, A. P., & Nosko, M., (2014). Effects of post-annealing on the microstructure and mechanical properties of friction stir processed Al-Mg-TiO2 nanocomposites. *Materials and Design, 63,* 30–41. doi: 10.1016/j.matdes.2014.05.065.
53. Zhang, Q., Xiao, B. L., Wang, Q. Z., & Ma, Z. Y., (2011). *In situ*, Al3Ti and Al2O3 nanoparticles reinforced Al composites produced by friction stir processing in an Al-TiO2 system. *Materials Letters, 65,* 2070–2072. doi: 10.1016/j.matlet.2011.04.030.
54. You, G. L., Ho, N. J., & Kao, P. W., (2013b). The microstructure and mechanical properties of an Al-CuO in-situ composite produced using friction stir processing. *Materials Letters, 90*, 26–29. doi: 10.1016/j.matlet.2012.09.028.
55. Lee, I. S., Kao, P. W., Chang, C. P., & Ho, N. J., (2013). Formation of Al-Mo intermetallic particle-strengthened aluminum alloys by friction stir processing. *Intermetallics, 35*, 9–14. doi: 10.1016/j.intermet.2012.11.018.
56. Anvari, S. R., Karimzadeh, F., & Enayati, M. H., (2013). Wear characteristics of Al-Cr-O surface nano-composite layer fabricated on Al6061 plate by friction stir processing. *Wear, 304*, 144–151. doi: 10.1016/j.wear.2013.03.014.
57. Michael, R. H. B., Dinaharan, I., Ramabalan, S., & Akinlabi, E. T., (2016). Influence of friction stir processing on microstructure and properties of AA7075/TiB2 in situ composite. *Journal of Alloys and Compounds, 657*, 250–260.
58. Ke, L., Huang, C., Xing, L., & Huang, K., (2010). Al-Ni intermetallic composites produced in situ by friction stir processing. *Journal of Alloys and Compounds, 503*, 494–499. doi: 10.1016/j.jallcom.2010.05.040.
59. Zhang, Q., Xiao, B. L., Wang, Q. Z., & Ma, Z. Y., (2011). *In situ*, Al3Ti and Al2O3 nanoparticles reinforced Al composites produced by friction stir processing in an Al-TiO2 system. *Materials Letters, 65,* 2070–2072. doi: 10.1016/j.matlet.2011.04.030.
60. Zhang, Q., Xiao, B. L., Wang, W. G., & Ma, Z. Y., (2012). Reactive mechanism and mechanical properties of in situ composites fabricated from an Al-TiO2 system by friction stir processing. *Acta Materialia, 60*, 7090–7103. doi: 10.1016/j.actamat.2012.09.016.
61. Zhang, Q., Xiao, B. L., & Ma, Z. Y., (2013). *In situ* formation of various intermetallic particles in Al-Ti-X(Cu, Mg) systems during friction stir processing. *Intermetallics, 40*, 36–44. doi: 10.1016/j.intermet.2013.04.003.

62. Ali, S., Seyed, F., Kashani, B., & Abbas, Z. H., (2013). Production of in-situ hard Ti/TiN composite surface layers on CP-Ti using reactive friction stir processing under nitrogen environment. *Surface & Coatings Technology, 218*, 62–70.
63. Yutao, Z., Xizhou, K., Gang, C., Weili, L., & Chunmei, W., (2016). Effects of friction stir processing on the microstructure and superplasticity of in situ nano-ZrB2/2024 Al composite. *Progress in Natural Science: Materials International, 26*, 69–77.
64. Basavarajappa, S., Chandramohan, G., Mahadevan, A., Thangavelu, M., Subramanian, R., & Gopalakrishnan, P., (2007). Influence of sliding speed on the dry sliding wear behavior and the subsurface deformation on hybrid metal matrix composite. *Wear, 262,* 1007–1012. doi: 10.1016/j.wear.2006.10.016.
65. Eskandari, H., Taheri, R., & Khodabakhshi, F., (2016). Friction-stir processing of an AA8026-TiB_{2}-$Al_{2}O_{3}$ hybrid nano-composite: Microstructural developments and mechanical properties. *Materials Science and Engineering A., 660*, 84–96.
66. Mostafapour, A. A., & Khandani, S. T., (2013). Role of hybrid ratio in microstructural, mechanical and sliding wear properties of the Al5083/Graphitep/Al2O3p a surface hybrid nanocomposite fabricated via friction stir processing method. *Materials Science and Engineering: A., 559*, 549–557. doi: 10.1016/j.msea.2012.08.140.
67. Alidokht, S. A., Abdullah-Zadeh, A., Soleymani, S., & Assadi, H., (2011). Microstructure and tribological performance of an aluminum alloy based hybrid composite produced by friction stir processing. *Materials and Design, 32*, 2727–2733. doi: 10.1016/j.matdes.2011.01.021.
68. Soleymani, S., Abdollah-Zadeh, A., & Alidokht, S. A., (2012). Microstructural and tribological properties of Al5083 based surface hybrid composite produced by friction stir processing. *Wear, 278*, 41–47. doi: 10.1016/j.wear.2012.01.009.
69. Devaraju, A., Kumar, A., & Kotiveerachari, B., (2013). Influence of rotational speed and reinforcements on wear and mechanical properties of aluminum hybrid composites via friction stir processing. *Materials and Design, 45*, 576–585. doi: 10.1016/j.matdes.2012.09.036.
70. Sahandi, Z. P., Khodabakhshi, F., Simchi, A., & Kokabi, A. H., (2016). Fatigue fracture of friction-stir processed Al–Al3Ti–MgO hybrid nanocomposites. *International Journal of Fatigue, 87*, 266–278.

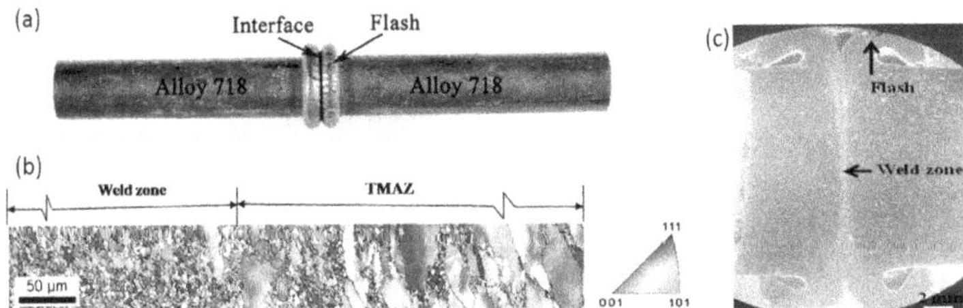

FIGURE 2.3 (a) Photograph showing friction welded 718 alloy (nickel-iron based superalloy widely used in aerospace and chemical plant industries), (b) EBSD map showing the microstructure at the weld zone and thermomechanical affected zone (TMAZ), and (c) microstructure of the interface. a) *Source:* Reprinted with permission from Damodaram et al., [2]. © 2013 Elsevier. c) *Source:* Reprinted with permission from Damodaram et al., [3]. © 2014 Elsevier.).

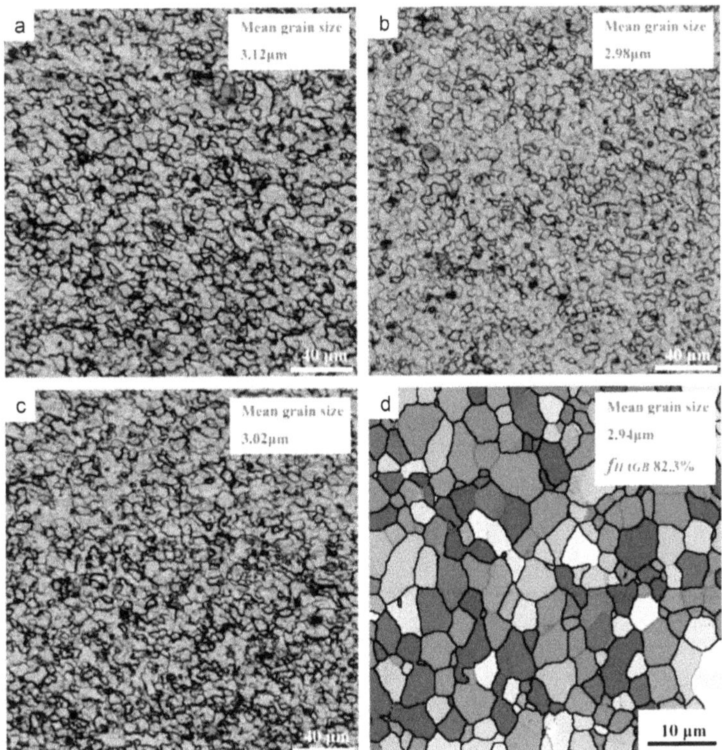

FIGURE 5.2 Optical microscope images of FSPed Al5083 alloy obtained at stir zone at 1200 tool rotational speed and 360 mm/min tool travel speed: (a) after 1st pass, (b) after 2nd pass (c) after 3rd pass and (d) EBSD map of sample after 3 FSP passes. (*Source:* Reprinted with permission from Chen et al., [7]. © 2016 Elsevier.)

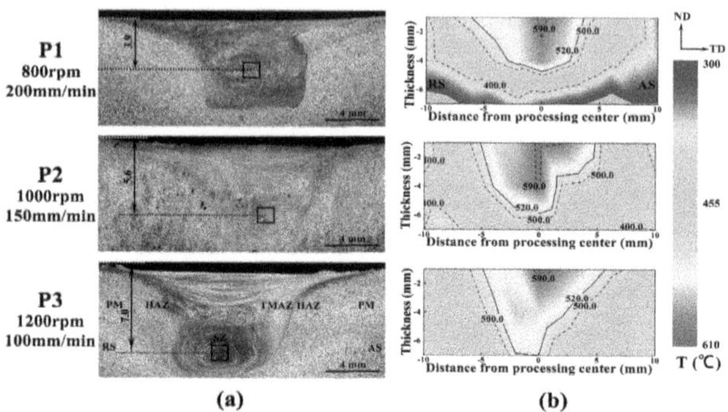

FIGURE 5.6 (a) Cross-sectional macrostructures of the FSPed Mg alloy, and (b) thermographs showing the temperature distribution. (*Source*: Reprinted with permission from Han et al., [24]. © 2016 Elsevier.)

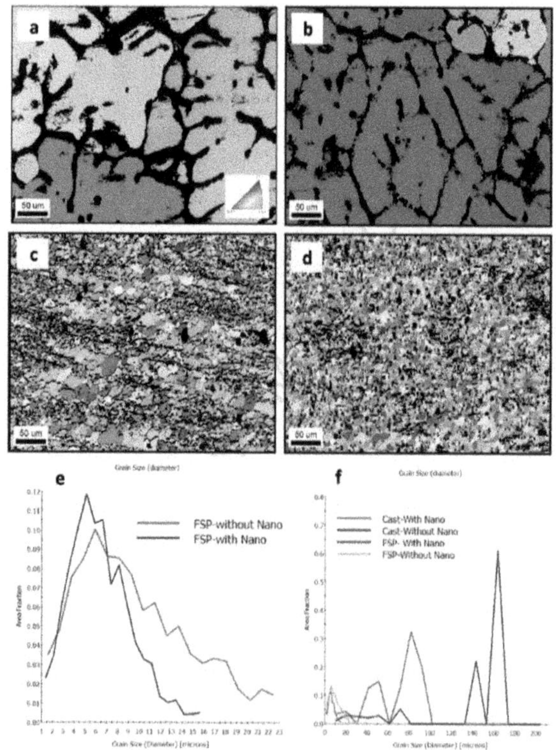

FIGURE 10.4 Orientation image maps with respect to FSP normal direction: (a) as cast without reinforcement, (b) as cast with reinforcement, (c) FSPed without reinforcement, (d) FSPed nanocomposite, (e) and (f) comparison of grain size distribution. (*Source:* Reprinted with permission Hoziefa et al. [18]. © 2016 Elsevier.)

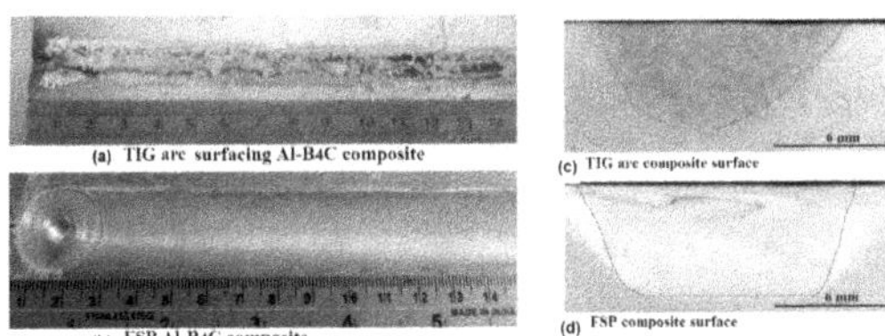

FIGURE 10.9 Comparison of the surface composites produced by two different methods: (a) photograph showing the surface composite produced by TIG welding, (b) surface composite by FSP, (c) cross-section comparison of composite produced by TIG, and (d) FSP. (*Source:* Reprinted with permission from Yuvraj et al., [29]. 2017 Elsevier.)

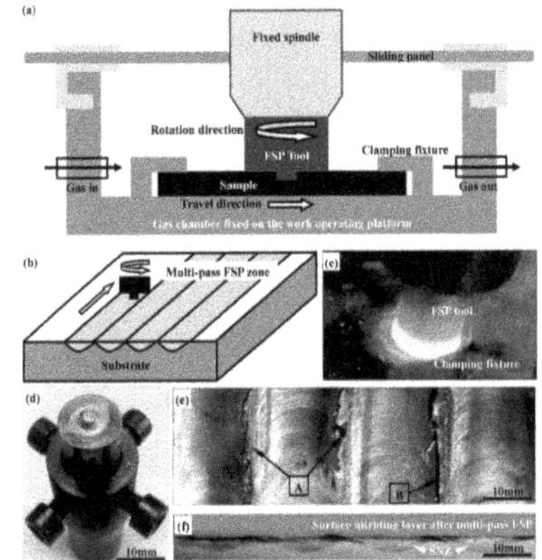

FIGURE 10.15 (a) Schematic diagram of special set-up developed to provide gas supply during FSP, (b) Procedure adopted for multipass FSP, (c) photograph showing FSP of Ti-6Al-4V, (d) photograph of FSP tool shoulder after one pass, (e) macrograph at the surface and microstructure at the cross section after surface nitriding by FSP. (*Source:* Reprinted with permission from Li et al., [84]. © 2013 Elsevier.)

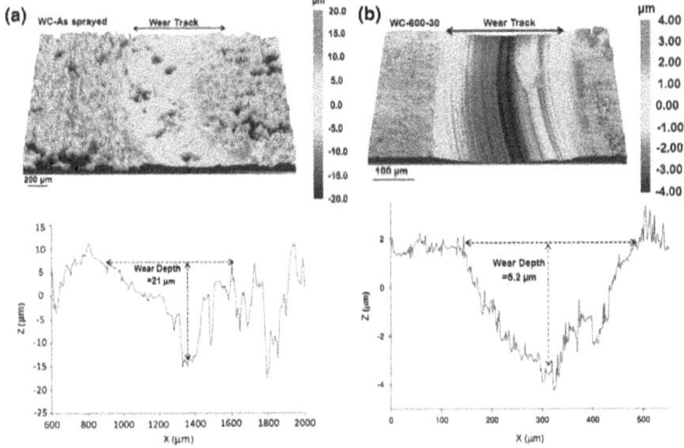

FIGURE 11.5 Wear topography of the surfaces: (a) WC surface coating, and (b) WC coating followed by FSP. (*Source:* Reprinted with permission from Rahbar-Kelishami et al., [3]. © 2015 Elsevier.)

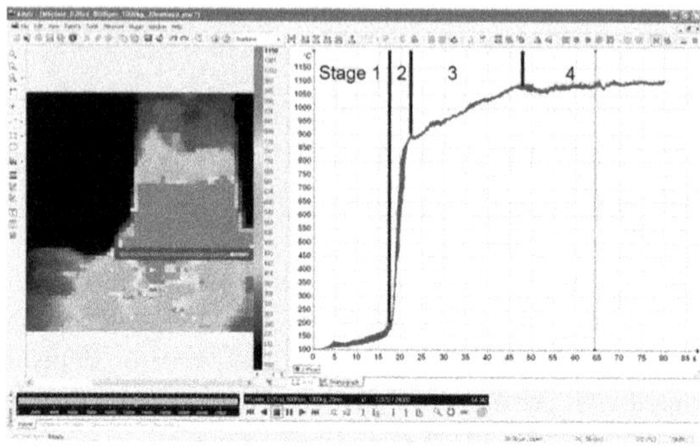

FIGURE 15.2 Thermal profile obtained during friction surfacing in which tool steel (high carbon and high chromium) D2 rod was used as a consumable rod to develop surface coating on steel substrate. (*Source:* Reprinted with permission from Rao et al., [7]. © 2012 Elsevier.)

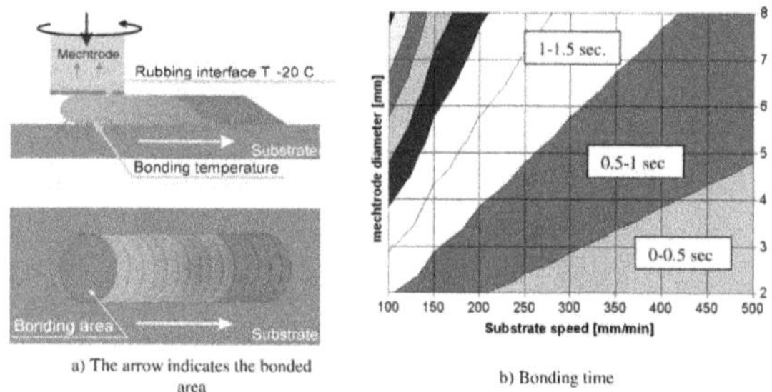

FIGURE 16.1 Bonding time influence on coating formation: (a) Schematic representation, and (b) relation between the mechtrode diameter and the substrate speed. (*Source:* Reprinted with permission from Vitanov & Voutchkov [4]. © 2005 Springer Nature.)

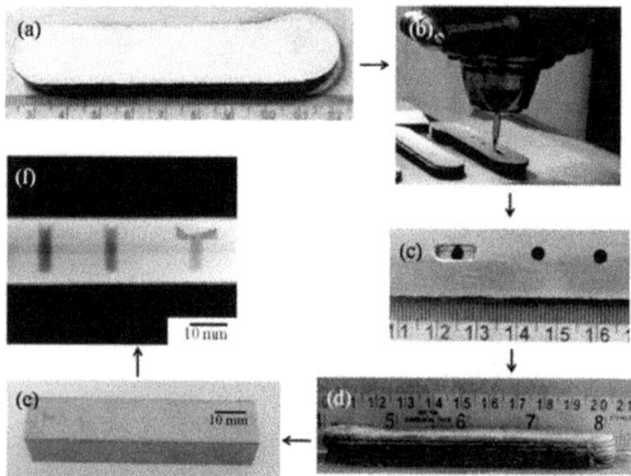

FIGURE 19.6 Photographs of a 3D part produced by friction deposition containing internal cavities: (a) initial surface layer, (b) producing holes by drilling, (c) top vies of the cavities and (d) surface layers deposition, (e) machined structure and (f) X-ray radiograph of the developed 3D structure. (*Source:* Reprinted with permission from Dilip et al., [7]. © 2013 Elsevier.)

CHAPTER 9

Effect of Processing Factors on the Composite Formation by FSP

For any process, a certain set of parameters play an important role and influence the success of the process. Understanding the influence of each parameter on the process is crucial before the processing. Certain parameters dictate the process feasibility which must be taken into serious consideration to achieve basic output. A few parameters influence the quality of the process which helps to achieve the structures with enhanced performance. All the factors or parameters which influence a manufacturing process can be grouped as: (i) process-based (ii) equipment-based, and (ii) material-based. Similar to other manufacturing processes, FSP is also governed by several processing parameters as discussed in this chapter.

9.1 FSP TOOL

Using appropriate FSP tool is crucial in successfully developing surface composites by FSP. The tool design including dimensions, shapes, and type of profile, relative ratios of the dimensions are crucial in FSP to develop surface MMCs. Appropriate tool design reduces the excess flash and defects and increases the material flow to introduce desired microstructural modifications. Unlike simple FSP, developing composites add additional factors such as the presence of secondary phase, amount of the secondary phase, method of introduction and the properties of the secondary phase powder. In Chapter 3 the use of different tools and important factors of tool design were generally discussed. Here, the influence of tool design on developing surface composites by FSP is discussed in connection with the incorporation of the secondary phase particles.

Tool geometry is crucial in dictating the thermal events which are developed at the vicinity of the stir zone. The distribution of secondary phase powder in matrix material depends on material flow in the nugget zone which is influenced by the amount of heat generation during FSP. Tool shoulder influences the amount of heat generation and controls the material flow at the surface. Different surface profiles are used on the shoulder such as flat, scroll, knurling, ridges and grooves to control the material flow at the surface. Tool pin influences the material flow within the stir zone. Several pin designs are used such as cross-sections of the square, triangular, cylindrical, and conical and threaded to introduce a different level of microstructure modification as schematically shown in Figure 9.1. The probe also contains some special flutes and threads to increase the rate of material flow. Simple cylindrical and cone probes with threads and without threads are the most widely used probe designs. By providing more sharp edges for example in the case of square and triangular probes, the amount of material which is stirred in the nugget zone is increased. By providing threads and flutes, the rate of material flow can be increased as the threads and flutes facilitate more material flow through the channels provided in the form of threads and flutes from the bottom of the workpiece towards the tool shoulder. Therefore, selecting the proper tool probe design is crucial in FSP. Threaded or fluted probes are optimum for welding applications where excellent material mixing is done from both base materials. Whereas, by adopting probes with more number of

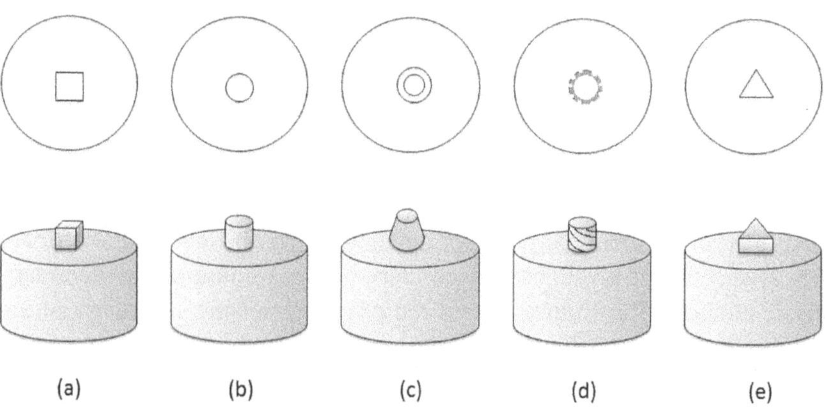

FIGURE 9.1 Schematic representation of different probes used in FSP: (a) square, (b) cylindrical, (c) cone, (d) threaded and (e) triangular.

sharp edges such as square design, more amount of material is plastically deformed, and a higher level of grain refinement also can be achieved. Figure 9.2 shows optical microscope images of the nugget zone of AZ91 Mg alloy processed with three different pin profiles. Among all, square pin profile has shown a higher level of grain refinement compared with simple taper and threaded taper profiles [1].

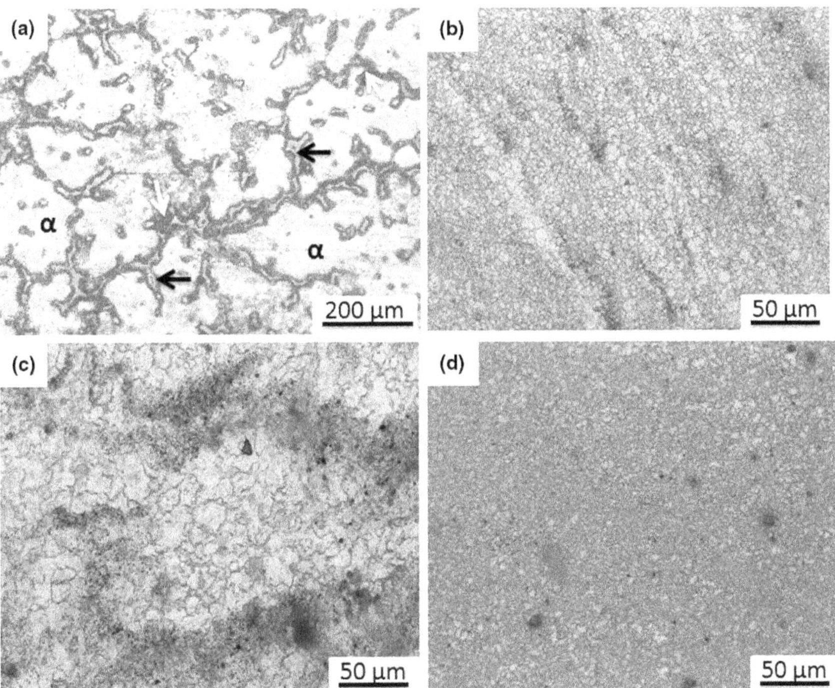

FIGURE 9.2 Optical microscope images of AZ91 Mg alloy processed with different tool pin profiles: (a) as a received condition, (b) simple taper pin, (c) threaded taper pin, and (d) square taper pin. (*Source:* Reprinted with permission from Hemender et al., [1]. © 2017 Springer Nature.)

Similar kind of observation was also reported by Elangovan et al., [2] while processing AA6061 alloy by FSP using different pin profiles. Among all, square pin profile has shown the highest grain refinement. Tool geometry includes the design aspects of the tool such as the shape of the pin, shoulder profile, ratio to the shoulder diameter to the pin diameter, additional features on the pin (threads, flutes, etc.). The volume of the workpiece material that undergoes mixing to disperse the incorporated secondary phase depends on the tool dimensions. The depth of the composite produced at the surface depends on the pin length, and the width of the composite produced at the surface depends on the dimensions of the tool shoulder. The volume fraction of the workpiece material that can be affected by the rotation of the tool is influenced by the dimensions of the pin. The rate of material flow during FSP is influenced by the special profiles developed on the tool pin. For example, threaded/flutes profiles result in a higher level of material mixing, but the width of the stir zone is decreased. Providing sharp edges increases the width of the stir zone but relatively give lower material flow. The pin profile influences the heat generation by which actually material undergoes plastic deformation at a lower amount of loads. Patel et al., [3] demonstrated the effect of tool pin profile on heat generation by using different polygonal pin profiles. Figure 9.3 shows the dimensions and photographs of three different tools with pins of the square, pentagon and hexagon profiles.

Figure 9.4 shows the thermal profiles obtained during FSP done by using different pin profiles. The temperature distribution during heating and cooling is compared among the three profiles. As the number of sides to the polygon pin is increased, the effective area of the shoulder that directly influences the heating mechanisms is affected. As the surface area associated with each pin profile is different, the ratio to the area of the shoulder and the area of the pin is crucial to understand the material flow if the shoulder design is unchanged. Similarly, Azaziah et al., [4] also demonstrated the effect of pin profile while developing AZ31-Al_2O_3 composites through FSP by using three different tool profiles. Compared with the threaded probe, pin profiles without thread produced defective stir zones.

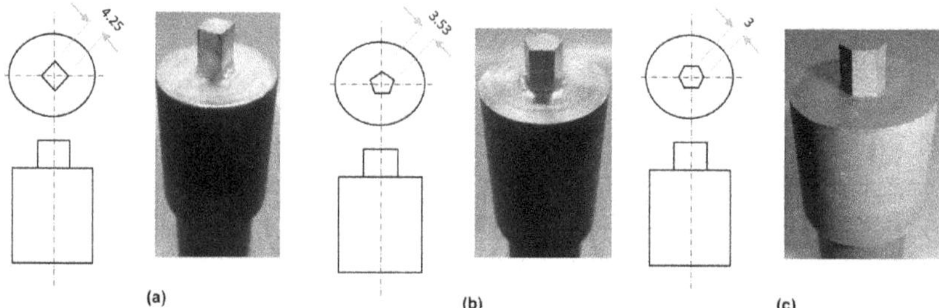

FIGURE 9.3 FSP tool pin dimensions and photographs: (a) square profile, (b) pentagon profile and (c) hexagon profile. (*Source:* Reprinted with permission from Patel et al., [3]. © 2017 Elsevier.)

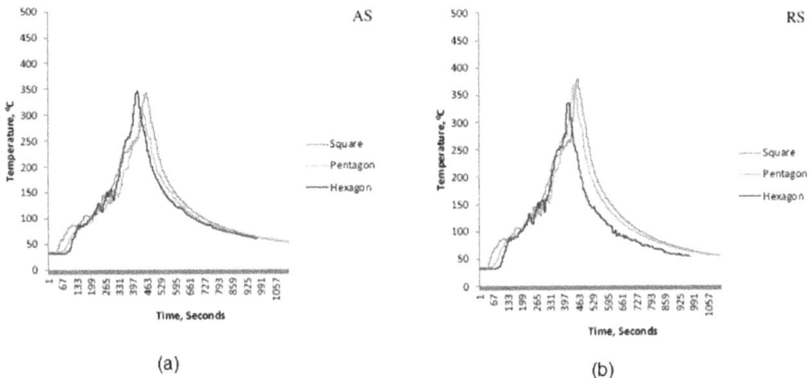

FIGURE 9.4 Thermal profiles during FSP done by using different tools pin profiles: (a) at the advancing side and (b) at the retreating side. (*Source:* Reprinted with permission from Patel et al., [3]. © 2017 Elsevier.)

9.2 PROCESSING VARIABLES

Tool rotational speed, tool travel speed, tool tilt angle, number of passes, and load are the important process parameters which influence the success of the process to develop surface MMCs by FSP. The optimized parameters which result in defect-free stir zone in developing microstructure modified surfaces by FSP may not be suitable to successfully produce composites even with the same metal is used as matrix material.The addition of the dispersing phase certainly alters the thermal events in the stir zone and affects the material flow behavior. The material at the stir zone becomes heterogeneous as the secondary phase added in the form of powder poses different mechanical and thermal properties compared with the base material. Therefore, optimizing processing parameters is essentially required while preparing the composites.

9.2.1 Tool Rotational Speed

Tool rotational speed is one of the important processing parameters which directly influences the success of composite formation. When the tool shoulder is in contact with the workpiece surface, heat is generated with tool rotation due to the friction that arose between the workpiece surface and the shoulder. The amount of heat generation depends on the tool rotational speed. Higher tool rotational speeds produce more heat and also better material flow. From the works of Jabbari et al., [5] while dispersing SiC particles into AZ31 Mg alloy, lower tool rotational speed was observed as optimum to uniformly dispersed the particles. The effect of tool rotational speed is coupled with the tool travel speed as observed in their results. The combination of lower tool rotational speed and tool travel speed produced a better composite. Azaziah et al., [4] demonstrated the effect of tool rotational speed on the material flow during the composite fabrication of AZ31/Al_2O_3.It was a clear observation that the increased speed produced defect free stir zones as shown in Figure 9.5. Compared with 800 and 1000 rpm, the stir zone obtained at 1200 rpm showed better material mixing close to the shoulder processed with the non-threaded probe.

In the process of developing composites by FSP, higher tool rotational speeds help to break the agglomerates of the reinforced particles. However, increased tool rotational speeds influence the level of grain refinement. As the generated heat is higher with the increased tool rotational speeds, the grain growth may lead to affect the grain refinement. In order to reduce the heat input in the stir zone and to achieve optimum grain refinement, suitable tool travel speed must be adopted. The objective behind selecting the optimum combination of tool rotation and travel speeds is to achieve better material mixing without defect and uniform distribution of the secondary phase. The role of higher tool rotational speeds on decreasing the grain refinement level was also demonstrated by Asadi et al., [6] in developing AZ91-SiC composites by FSP at different tool rotational speeds. By increasing the tool travel speed, the stir zone was exposed to a shorter period of time, and hence, fine grains were produced.

Again Asadi et al., [7] demonstrated the effect of tool higher tool rotational speeds on breaking the size of Al_2O_3/AZ31 clusters in the surface composite produced by FSP. By increasing the rotational speed, the level of distribution of the dispersing phase is improved as reported by Kurt et al., [8]. SiC particles were uniformly distributed in AA1050 alloy due to increased rotational speeds. Furthermore, the strength of the bond between the substrate and surface composite layer was observed as poor with increased rotational speeds. Salehi et al., [9] adopted optimization

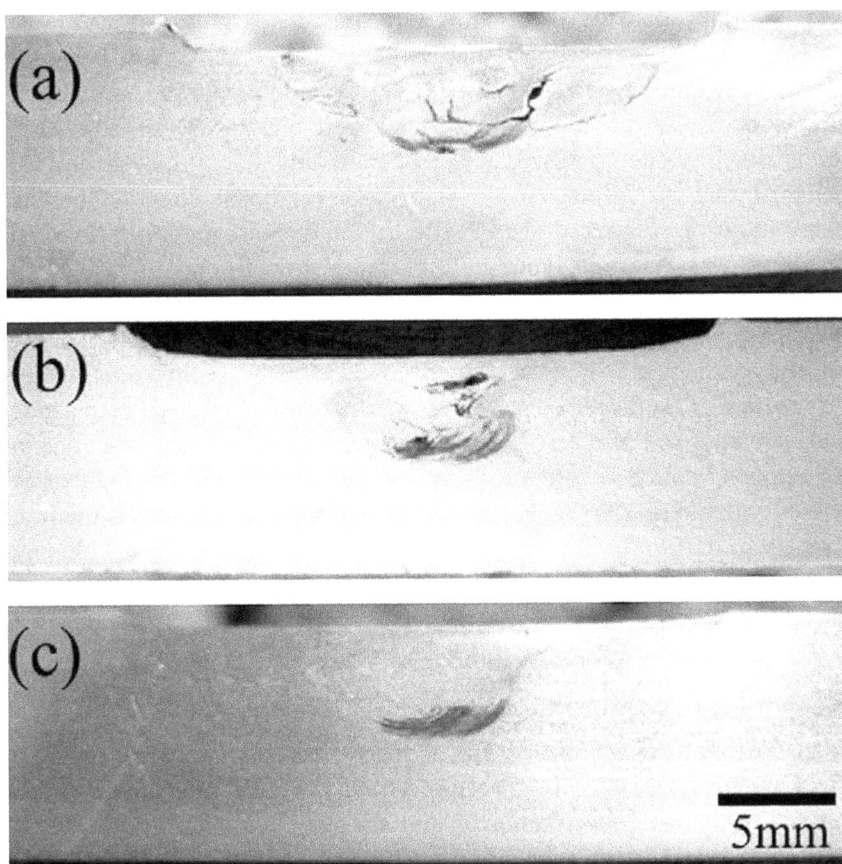

FIGURE 9.5 Optical macroscopic view of the cross-section of the composites produced at different tool rotational speeds with non-threaded probe: (a) 800 rpm, (b) 1000 rpm and (c) 1200 rpm. (*Source:* Reprinted with permission from Azaziah et al., [4]. © 2011 Elsevier.)

techniques (Taguchi analysis) and investigated the contribution of different process parameters on mechanical properties of AA6061-SiC composite as produced by FSP. As per the analysis, ultimate tensile strength (UTS) was influenced by the rotational speed by 43.7% contribution and 33.79%, 11.22%, and 4.21% by tool travel speed, tool pin profile, and penetration, respectively.

In FSP, the rate of material flow is different in the advancing side and retreating side. When the secondary phase powder is added to develop surface composites, the distribution of the powder particles is found to be different in the advancing side and retreating side. This kind of behavior was demonstrated by Dolatkhah et al., [10] while preparing Al5052-SiC composite. Relatively more SiC particles were observed as distributed in the advancing side compared with the retreating side. Asadi et al. [11] demonstrated the effectiveness of the tool rotational direction on the distribution of secondary phase. After a single pass, the tool rotational direction was reversed, and uniform microstructure was observed at the advancing and retreating side which was reflected in the hardness measurements. Similar kind of observation was also made by Rejil et al., [12] by conducting FSP in the opposite direction in the second pass. Better distribution of the dispersed phase was observed in both the advancing side and retreating side after two passes in which, the second pass was done in the opposite direction to that of the first pass. Increasing the number of passes with altering the tool rotational direction for every pass is the best strategy to uniformly disperse the secondary phase powder in developing surface composites by FSP.

9.2.2 Tool Travel Speed

Next, to tool rotational speed, travel speed occupies important space in dictating the success of the process. Usually, higher tool travel speeds result lower heat concentration in the stir zone and fine grain size. If the travel speed is too high, insufficient heat generation results, defect stir zones due to insufficient material flow. At lower travel speeds, more amount of heat input results in grain growth in the stir zone. Kurt et al., [8] observed the better distribution of SiC particles in AA1050 alloy at higher tool travel speeds. Khayyamin et al. [13] produced AZ91-SiO_2 composite by FSP at three different travel speeds (20, 40 and 63 mm/min). It was demonstrated that the increased travel speed produced refined microstructure which helped to increase the hardness of the composite. It is not always true that increased travel speeds produce composites with better particle distribution. The material type also influences the particle dispersion. Hence optimizing the combination of tool rotation and travel speeds is crucial for every new material system. Morisada et al., [14] dispersed multi-walled carbon nanotubes (MWCNT) into AZ31 Mg alloy by fixing the tool rotational speed as 1500 rpm and altering the tool travel speed from 25 to 100 mm/min. At higher tool travel speeds, the dispersion of MWCNT was observed as nonuniform due to insufficient material flow. In the composite produced at 25 mm/min, the distribution of MWCNT was observed relatively better due to more viscosity in the base material at lower tool travel speeds.The level of MWCNT distribution at different tool travel speeds can be seen in Figure 9.6.

Formation of secondary phase particle during FSP is another kind of variant in which the developed new phases act as secondary phase. In order to develop secondary phase due to an in-situ chemical reaction during FSP, a sufficient amount of heat is required so that the reaction between the chemical species will lead to develop new compounds. Higher tool travel speeds limit the exposure of the stirring material to a high temperature which is necessarily required to develop desired compounds. Low tool travel speeds may result in a more aggressive chemical reaction and the formation of the secondary phases. Therefore, optimum tool travel speed is required to be adopted while planning for in-situ chemical reactions. The role of tool travel speed on the formation of secondary phase was

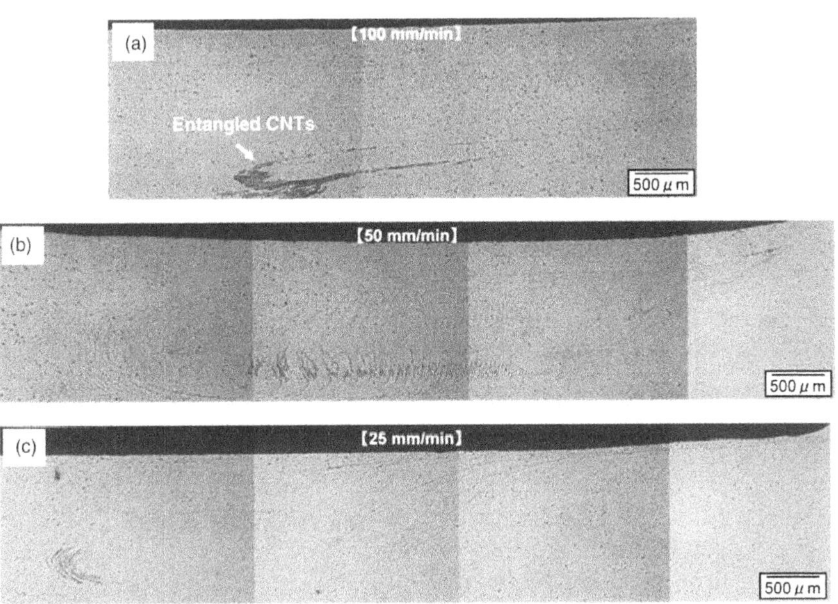

FIGURE 9.6 Optical microscope images of AZ31-MWCNT composite developed at a constant tool rotational: the speed with three different tool travel speeds: (a) 100mm/min, 50 mm/min and 25 mm/min. (*Source:* Reprinted with permission from Morisada et al., [14]. © 2006 Elsevier.)

clearly demonstrated by Chen et al., [15] while processing Al-CeO_2 system by FSP. Lower tool travel speeds allowed sufficient time and temperature to develop very fine in-situ particles uniform in size and distribution. However, the particle developed at higher tool travel speeds were observed as large and also the distribution was found to be non-uniform. This behavior is due to the shorter time allowed for the chemical reaction to be taken at higher tool travel speeds. This is similar to what You et al., [16] observed while processing Al-SiO_2 by FSP due to longer processing time associated with slower tool travel speeds.

Zhang et al., [17] also processed Al-TiO_2 system to develop in-situ surface composites and demonstrated the significant role of tool travel speeds on in-situ reaction compared with tool rotational speed. The formation of Al_3Ti and Al_2O_3 phases was observed as more at lower tool travel speeds. The effect of rotational speed was relatively lower on the size and distribution of the developed particles. Increased diffusion due to more available time was observed as the main reason behind the increased formation of in-situ formation of phases at lower tool travel speeds. Barmouz et al., [18] process Cu by adding SiC particles and without the addition of SiC particles at different tool rotational and travel speeds. Figure 9.7 shows the average grain size with respect to traverse speed.

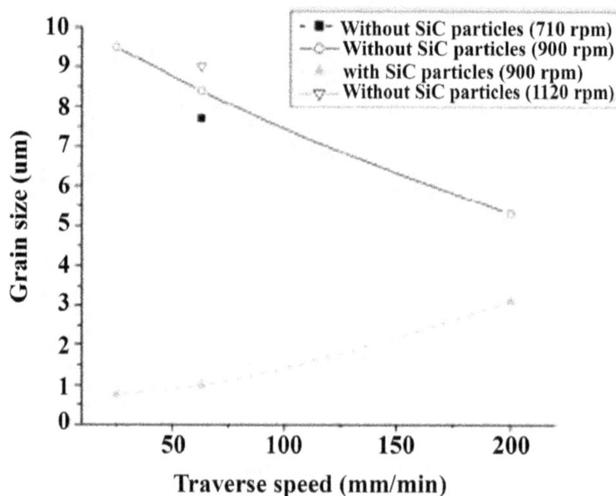

FIGURE 9.7 Effect of SiC particles addition on grain size at different tool rotational speeds and travel speeds as reported by Barmouz et al. (*Source:* Reprinted with permission from Barmouz et al., [18]. © 2011 Elsevier.)

Without SiC particles, the grain size was observed as increased with decreasing the tool travel speeds. Interestingly, after SiC addition, grain size was observed as decreased compared with FSPed Cu without SiC. The decreased grain size in the composite was attributed to the presence of SiC which introduced pinning effect to sustain the grain refinement achieved by FSP.The combined effect of tool rotational speed and travel speed on composite formation occupies prime space in the optimization of process parameters in FSP. Table 9.1 list different combinations of materials and the corresponding tool rotational and travel speeds adopted in developing the surface composites.

TABLE 9.1 Summary of the Work Done in Developing Surface Composites At the Different Best Combination of Tool Rotational and Travel Speeds.

Surface composite	Rotational speed(rpm)	Traverse speed (mm/min.)	Reference
$(SiC+MoS_2)/A356$	1600	50	Alidokht et al. [20]
SiC/AZ91	900	63	Asadi et al. [6]
$Al_2O_3/AZ31$	800	45	Azizieh et al. [4]
SiC/Cu	900	40	Barmouz et al. [20]
Nano clay/polymer	900	160	Barmouz et al. [21]
TiC/Al	1000	60	Bauri et al. [22]
SiC/AA5052	1120	80	Dolatkhah et al. [10]
$(SiC + Al_2O_3)/AA1050$	1500	100	Mahmoud et al. [23]
$Al_2O_3/AZ91$	1600	31.5	Faraji et al. [24, 66]
HA/Ti-6Al-4V	250	16	Farnoush et al. [25]
TiC/Steel	1120	31.5	Ghasemi-Kahrizsangi and Kashani-Bozorg [26]
$SiO_2/AZ91$	1250	63	Khayyamin et al. [27]
SiC/AA1050	1000	15	Kurt et al. [8]
MWCNT/AA1016	950	30	Liu et al. [28]
Ni/AA1100	1180	60	Qian et al. [29]
(TiC + B4C)/AA6360	1600	60	Rejil et al. [12]
SiC/AA6061	1600	40	Salehi et al. [30]
B_4C/Cu	1000	40	Sathiskumar et al. [31]
SiC/Ti	800	45	Shamsipur et al. [32]

(*Source:* Reprinted with permission from Sharma et al., [19]. © 2015 Elsevier.)

TABLE 9.1 *(Continued)*

Surface composite	Rotational speed(rpm)	Traverse speed (mm/min.)	Reference
$(SiC+MoS_2)/AA5083$	1250	50	Soleymani et al. [33]
$Al_2O_3/AA2024$	800	25	Zahmatkesh and Enayati [34]
$Al_2O_3/AA6082$	1000	135	Shafiei-Zarghani et al. [35]
Cu/AA5083	750	25	Zohoor et al. [36]

9.2.3 Applied Load

Axial load that is applied to the FSP tool during FSP influences the material flow. The heat due to the appropriate combination of tool rotational and travel speed plasticize the material locally in the vicinity of the tool pin. The volume of the material that is stirred along with the pin rotation is decreased in the thickness direction of the workpiece. This leads to have decreased width of the stir zone in the thickness direction of the workpiece usually in FSP. Along with tool pin profile, axial load that is applied on the FSP tool influence the amount of material that is plastically deformed in the stir zone. Figure 9.8 schematically compares the influence of applied axial load on material flow behavior. Insufficient load may

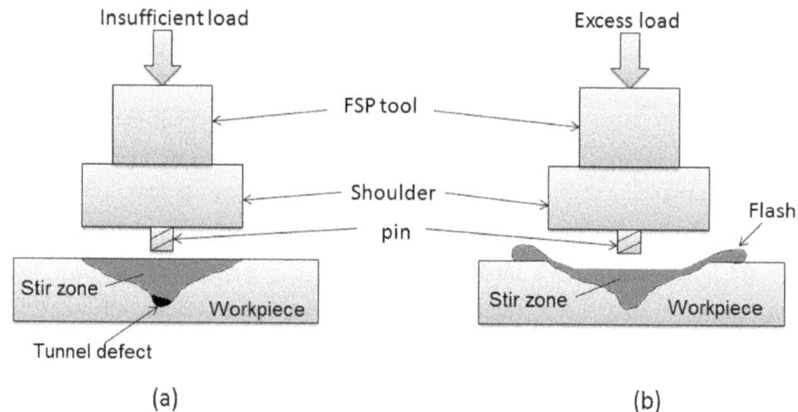

FIGURE 9.8 Schematic illustration of defects resulted due to (a) insufficient axial load and (b) excess axial load applied on the FSP tool.

cause a tunneling defect due to insufficient material filling even though appropriate tool rotational and travel speeds are adopted. If the applied axial load is higher, an excess flash may result which is flushed away from the processed region. Hence, adopting appropriate axial load is also crucial to get optimum material flow behavior to avoid defects resulted due to the poor material filling.

9.2.4 Tool Tilt Angle

Tool tilt angle (θ) is another important factor in FSP which affects the material flow in the stir zone. A slight tilt is given to the spindle or tool with respect to the workpiece surface. A slight inclination of the axis of the FSP tool usually up to 3° in the trailing direction helps to improve the material flow from the front region of the rotating FSP tool to the back side cavity produced by the traveling tool pin. A localized reservoir of stirred material is created with tool tilt. Figure 9.9 schematically illustrates the tool tilt during FSP. The tool penetration depth is directly affected by the tool tilt angle. Higher penetration is associated with lower tool tilt, and the penetration depth is increased with increased tool tilt.

Asadi et al., [6] demonstrated the role of tool tilt angle on obtaining defect free stir zones while developing AZ91/SiC composites. At 1120 rpm and 63mm/min tool travel speed, the optimum penetration depth of 0.4, 0.3 and 0.33 were observed with a tool tilt angle of 3.5, 3 and 2.5°, respectively. The contact area between the tool shoulder and the workpiece is decreased if the tilt angle is increased in the case of constant penetration depth. At lower tilt angles, the material may deposit back to the shoulder as a coating. If the tilt angle is increased more than 3.5°, tunnel defect was noticed. This kind of tool tilt angle effect is common in all material systems in developing surface composites.

9.2.5 Tool Penetration Depth

Penetration depth is another influencing parameter in FSP which governs the heat generation and material flow. Insufficient penetration depth gives poor contact between the shoulder surface and the workpiece and therefore, insufficient heat generation results poor material flow.This kind of material flow results longitudinal cracks and tunneling defects as reported by Asadi et al., [6]. As per the works of Salehi et al. [9], tool penetration depth occupies 4th important role next to tool rotational speed, travel speed and pin profile in successfully develop surface composites. If tool tilt is altered to different angles, as shown in Figure 9.9, penetration depth also altered.

9.2.6 *Number of Passes*

Due to the heterogeneity in the material flow rate from the advancing side to the retreating side in FSP, the added particles also are distributed at

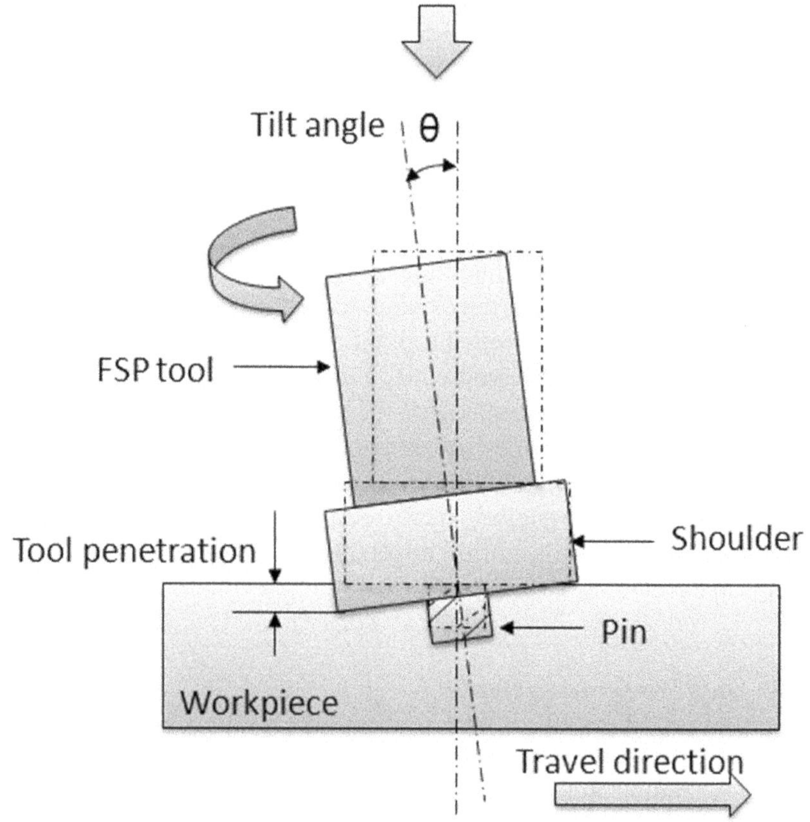

FIGURE 9.9 Schematic illustration of tool tilt during FSP.

different scales in the stir zone. After a single pass, a considerable level of grain refinement can be achieved. Multiple passes may help to further modify the microstructure, in particular, the distribution of the secondary phase. If the secondary phase is in the size of the nano-meter level, agglomeration is a common problem associated with the high energy nature of nanoparticles. Multi-pass FSP helps to uniformly distribute the nanoparticles in the matrix. Eftekharinia et al., [37] developed AA6061/SiC surface composite by FSP and higher level of grain refinement was achieved as the number of passes were increased to four. This is similar to what observed by Asadi et al., [7] while developing AZ91-SiC composite by FSP. Ke et al., [38] applied for three FSP passes and successfully demonstrated increased Al-Ni reaction in developing intermetallic composites by FSP. Similarly, Qian et al., [39] observed the formation of Al_3Ni particles during FSP while developing AA100-Ni composite. The fraction of these particles was observed as increased with increased number of passes. Prolonged time due to the more number of passes was the reason behind the formation of a higher amount of intermetallic particles. Increased hardness and ultimate tensile strength were also observed due to the uniform distribution of the particles after six FSP passes.

Lee et al., [40] produced AZ61-SiO_2 nano-composites by FSP.After a single pass, the agglomerated cluster size of SiO_2 was measured as ranging from 0.1 to 3 μm. The cluster size was measured as reduced to 150 nm, and a grain size of 0.8 μm was measured after four passes of FSP. Asadi et al., [7] observed the decrease of cluster size after six FSP passes to 500 nm compared with the size of cluster (7 μm) after single pass FSP. Sharifitabar et al., [41] also produced composites of AA5052- Al_2O_3 by FSP and clustering of nanopowder was observed with a size of 650 nm after one FSP pass. The same composite has shown the size of clusters around 70 nm after four passes. Barmouz and Givi [42] developed Cu-SiC composite by FSP and observed a poor bonding between the matrix and the dispersed particles after one FSP pass. In addition, presence of pores was also observed in the vicinity of the SiC particles. These pores fraction was observed as 19% compared with the matrix material. Interestingly after 8 passes, the porosity was measured as decreased to 5% due to enhanced bonding between the matrix and the dispersed SiC.

Dolatkhah et al., [10] produced Al5052-SiC composite by using SiC powder of two different sizes (5 μm and 50 nm). FSP was carried out up to four passes. FSP was also carried out without adding SiC particles. As the number of passes were increased to four and by adopting change of direction between passes, uniform distribution of SiC particles was achieved. Figure 9.10 shows the optical microscope images of as

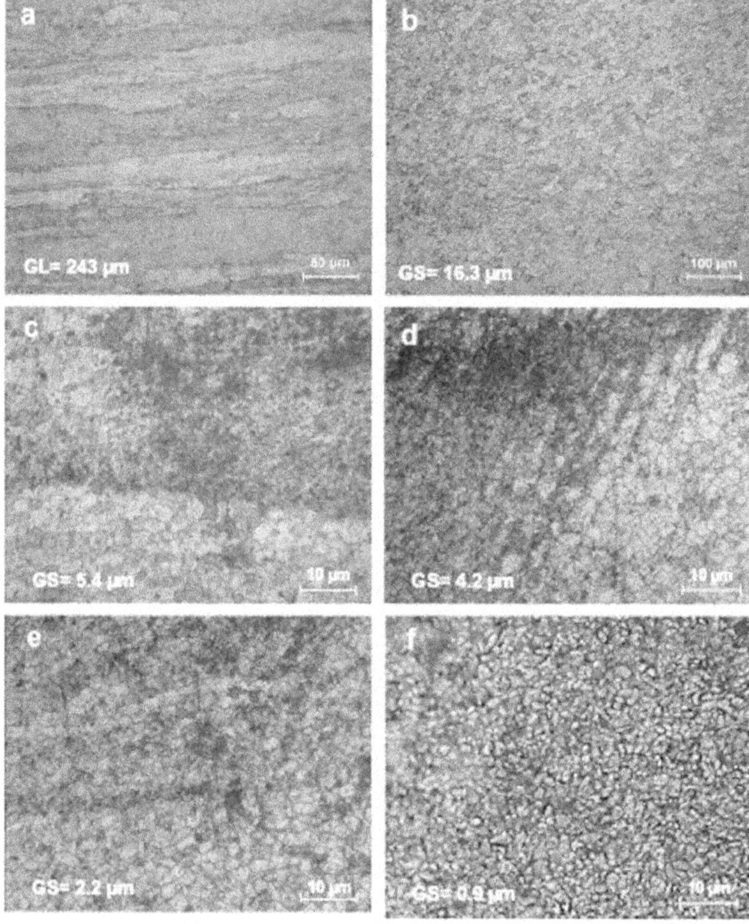

FIGURE 9.10 Optical microscope images of Al5052/SiC composite produced by FSP: (a) base material, (b) FSP one pass without SiC powder, (c) FSP one pass with 5 µm size SiC powder, (d) FSP four passes with µm SiC powder without change in rotational direction between passes, (e) FSP four passes with 5 µm SiC powder with change of rotational direction between passes, and (f) FSP four passes with 50 nm SiC powder with change of rotational direction between passes. (*Source:* Reprinted with permission from Dolatkhah et al., [10]. © 2012 Elsevier.)

received, FSPed Al5052 and FSPed Al5052-SiC composite. Due to the use of nano-SiC powder, changing the direction of processing between FSP passes resulted excellent grain refinement from a starting grain size of 243 μm to 0.9 μm. Further, the nano-composite exhibited an increase in hardness up to 55% and a decrease in wear resistance by 9.7 times compared with unprocessed Al5052 alloy. From the available literature, it is true that the number of FSP passes introduces uniform distribution of the dispersed phase and also results better grain refinement. However, obtaining optimum pass number is crucial as each pass is associated with energy consumption.

9.2.7 Tool Material

FSP tool is the most important element in producing composites by FSP. The tool is subjected to higher amounts of heat produced during FSP and loads during the process. The strength and hardness of the tool is affected if the temperature is increased beyond certain limits. The addition of hard secondary phase in the form of powder/particles increase mechanical wear at the tool shoulder and pin which decreases the tool life and alter the tool actual dimensions. Tool steels such as H13 are used as material to fabricate FSP tools to process relatively soft materials such as aluminum, magnesium and their alloys. Whereas for hard workpiece materials; tungsten carbide, cermets, poly cubic boron nitride (PCBN), etc. are used as tool materials. The following important factors must be considered while selecting a material to fabricate FSP tool to develop composites.

i) The material should withstand the load at the elevated temperatures
ii) The material should possess high strength, hardness, wear resistance and toughness
iii) The material can be easily re-machined/producedafter worn out
iv) The chemical elements must not diffuse from the tool material into the workpiece and vice-versa.
v) The potential to participate in chemical reaction with oxygen at high temperature must be low.
vi) The material should be relatively of low cost and high durability.

Prado et al., [43] developed AA 6061 – 20%SiC metal matrix composite (MMC) by using a pin with threaded profile. Due to the tool wear, the threaded profile has turned into a smooth profile. From their observations, it can be suggested that tool wear can be greatly reduced even for MMCs when using the optimized tool shape. Formation of new phases due to a chemical reaction between the tool material and workpiece material is another complication need to understand while selecting a suitable material for FSP tool. Feng et al., [44] observed the formation of Cu_2FeA_{17} phase due to wear of tool steel in FSW of SiC/AA2009 bulk composite.Due to formation of these phases, the strength of the weld joint was observed as decreased.

Using materials of high strength and hardness have shown some promising results while processing high melting alloys with hard dispersing phases to fabricate surface composites by FSP. Morisada et al., [45] used a sintered WC-Co tool for doing FSP to produce (WC-CrC-Ni) layer on SKD61 steel substrate without any defect. The hardness of the surface layer was measured as 1.5 times higher compared with that of starting condition. Similarly, no tool wear was observed by Swaminathan et al., [46] during FSP of NiAl bronze material by using tungsten alloy Densimet (W-7% Ni, Fe) as tool material.Grewal et al., [47]also observed no tool wear in FSP of 13Cr-4Ni steel by using WC as tool material. London et al., [48]also observed no wear when PcBN and W-RE materials were used for tool while doing FSP of Nitinol (49.2 % Ti, 50.8 % Ni).

Rai et al. [49] explained the potential of tungsten, molybdenum, and iridium as tool materials as they exhibit high hardness, strength and low chemical affinity with oxygen at elevated temperatures. Also, they suggested that the properties of these materials can be further improved by adding alloying elements or by providing hard and wear-resisting surface coatings. However, it is true that the cost of tool fabrication is increased with these alloys or surface coatings. The authors demonstrated lower tool wear while doing FSW of titanium and high strength steel by using tungsten-rhenium (W-25% Re) alloys. Thompson and Babu [50] used W-1% La_2O_3, W-25% Re, and W-20%Re-10% HfC alloys as materials for tool to join high strength steels. Among all the tool materials, W with HfC exhibited higher resistance to tool degradation. This behavior was attributed to presence of HfC particles as they prevent softening of the tool material. Miyazawa et al. [51] used iridium alloy as tool material to carry out FSW of SUS304 stainless steel. The recrystallization temperature was observed as higher for Ir-1 At.% Re alloy compared with other alloys. Addition of Ta, W, Nb, and Mo was suggested as alloying elements to introduce high hardness and compressive strength. Buffa et al. [52] used different tungsten based alloys such as WC-4.2%Co, WC-12%Co and a W-25%Re for FSW of Ti-6Al-4V alloy. Among all combinations, WC-4.2% Co demonstrated high hardness. However, the tool experienced a fracture failure even before the expected lifetime. When the Co% was increased to 12% (WC-12%Co), material adhesive wear was noticed. FSW tool made of W-25%Re material showed lowest wear and a higher life span. Sato et al. [53] used Co-based alloy as tool material to perform FSW steel and titanium. The Co-alloy was strengthened by precipitation hardening method which served to resist the loss in mechanical properties at high temperature. The tool wear was observed as lower, and a sound weld joint was achieved.

Other strategies such as providing cooling to the tool or developing surface coatings on the tool to increase the life span of tool also applied as observed from the reported literature. Xue et al. [54] used a common tool steel (42) and successfully done FSP of nickel (Ni) by using water-cooling. The processed zone was defect free, and the FSP tool has shown lower tool wear. Without water-cooling, the tool has failed as soon the shoulder touched the workpiece due to higher amount of generated heat. Developing wear resisting surface coatings on FSP tools by adopting different vapor deposition methods such as chemical vapor deposition (CVD) and physical vapor deposition (PVD) offers some advantages such as improved wear resistance, decreased adhesion of the workpiece material to the tool shoulder and decreased diffusion at the process zone. Batalha et al. [55] developed coatings of Al, Cr, and Ni by PVD method onWC-Co tool for FSW ofTi alloy sheets. But the coating has disappeared on the FSP tool after processing and also tool wear was observed. Lakshminarayanan et al. [56] developed various refractory ceramic-based composite coatings by using Inconel 738 alloy used as FSW tool for AISI 304 grade austenitic stainless steel. The tool with 60% B_4N coating by plasma transferred arc process has exhibited better wear characteristics compared with the other tools. Similarly, Kahrizsangi and Bozorg [26] developed steel-TiC composite by using WC tool and a significant tool wear (20%) was measured after completing 400 mm of processing. Farias et al. [57] also reported surface mechanical degradation due to severe tool wear in FSW of Ti-6Al-4V. From the overall observations, it is understood that the selection of tool material is crucial to successfully develop a surface composite by FSP. For materials such as aluminum, magnesium and their alloys, using tool steels may be appropriate. For high hard materials such as steels, nickel, copper, etc., adopting carbide-based tools which poses excellent hardness and wear resistance is optimum. Therefore, depending on the workpiece material appropriate material is selected to fabricate FSP tool.

9.2.8 Post FSP Cooling Rate

Cooling rate after FSP is another influencing factor which particularly influences the microstructure. Due to the decreased temperature by providing follow up cooling process, tool wear is decreased, and the tool life is increased. Chabok and Dehghani [58] FSPed interstitial free steel and follow up quenching was also carried out. From the microstructural studies, formation of a nano-grain layer of 150 μm thickness with an average grain size from 50–100 nm was achieved. Special arrangements to provide higher cooling rates after FSP suppress the grain growth and results nanostructure. Submerged FSP is another variation in FSP to introduce different cooling rates in which the process is done underwater so that the cooling rate is relatively rapid compared with FSP done in open air. Hofmann and Vecchio [59] performed submerged FSP of AA6061-T6 alloy and a grain refinement from 50 um to 100–200 nm was achieved. Instead of performing FSP under water, applying cooling while FSP is carried out is another way of altering the cooling rate. Najafi et al., [60] developed AZ31-SiC composite by single FSP pass by applying mixture of methanol and dry ice. A grain refinement up to 800 nm was successfully achieved in the composite. This was relatively smaller compared with 1 μm observed in FSPed AZ31 without SiC particles. Another way of applying coolant is by arranging coolant circulation below the workpiece while FSP is carried out. In such conditions, Arora et al., [61] developed AZ91-Al_2O_3 and TiC hybrid composites and demonstrated increased hardness up to 100 %. The grain refinement was observed as 2.4 μm compared with the grain size (5.1 μm) achieved in FSP carried out at open air.Zhang et al. [62] developed in-situ Al-based composites by using Al_3Ti and Al_2O_3 nanoparticles. FSP was carried out up to four passes in open air and also two passes in flowing water. Interestingly after two passes in flowing water, the grain size was measured as 602 nm and 1285 nm after four passes in open air. These findings clearly demonstrate the effect of cooling rate achieved by applying suitable coolant in the form of complete immersion, spraying on the surface or providing beneath the workpiece. However, formation of oxide layers is an issue associated with rapid cooling after FSP. Particularly, if more number of FSP passes are planned to continue on the same processed region, presence of these oxide layers may entrapped into the stir zone in the subsequent passes as observed by Asadi et al., [11] while processing AZ91 Mg alloy.

9.3 METHOD OF SECONDARY PHASE INTRODUCTION INTO THE MATRIX BY FSP

Incorporating the secondary phase and control over the distribution of secondary phase is crucial while using FSP as the processing method to develop composites. The martial flow during FSP is complex in nature [63]. However, optimizing the process parameters can address the distribution of the secondary phase during FSP. Along with the other influencing factors such as tool design and processing parameters, method of secondary phase incorporation is also found to have a major role in distributing the powder particles and producing a successful composite [64, 65]. The thickness of the composite layer produced using FSP depends on the method of secondary phase incorporation. Several methods have been demonstrated by different authors to incorporate the secondary phase powder into the matrix material which brings advantages as well as limitations to produce the composites. The following sections give a detailed discussion on the methods and compare them with all aspects.

9.3.1 Groove Filling Method

Groove filling method was first introduced by Mishra et al. [66] while developing 7075-SiC composite by FSP. In groove filling method, a narrow groove is produced on the surface of the sheet or plate before FSP. Then the grove is completely filled with the secondary phase particles intended to be embedding in the substrate. Now FSP is done on the powder filled groove to develop surface MMCs [67–69]. Figure 9.11(a) schematically explains the groove filling method. Afterward, a slight modification was proposed to the groove filling method by adding a small additional step before doing actual FSP. The powder filled groove is closed by using a modified non-consumable tool in which the shoulder does not contain any pin or probe as explained by Lee et al. [40] as shown in Figure 9.11(b). During the first step, sufficient amount of heat is generated due to the friction between the flat surface of the tool shoulder and the surface of the workpiece which plasticize the material around the groove and helps to cover the groove completely as a closing surface layer. Then the groove filled with secondary phase powder is closed on the surface, and the hidden cavity contains the powder particles. This helps to arrest the secondary phase particle within the groove, and the particles do not fly away or escape from the groove during FSP. This method can be referred as groove filling and closing.

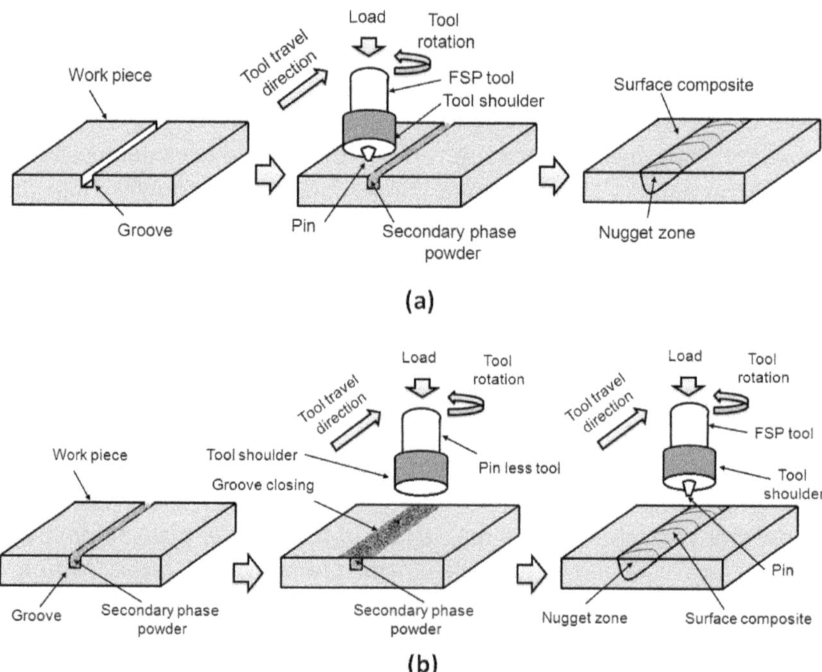

FIGURE 9.11 Schematic representation of composite fabrication using (a) groove filling and (b) groove filling and closing method (*Source:* B. Ratna Sunil [69]; http://creativecommons.org/licenses/by/4.0/).

9.3.2 Holes Filling Method

In holes filling method, tiny holes are produced on the surface of the sheet to incorporate the secondary phase particles into the substrate during FSP [70, 71] as schematically shown in Figure 9.12(a). Later on, a few works have been carried out similar to groove filling and closing method, which requires two processing tools; one without pin and second with a designed pin at the end of the shoulder [72]. The holes are initially filled with powder and then closed with the help of pin-less tool as schematically shown in Figure 9.12(b). Then the same region is processed with FSP tool that contains a pin to produce surface MMCs.

9.3.3 Sandwich Method

Mertens et al. [73] demonstrated another method of secondary phase introduction into the matrix by FSP as schematically illustrated in Figure 9.13. This can be known as sandwich method, in which secondary phase is arranged in the form of a layer or lamina between the sheets or plates of matrix material. Then FSP is done such a way that the FSP tool pin is penetrated through the plates and the secondary phase layer or lamina. Due to the stirring action and traverse motion of the FSP tool, the secondary phase layer or lamina is broken into small particles or fibers, and distributed into the matrix phase and results a composite. By providing more number of layers in between the matrix sheets, the amount of secondary phase that is distributed into the matrix can be increased. Uniform distribution of secondary phase can be achieved by increasing the number of FSP passes.

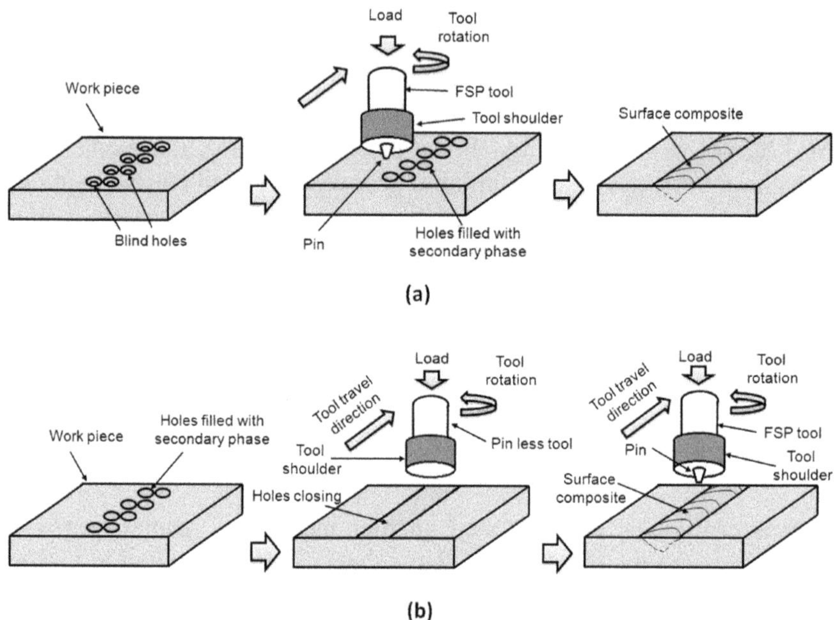

FIGURE 9.12 Schematic representation of composite fabrication by (a) holes filling method and (b) holes filling and closing method (*Source:* Ratna Sunil [69]; http://creativecommons.org/licenses/by/4.0/).

9.3.4 Direct Friction Stir Processing (DFSP) Tool Method

Huang et al. [74] demonstrated another new kind of FSP tool that facilitates introducing secondary phase powder directly during the process through the tool itself. The specially designed FSP tool contains a hole along the longitudinal axis of the tool through which secondary phase powder is supplied to the processing zone during FSP. This tool was named as direct friction stir processing (DFSP) tool. Figure 9.14 shows the schematic illustration of DFSP process. Figure 9.15 shows the typical photographs of DFSP tool. In DFSP method, secondary phase particles are supplied through a continuous hole provided in the DFSP tool.

Huang et al. [74] explained the mechanism behind the composite formation using this modified tool. Unlike in FSP, the reinforcement particles are not introduced into the surface of the base metal before processing but supplied through the hole which is designed within the DFSP tool. As

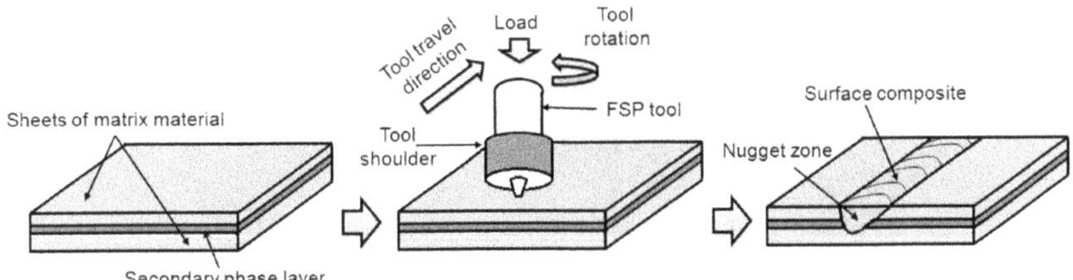

FIGURE 9.13 Schematic representation of sandwich method to produce composites using FSP (*Source:* Ratna Sunil [69]; http://creativecommons.org/licenses/by/4.0/).

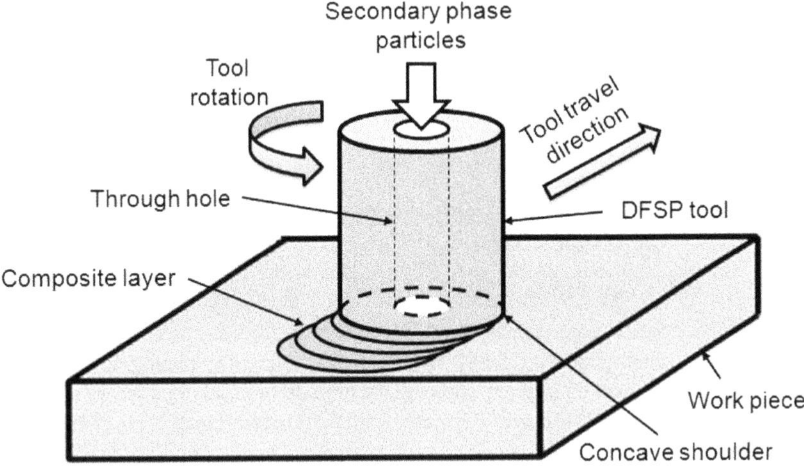

FIGURE 9.14 Schematic representation of DFSP tool method as explained by Huang et al. (*Source:* Reprinted with permission Huang et al., [74]. © 2014 Elsevier.)

the tool is moved in the traverse direction at the surface of the workpiece, the reinforcement particles are directly placed within the space formed between the shoulder of the DFSP tool and the workpiece. Therefore, the particles are not escaped during the process but entrapped between the concave space of the shoulder and the surface of the workpiece. Then, these entrapped particles are stirred and pressed into the workpiece uniformly to produce the surface composite. As described by the authors, in a single pass, more amount of secondary phase can be introduced into the matrix using DFSP tool compared with that of FSP tool.

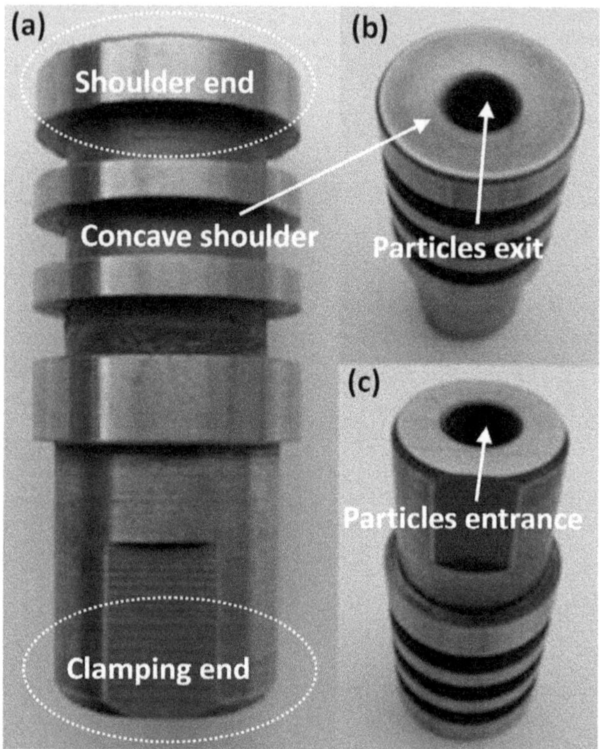

FIGURE 9.15 Photographs of DFSP tool: (a) front view, (b) photographs showing concave shoulder and (c) photograph showing the through hole provided on the top of the tool to supply the particles. (*Source:* Reprinted with permission Huang et al., [74]. © 2014 Elsevier.)

9.3.5 Surface Coating Followed by FSP Method

Providing the surface coatings on the material before FSP is another strategy to incorporate the secondary phases into the matrix material [75, 76]. Secondary phase is coated on the surface by any suitable coating technique before FSP. Then the coated sheet is processed using an FSP tool as shown in Figure 9.16 to produce the surface composite. Due to the plastic deformation and material flow, as the rotating FSP tool travels across the coated surface, matrix material and secondary phase are properly mixed and results a composite layer.

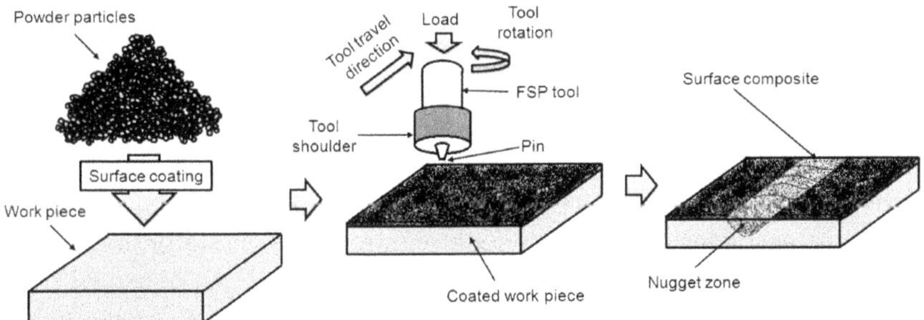

FIGURE 9.16 Schematic representation of composite fabrication by surface coating followed by FSP method (*Source:* Ratna Sunil [69]; http://creativecommons.org/licenses/by/4.0/).

9.3.5.1 Comparison of Different Methods

Table 9.2 briefly summarizes the maximum thickness of the surface composite layer developed on different material systems by adopting different strategies of secondary phase introduction during FSP. Among all the methods, groove filling method was the first one appeared in the literature in developing surface MMC by FSP. This method is very simple and requires less machining. A groove is machined on the surface of the workpiece by using a milling cutter. Based on the studies done by Gandra et al. [64], the position of the FSP tool pin during the composite formation is found to be crucial. In their work, more powder distribution was observed when the position of the groove was placed under the pin compared with placing the groove away from the pin interaction area (advancing or retrieving). Heydarian et al. [77] proposed another variation in which parallel grooves with gradient groove depths are produced. Compared with providing a single groove on the surface, providing parallel gradient grooves has given uniform distribution of the dispersing powder. Groove filling and closing method requires additional non-consumable pin-less tool. Furthermore, the process is done in two steps; first to close the groove, and in the second to produce the composite. Groove filling and closing method helps to introduce more amount of secondary phase into the workpiece compared with simple groove filling method as the former one prevents the escape of filled powder from the groove during FSP.

More amount of powder can be introduced by adopting holes filling method compared with the groove filling method. In holes filling method, the escape of the powder during FSP is also lower compared with that of groove filling method. Furthermore, holes filling and closing method gives more control on the particle incorporation compared with groove filling and closing method. However, additional operations are required compared with the groove filling or holes filing methods. For some applications, slight modification at the surface is sufficient and may not demand higher amount of secondary phase. For such requirement, simple groove filling or groove filling with closing methods can be suitable. As observed from the previous works, the groove filling method was observed as appropriate to produce a thick composite layer compared with the other processes as listed in Table 9.2. Of course, the thickness of a composite layer also depends on the FSP tool pin dimensions and processing parameters.

TABLE 9.2 Maximum Thickness of Surface Composite Layer Achieved in Different Material Systems Using FSP and the Corresponding Method Used To Incorporate the Secondary Phase Into the Matrix (*Source:* B. Ratna Sunil [69]; http://creativecommons.org/licenses/by/4.0/)

Material system	Secondary phase	Method of secondary phase incorporation	Composite layer thickness (mm)	Reference
Aluminum/ it alloys	SiC	Groove filling	2	Wang et al., [78]
	Al_2O_3	Groove filling and closing	4	Shafiei-Zarghani et al.,[35]
	MWCNTs	Groove filling	2.2	Lim et al.,[79]
	Al_2O_3	Holes filling	3	Yang et al.,[70]
	SiC and MoS_2	Groove filling	3.5	Alidokht et al.,[19]
	Ni particles	Groove filling	0.15	Devinder Yadav and Ranjit Bauri [80]
	Al_2O_3	Groove filling and closing	3.8	Sharifitabar et al., [81]
	TiC B_4C	Groove filling	6	Maxwell Rejil et al.,[12]
	Graphite, Al_2O_3, and SiC	Groove filling	3.5	Anvari et al.,[82]
	SiC and Al_2O_3	Surface coating and FSP	0.204	Miranda et al.,[65]
	Al_2O_3	Surface coating and FSP	0.1	Mazaheri et al.,[83]

TABLE 9.2 *(Continued)*

Material system	Secondary phase	Method of secondary phase incorporation	Composite layer thickness (mm)	Reference
Magnesium/ its alloys	SiO_2	Groove filling and closing	3–3.5	Lee et al., [40]
	CNTs	Groove filling	2	Morisada et al., [14]
	SiC	Groove filling and closing	2.5	Asadi et al.,[7]
	Al_2O_3	Groove filling	5–6	Azizieh et al.,[4]
	Al_2O_3	Groove filling	2–2.5	Faraji et al.,[24]
	Carbon fibers	Sandwich method	2.7	Mertens et al., [73]
	Sic and B_4C	Holes filling and closing	3	Madhusudan Reddy et al., [72]
	SiC	Direct friction stir processing tool	0.15	Huang et al., [74]
	Hydroxyapatite	Groove filling	2	Ratna Sunil et al., [67]
Copper/ its alloys	SiC	Groove filling	2	Barmouz, et al., [20]
	Graphite	Groove filling and closing	Not reported	Sarmadi et al., [84]
	B_4C	Groove filling and closing	5	Sathiskumar et al., [31]
	TiC	Holes filling and closing	Not reported	Sabbaghian et al., [85]
	SiC	Holes filling	Not reported	Akramifard et al., [71]
Titanium/ its alloys	Hydroxyapatite	i) Groove filling ii) Surface coating	0.16	Farnoush et al., [86]
	Hydroxyapatite	Holes filling	0.16	Farnoush et al., [87]
Steels	TiC	iii) Groove filling iv) Surface coating	1.5–2	Ghasemi-Kahrizsang and Kashani-Bozorg [26]

For the applications which require more amount of secondary phase at the surface, holes filling or holes filling and closing method can be chosen. Unlike groove filling and hole filling methods, sandwich method does not require any additional machining processes before FSP. Also, it reduces the number of processing steps and the necessity of using additional tool (without pin). However, the distribution of secondary phase is observed as uniform in sandwich method as reported by Faraji et al. [24]. However, obtaining the secondary phase in the form of layers or laminas is difficult particularly with hard and brittle ceramic materials. Also, the secondary phase may distribute in the matrix material in the form of large particles, or the regions of high concentration with secondary phase after FSP may result as there is no control over the breakdown of the layers during the process. It is true that if the dispersing particle size is reduced, the surface properties were found to be improved. Therefore, in sandwich method, control from the FSP within the matrix is very poor and completely depends on the processing parameters and the design of the tool.

Using DFSP tool as proposed by Huang et al., [74] is another interesting development in FSP. All additional machining processes associated with the previous methods can be completely eliminated by using DFSP tool. By using DFSP tool, the secondary phase particles can be effectively introduced into the surface of the matrix material. But the penetration depth, which directly influences the thickness of the surface composite play important role, is the main limitation in DFSP method. The thickness of surface composite layer produced by DFSP is relatively lower compared with the surface composite produced by other methods and also DFSP tool fabrication is complex. However, the level of uniformity in the distribution of the secondary phase is superior in the case of DFSP method compared with other methods. DFSP tool design showed interesting results in developing surface composites by FSP. However, no significant work was reported using this kind of tools to develop metallic surface composites. Providing surface coating on the workpiece by different means before FSP has also shown promising results in successfully developing the surface composites. More amount of secondary phase can be introduced into the surface, but the thickness of the produced composite layer is less compared with the composites produced by the other methods. Similar to DFSP tool method, surface coating followed by FSP method gives the modified surface depth limited to a few hundreds of micrometers from the surface. Also, pre-coating requires additional processing which increase the overall work to produce a composite.

Based on the required thickness and the amount of secondary phase that is required to be dispersed at the surface of the workpiece, appropriate method can be chosen in developing surface composites using FSP. Among all of the methods proposed in the literature, one common limitation is producing composites with tailored composition. The composition of the composites produced by FSP can be approximately assessed by assuming that the dispersion is uniform. However, exact amount of secondary phase that is distributed in the matrix material is not the same throughout the processed region. Variation in the distribution of secondary phase also influences the material properties if not macroscopic but microscopic level. Majority of the work is being carried out on developing different composite systems but focus on developing strategies to address these issues is inferior. Developing new strategies to increase the level of incorporation, distribution of secondary phase and increasing the thickness of the affected surface layer is in a great need to utilize the high potential of FSP technique in producing surface composites. For example, Asadi et al., [6, 7] studied the effect of penetration depth (PD) of FSP tool and demonstrated that the PD also has a great influence on composite formation. Further, the authors have shown the effect of PD on composite formation again depends on tool travel and rotational speeds. But, the composite layer thickness was not clearly reported. Investigating the combined effect of method of particles incorporation, process parameters, and type of tool on successful fabrication of surface composites by FSP is also enhancing the knowledge to address challenges involved in this area. Composite fabrication using FSP has wide industrial applications and also the process does not require huge capital investments. Existing CNC machines can be modified with appropriate attachments, and FSP can be carried out. Therefore, industries can easily and rapidly adapt the process.

KEYWORDS

- **metal matrix composite**
- **multi-walled carbon nanotubes**
- **tool material**
- **tool rotational speed**
- **tool tilt angle**
- **tool travel speed**
- **ultimate tensile strength**

REFERENCES

1. Patle, H., Ravikumar Dumpala, & Ratna Sunil, B. (2018). Machining Characteristics and Corrosion Behavior of Grain Refined AZ91 Mg Alloy Produced by Friction Stir Processing: Role of Tool Pin Profile, *Trans Indian Inst Met*, https://doi.org/10.1007/s12666-017-1250-3.
2. Elangovan, K., Balasubramanian, V., & Valliappan, M., (2007). Influences of tool pin profile and axial force on the formation of friction stir processing zone in AA6061 aluminum alloy. *The International Journal of Advanced Manufacturing Technology 38,* 285–295. doi:10.1007/s00170-007-1100-2.
3. Vivek V. Patel, Vishvesh Badheka, & Abhishek Kumar, (2017). Effect of polygonal pin profiles on friction stir processed superplasticity of AA7075 alloy, *Journal of Materials Processing Technology 240,* 68–76.
4. Azizieh, M. Kokabi, A. H. & Abachi, P. (2011). Effect of rotational speed and probe profile on microstructure and hardness of AZ31/Al_2O_3 nanocomposites fabricated by friction stir processing, *Materials and Design 32,* 2034–2041.
5. Jabbari, A. H., Sedighi, M., Vallant, R., Huetter, A., & Sommitsch, C. (2015). Effect of Pass Number, Rotational and Traverse Speed on Particle Distribution and Microstructure of AZ31/SiC Composite Produced by Friction Stir Processing, *Key Engineering Materials,* 651-653, 765–770.
6. Asadi, P., Faraji, G., Besharati, M. K., (2010). Producing of AZ91/SiC composite by friction stir processing (FSP). *The International Journal of Advanced Manufacturing Technology 51*, 247–260. doi:10.1007/s00170-010-2600-z.
7. Asadi, P., Faraji, G., Masoumi, A., Besharati Givi, M. K., (2011). Experimental investigation of magnesium-base nanocomposite produced by friction stir processing: Effects of particle types and number of friction stir processing passes. *Metallurgical and Materials Transactions A 42*, 2820–2832. doi:10.1007/s11661-011-0698-8.
8. Kurt, A., Uygur, I., & Cete, E., (2011). Surface modification of aluminum by friction stir processing. *Journal of Materials Processing Technology 211,* 313–317. doi:10.1016/j.jmatprotec.2010.09.020.
9. Salehi, M., Saadatmand, M., & Aghazadeh Mohandesi, J., (2012). Optimization of process parameters for producing AA6061/SiC nanocomposites by friction stir processing. *Transactions of Nonferrous Metals Society of China 22,* 1055–1063. doi:10.1016/s1003-6326(11)61283-1.
10. Dolatkhah, A., Golbabaei, P., Givi, M. K. B., & Molaiekiya, F., (2012). Investigating effects of process parameters on microstructural and mechanical properties of Al5052/SiC metal matrix composite fabricated via friction stir processing. *Materials and Design 37,* 458–464. doi:10.1016/j.matdes.2011.09.035.
11. Asadi, P., Givi, M.K.B., Parvin, N., Araei, A., Taherishargh, M., & Tutunchilar, S., (2012). On the role of cooling and tool rotational direction on microstructure and mechanical properties of friction stir processed AZ91. *The International Journal of Advanced Manufacturing Technology 63,* 987–997. doi:10.1007/s00170-012-3971-0.
12. Rejil, C. M., Dinaharan, I., Vijay, S. J., & Murugan, N., (2012). Microstructure and sliding wear behavior of AA6360/ (TiC+B4C) hybrid surface composite layer

synthesized by friction stir processing on aluminum substrate. *Materials Science and Engineering: A 552,* 336–344. doi:10.1016/j.msea.2012.05.049.
13. Khayyamin, D., Mostafapour, A., & Keshmiri, R., (2013). The effect of process parameters on microstructural characteristics of AZ91/SiO_2 composite fabricated by FSP. *Materials Science and Engineering: A 559,* 217–221. doi:10.1016/j.msea.2012.08.084.
14. Morisada, Y., Fujii, H., Nagaoka, T., & Fukusumi, M., (2006). MWCNTs/AZ31 surface composites fabricated by friction stir processing. *Materials Science and Engineering: A 419,* 344–348. doi:10.1016/j.msea.2006.01.016.
15. Chen, C. F., Kao, P. W., Chang, L. W., & Ho, N. J., (2009). Effect of processing parameters on microstructure and mechanical properties of an Al-Al11Ce_3-Al_2O_3 in-situ composite produced by friction stir processing. *Metallurgical and Materials Transactions A 41,* 513–522. doi:10.1007/s11661-009-0115-8.
16. You, G. L., Ho, N. J., & Kao, P. W., (2013a). In-situ formation of Al_2O_3 nanoparticles during friction stir processing of AlSiO_2 composite. *Materials Characterization 80,* 1–8. doi:10.1016/j.matchar.2013.03.004.
17. Zhang, Q., Xiao, B. L., Wang, Q. Z., & Ma, Z. Y., (2014). Effects of processing parameters on the microstructures and mechanical properties of in situ (Al_3Ti + Al_2O_3)/Al composites fabricated by hot pressing and subsequent friction-stir processing. *Metallurgical and Materials Transactions A 45,* 2776–2791. doi:10.1007/s11661-014-2221-5.
18. Vipin Sharma, Ujjwal Prakash, B. V., & Manoj Kumar, (2015). Surface composites by friction stir processing: A review, *Journal of Materials Processing Technology, 224,* 117-134.
19. Alidokht, S. A., Abdollah-Zadeh, A., Soleymani, S., & Assadi, H., (2011). Microstructure and tribological performance of an aluminum alloy based hybrid composite by friction stir processing. *Materials and Design 32,* 2727-2733. doi:10.1016/j.matdes.2011.01.021.
20. Barmouz, M., Givi, M. K. B., & Seyfi, J., (2011a). On the role of processing parameters in producing Cu/SiC metal matrix composites via friction stir processing: Investigating microstructure, microhardness, wear, and tensile behavior. *Materials Characterization 62,* 108–117. doi:10.1016/j.matchar.2010.11.005.
21. Barmouz, M., Seyfi, J., Givi, M. K. B., Hejazi, I., & Davachi, S. M., (2011). A novel approach for producing polymer nanocomposites by in-situ dispersion of clay particles via friction stir processing. *Materials Science and Engineering: A 528,* 3003–3006. doi:10.1016/j.msea.2010.12.083.
22. Bauri, R., Yadav, D., & Suhas, G., (2011). Effect of friction stir processing (FSP) on microstructure and properties of Al-TiC in situ composite. *Materials Science and Engineering: A 528,* 4732–4739. doi:10.1016/j.msea.2011.02.085.
23. Mahmoud, E. R. I., Takahashi, M., Shibayanagi, T., & Ikeuchi, K., (2009b). Fabrication of Surface-Hybrid-MMCs Layer on Aluminum Plate by Friction Stir Processing and Its Wear Characteristics. *Materials Transactions 50,* 1824–1831. doi:10.2320/matertrans.M2009092.
24. Faraji, G., Dastani, O., & Akbari Mousavi, S. A. A. (2011). Effect of process parameters on microstructure and micro-hardness of AZ91/Al_2O_3 surface composite produced by FSP. *J Mater Eng Perform 20,* 1583–1590.

25. Farnoush, H., Sadeghi, A., Abdi Bastami, A., Moztarzadeh, F., & Aghazadeh Mohandesi, J., (2013). An innovative fabrication of nano-HA coatings on Ti-CaP nanocomposite layer using a combination of friction stir processing and electrophoretic deposition. *Ceramics International 39*, 1477–1483. doi:10.1016/j.ceramint.2012.07.092.
26. Ghasemi-Kahrizsangi, A., Kashani-Bozorg, S. F., (2012). Microstructure and mechanical properties of steel/TiC nano-composite surface layer produced by friction stir processing. *Surface and Coatings Technology 209*, 15–22. doi:10.1016/j.surfcoat.2012.08.005.
27. Khayyamin, D., Mostafapour, A., & Keshmiri, R., (2013). The effect of process parameters on microstructural characteristics of AZ91/SiO2 composite fabricated by FSP. *Materials Science and Engineering: A 559,* 217–221. doi:10.1016/j.msea.2012.08.084.
28. Liu, Q., Ke, L., Liu, F., Huang, C., & Xing, L., (2013). Microstructure and mechanical property of multi-walled carbon nanotubes reinforced aluminum matrix composites fabricated by friction stir processing. *Materials and Design 45*, 343–348. doi:10.1016/j.matdes.2012.08.036.
29. Qian, J., Li, J., Xiong, J., Zhang, F., & Lin, X., (2012). In situ synthesizing Al3Ni for fabrication of intermetallic reinforced aluminum alloy composites by friction stir processing. *Materials Science and Engineering: A 550,* 279–285. doi:10.1016/j.msea.2012.04.070.
30. Salehi, M., Saadatmand, M., & Aghazadeh Mohandesi, J., (2012). Optimization of process parameters for producing AA6061/SiC nanocomposites by friction stir processing. *Transactions of Nonferrous Metals Society of China 22,* 1055–1063. doi:10.1016/s1003-6326(11)61283-1.
31. Sathiskumar, R., Murugan, N., Dinaharan, I., & Vijay, S. J., (2013). Characterization of boron carbide particulate reinforced in situ copper surface composites synthesized using friction stir processing. *Materials Characterization 84,* 16–27. doi:10.1016/j.matchar.2013.07.001.
32. Shamsipur, A., Kashani-Bozorg, S. F., & Zarei-Hanzaki, A., (2011). The effects of friction-stir process parameters on the fabrication of Ti/SiC nano-composite surface layer. *Surface and Coatings Technology 206*, 1372–1381. doi:10.1016/j.surfcoat.2011.08.065.
33. Soleymani, S., Abdollah-Zadeh, A., & Alidokht, S. A., (2012). Microstructural and tribological properties of Al5083 based surface hybrid composite produced by friction stir processing. *Wear 278,* 41–47. doi:10.1016/j.wear.2012.01.009.
34. Zahmatkesh, B., & Enayati, M. H., (2010). A novel approach for development of surface nanocomposite by friction stir processing. *Materials Science and Engineering: A 527,* 6734–6740. doi:10.1016/j.msea.2010.07.024.
35. Shafiei-Zarghani, A., Kashani-Bozorg, S. F., & Zarei-Hanzaki, A., (2009). Microstructures and mechanical properties of Al/Al_2O_3 surface nano-composite layer produced by friction stir processing. *Materials Science and Engineering: A 500,* 84–91. doi:10.1016/j.msea.2008.09.064.
36. Zohoor, M., Besharati Givi, M. K., & Salami, P., (2012). Effect of processing parameters on fabrication of Al-Mg/Cu composites via friction stir processing. *Materials and Design 39,* 358–365. doi:10.1016/j.matdes.2012.02.042.

37. Hamidreza Eftekharinia, Ahmad Ali Amadeh, Alireza Khodabandeh, & Moslem Paidar, (2016). Microstructure and wear behavior of AA6061/SiC surface composite fabricated via friction stir processing with different pins and passes, *Rare Met,* doi: 10.1007/s12598-016-0691-x.
38. Ke, L., Huang, C., Xing, Li, & Huang, K. (2010). Al–Ni intermetallic composites produced in situ by friction stir processing. *Journal of Alloys and Compounds 503*(2), 494–499
39. Qian, J., Li, J., Xiong, J., Zhang, F., & Lin, X., (2012). In situ synthesizing Al_3Ni for fabrication of intermetallic reinforced aluminum alloy composites by friction stir processing. *Materials Science and Engineering: A 550,* 279–285. doi:10.1016/j.msea.2012.04.070.
40. Lee, C., Huang, J., & Hsieh, P., (2006). Mg-based nano-composites fabricated by friction stir processing. *Scripta Materialia 54,* 1415–1420. doi:10.1016/j.scriptamat.2005.11.056.
41. Sharifitabar, M., Sarani, A., Khorshahian, S., & Shafiee Afarani, M., (2011). Fabrication of 5052Al/Al_2O_3 nanoceramic particle reinforced composite via friction stir processing route. *Materials and Design 32,* 4164–4172. doi:10.1016/j.matdes.2011.04.048.
42. Barmouz, M., & Givi, M. K. B., (2011b). Fabrication of in situ Cu/SiC composites using multi-pass friction stir processing: Evaluation of microstructural, porosity, mechanical and electrical behavior. *Composites Part A: Applied Science and Manufacturing 42,* 1445–1453. doi:10.1016/j.compositesa.2011.06.010.
43. Prado, R. A., Murr, L. E., Soto, K. F., & McClure, J. C., (2003). Self-optimization in tool wear for friction-stir welding of Al 6061+20% Al_2O_3 MMC. *Materials Science and Engineering: A 349,* 156–165. doi: 10.1016/s0921- 5093(02)00750-5.
44. Feng, A., Xiao, B., & Ma, Z.Y., (2008). Effect of microstructural evolution on mechanical properties of friction stir welded AA2009/SiCp composite. *Composites Science and Technology 68,* 2141–2148. doi:10.1016/j.compscitech.2008.03.010
45. Morisada, Y., Fujii, H., Mizuno, T., Abe, G., Nagaoka, T., & Fukusumi, M., (2010). Modification of thermally sprayed cemented carbide layer by friction stir processing. *Surface and Coatings Technology 204,* 2459–2464. doi:10.1016/j.surfcoat.2010.01.021.
46. Swaminathan, S., Oh-Ishi, K., Zhilyaev, A. P., Fuller, C. B., London, B., Mahoney, M. W., & McNelley, T. R., (2009). Peak Stir Zone Temperatures during Friction Stir Processing. *Metallurgical and Materials Transactions A 41,* 631–640. doi:10.1007/s11661-009-0140-7.
47. Grewal, H. S., Arora, H. S., Singh, H., & Agrawal, A., (2013). Surface modification of hydro turbine steel using friction stir processing. *Applied Surface Science 268,* 547–555. doi:10.1016/j.apsusc.2013.01.006.
48. London, B., Fin, J., Pelton, A., Fuller, C., & Mahoney, M., (2005). Friction stir processing of Nitinol, In: Jata, K. V., Mahoney, M. W., Mishra, R. S., & Lienert, T. J. (Eds.), *Friction Stir Welding and Processing III.* TMS, Warrendale, PA, pp. 67–74.
49. Rai, R., De, A., Bhadeshia, H. K. D. H., & DebRoy, T., (2011). Review: friction stir welding tools. *Science and Technology of Welding and Joining 16,* 325–342. doi:10.1179/1362171811y.0000000023

50. Thompson, B., & Babu, S. S., (2010). Tool degradation characterization in the friction stir welding of hard metals. *Welding Journal 89*, 256–261
51. Miyazawa, T., Iwamoto, Y., Maruko, T., & Fujii, H., (2011). Development of Ir based tool for friction stir welding of high-temperature materials. *Science and Technology of Welding and Joining 16,* 188–192. doi:10.1179/1362171810y.0000000025
52. Buffa, G., Fratini, L., Micari, F., & Settineri, L. (2012). On the choice of tool material in friction stir welding of titanium alloys. In: *Transaction of North American Manufacturing Research Conference of SME June 4–8, 2012 Notre Dame*, Indiana, USA, pp. 785–794.
53. Sato, Y., Miyake, M., Kokawa, H., Omori, T., Ishida, K., Imano, S., Park, S., & Hirano S., (2011). Development of a cobalt-based alloy FSW tool for high-softening-temperature materials, In: Mishra, R. S., Mahoney, M. W., Sato, Y., Hovanski, Y., Verma, R. (Eds.), *Friction Stir Welding and Processing VI.* TMS, San Diego, California, USA, pp. 3–9.
54. Xue, P., Xiao, B. L., & Ma, Z. Y., (2014). Achieving ultrafine-grained structure in a pure nickel by friction stir processing with additional cooling. *Materials and Design 56,* 848–851. doi:10.1016/j.matdes.2013.12.001.
55. Batalha, G. F., Farias, A., Magnabosco, R., Delijaicov, S., Adamiak, M., & Dobrzański, L. A., (2012). Evaluation of an AlCrN coated FSW tool. *Journal of Achievements in Materials and Manufacturing Engineering 55,* 607–615.
56. Lakshminarayanan, A. K., Ramachandran, C. S., & Balasubramanian, V., (2014). Feasibility of surface-coated friction stir welding tools to join AISI304 grade austenitic stainless steel. *Defence Technology 10,* 360–370. doi:10.1016/j.dt.2014.07.003
57. Farias, A., Batalha, G. F., Prados, E. F., Magnabosco, R., & Delijaicov, S., (2013). Tool wear evaluations in friction stir processing of commercial titanium Ti-6Al-4V. *Wear 302,* 1327–1333. doi:10.1016/j.wear.2012.10.025
58. Chabok, A., & Dehghani, K., (2012). Effect of processing parameters on the mechanical properties of interstitial free steel subjected to friction stir processing. *Journal of Materials Engineering and Performance 22,* 1324–1330. doi:10.1007/s11665–012–0424–8.
59. Hofmann, D. C., & Vecchio, K. S., (2005). Submerged friction stir processing (SFSP): An improved method for creating ultra-fine-grained bulk materials. *Materials Science and Engineering: A 402,* 234–241. doi:10.1016/j.msea.2005.04.032
60. Najafi, M., Nasiri, A. M., & Kokabi, A. H., (2008). Microstructure and hardness of friction stir processed AZ31 with SiCP. *International Journal of Modern Physics B 22,* 2879–2885. doi:10.1142/S0217979208047717.
61. Arora, H. S., Singh, H., Dhindaw, B. K., & Grewal, H. S., (2012). Some investigations on friction stir processed zone of AZ91 alloy. *Transactions of the Indian Institute of Metals 65,* 735–739. doi:10.1007/s12666-012-0219-5.
62. Zhang, Q., Xiao, B.L., Wang, Q.Z., Ma, Z.Y., (2011). In situ, Al_3Ti and Al_2O_3 nanoparticles reinforced Al composites produced by friction stir processing in an $Al\text{-}TiO_2$ system. *Materials Letters 65,* 2070–2072. doi:10.1016/j.matlet.2011.04.030.
63. Arbegast, W. J. (2008). A flow-partitioned deformation zone model for defect formation during friction stir welding. *Scripta Mater 58,* 372–376.

64. Gandra, J., Miranda, R., Vilac, P., Velhinho, A., & Pamies Teixeira, J. (2011). Functionally graded materials produced by friction stir processing. *J Mater Process Technol 211,* 1659–1668.
65. Miranda, R. M., Telmo, G., Santos, Gandra, J., Lopes, N., & Silva, R. J. C. (2013). Reinforcement strategies for producing functionally graded materials by friction stir processing in aluminum alloys. *J Mater Process Technol 213,* 1609–1615.
66. Mishra, R. S., Ma, Z. Y., & Charit, I. (2003). Friction stir processing: a novel technique for fabrication of surface composite. *Mater Sci Eng A 341,* 307–310.
67. Ratna Sunil, B., Sampath Kumar, T. S., Uday Chakkingal, Nandakumar, V., & Doble, M. (2014). Friction stir processing of magnesium–nanohydroxyapatite composites with controlled in vitro degradation behavior. *Mater Sci Eng C 39*, 315–324.
68. Ratna Sunil, B., Sampath Kumar, T. S., Uday Chakkingal, Nandakumar, V., & Doble, M. (2014). Nano-hydroxyapatite reinforced AZ31 magnesium alloy by friction stir processing: A solid-state processing for biodegradable metal matrix composites. *J Mater Sci: Mater Med 25,* 975–988.
69. Ratna Sunil, B. (2016). Different strategies of secondary phase incorporation into metallic sheets by friction stir processing in developing surface composites, *International Journal of Mechanical and Materials Engineering 11,* 12. https://doi.org/10.1186/s40712-016-0066-y,
70. Yang, M., Xu, C., Wu, C., Kuo-Chi Lin, Yuh, J.,& Chao, Linan (2010). An Fabrication of AA6061/Al_2O_3 nano-ceramic particle reinforced composite coating by using friction stir processing. *J Mater Sci 45,* 4431–4438.
71. Akramifard, H. R., Shamanian, M., Sabbaghian, M., & Esmailzadeh, M. (2014) Microstructure and mechanical properties of Cu/SiC metal matrix composite fabricated via friction stir processing. *Mater Design 54,* 838–844
72. Madhusudhan Reddy, G., Sambasiva Rao, A., & Srinivasa Rao, K. (2013). Friction stir processing for enhancement of wear resistance of ZM21 magnesium alloy. *Trans Indian Inst Met 66*(1), 13–24.
73. Mertens, A., Simar, A., Montrieux, H. M., Halleux, J., Delannay, F., &Lecomte-Beckers, J. (2012).Friction stir processing of magnesium matrix composites reinforced with carbon fibers: influence of the matrix characteristics and of the processing parameters on microstructural developments. In: Poole, W. J., & Kainer, K. U., (ed). *Proceedings of the 9th International Conference on Magnesium Alloys and Their Applications.* Vancouver (Canada), 845–850. http:// hdl.handle.net/2268/120134.
74. Huang, Y., Wang, T., Guo, W., Wan, L., & Lv, S. (2014). Microstructure and surface mechanical property of AZ31 Mg/SiC surface composite fabricated by direct friction stir processing. *Materials and Design, 59*, 274–278.
75. Mazaheri, Y., Karimzadeh, F., &Enayati, M. H. (2014). Tribological Behavior of A356/Al_2O_3 Surface nanocomposite prepared by friction stir processing. *Metall Mater Trans A 45*(a), 2250–2259.
76. Kurt, A., Uygur, I., & Cete, E. (2011). Surface modification of aluminum by friction stir processing. *J Mater Process Technol 211*, 313–317
77. Heydarian, A., Dehghani, K., & Slamkish, T. (2014). Optimizing powder distribution in production of surface nano-composite via friction stir processing. *Metall Mater Trans B 45*(b), 821–826.

78. Wang, W., Shi, Q.-Y., Liu, P., Li, H.-K., & Li, T. (2009). A novel way to produce bulk SiCp reinforced aluminum metal matrix composites by friction stir processing. *Journal of Material Processing Technology, 209,* 2099–2103.
79. Lim, D. K., Shibayanagi, T., & Gerlich, A. P. (2009). Synthesis of multi-walled CNT reinforced aluminum alloy composite via friction stir processing. *Materials Science and Engineering A, 507*, 194–199.
80. Devinder, Y., & Bauri, R. (2011). Processing, microstructure and mechanical properties of nickel particles embedded aluminum matrix composite. *Materials Science and Engineering A, 528,* 1326–1333.
81. Sharifitabar, M., Sarani, A., Khorshahian, S., & Shafiee Afarani, M. (2011). Fabrication of 5052Al/Al2O3 nanoceramic particle reinforced composite via friction stir processing route. *Materials and Design, 32,* 4164–4172.
82. Anvari, S. R., Karimzadeh, F., & Enayati, M. H. (2013). A novel route for development of Al–Cr–O surface nano-composite by friction stir processing. *Journal of Alloys and Compounds, 562,* 48–55.
83. Mazaheri, Y., Karimzadeh, F., & Enayati, M. H. (2014). Tribological Behavior of A356/Al_2O_3 Surface nanocomposite prepared by friction stir processing. *Metallurgical and Materials Transactions A, 45*(a), 2250–2259.
84. Sarmadi, H., Kokabi, A. H., & Seyed Reihani, S. M. (2013). Friction and wear performance of copper–graphite surface composites fabricated by friction stir processing (FSP). *Wear, 304*, 1–12.
85. Sabbaghian, M., Shamanian, M., Akramifard, H. R., & Esmailzadeh, M. (2014). Effect of friction stir processing on the microstructure and mechanical properties of Cu–TiC composite. *Ceram Int* 12969–123976.
86. Farnoush, H., Sadeghi, A., Bastami, A. A., Moztarzadeh, F., & Mohandesi, J. A. (2013). An innovative fabrication of nano-HA coatings on Ti-CaP nanocomposite layer using a combination of friction stir processing and electrophoretic deposition. *Ceramics International, 39,* 1477–1483.
87. Farnoush, H., Bastami, A. A., Sadeghi, A., Mohandesi, J. A., & Moztarzadeh, F. (2013). Tribological and corrosion behavior of friction stir processed Ti-CaP nanocomposites in simulated body fluid solution. *Journal of the Mechanical Behavior of Biomedical Materials, 20,* 90–97.

CHAPTER 10

Different Material Systems

Several pure metals and alloys in particular non-ferrous metals have been used as the matrix material to develop surface MMCs by FSP. Different external secondary phases were added to the matrix to develop the composites. Further, with the help of the heat generated during FSP, in-situ composites were also developed by accelerating the chemical reaction in the matrix material. Hybrid composites were also developed by adding two or more secondary phases through FSP. This chapter summarizes all material systems which were processed by FSP in developing surface MMCs.

10.1 ALUMINUM

Aluminum and its alloys are the first group of materials widely used as matrix materials. The first report on developing surface MMCs by FSP was appeared on developing Al5083-SiC composite by Mishra et al., [1]. Figure 10.1 shows microstructure of the surface composite obtained at 300 rpm with 101.6 mm/min feed. Later on, many aluminum alloys were selected as the matrix material to developed MMCs. Table 10.1 lists different aluminum alloys and their commercial identification, alloying elements and processing/treatment to improve mechanical and other basic properties. Commercial aluminum alloys are designated with 4 digits. The first digit gives the major constituting alloying element. The second digit indicated the modification of the alloy from its starting condition. Third and fourth digits are given to differentiate a particular alloy in the corresponding series.

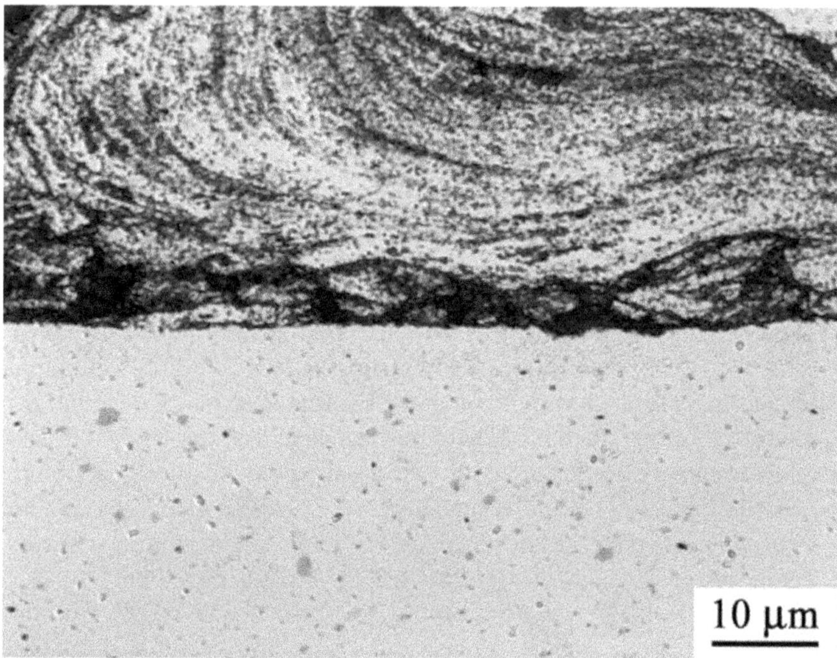

FIGURE 10.1 Microstructure of Al5083-SiC composite produced by FSP at 300 tools rotational speed and 101.6 travel speed Mishra et al. (Source: Reprinted with permission Mishra et al., [1]. © 2003 Elsevier.)

TABLE 10.1 List of Different Commercial Aluminum Alloys Used in the Manufacturing Industry

S. No.	Aluminum alloy series	Major alloying elements	Processing/treatment
1	1xxx	Pure aluminum (99%)	Work hardening
2	2xxx	Copper	Precipitation hardening
3	3xxx	Manganese	Work hardening
4	4xxx	Silicon	Work hardening
5	5xxx	Magnesium	No heat treatment
6	6xxx	Magnesium and silicon	Precipitation hardening
7	7xxx	Zinc	Precipitation hardening
8	8xxx	Other elements	Hardening depends on the alloying elements

10.1.1 1xxx Series

Pure aluminum with the presence of other elements at an insignificant level is categorized as 1xxx series. A considerable amount of work has been done using 1xxx series as matrix materials. Kurt et al., [2] developed aluminum – SiC composite by providing a surface layer of SiC/methanol paste on AA1050 base material followed by FSP. Good interfacial bonding was observed between the SiC particles and the matrix. The distribution of SiC was observed as uniform with higher tool rotational speed and travel speed. However, the bond between the composite layer and the substrate was observed as a week with increased tool travel speed. Increased hardness was observed with increase in tool rotational speed in the composite. Decreased formability was observed for the composite due to the decreased ductility from the three-point bending test as shown in Figure 10.1. Liu et al., [3] developed a composite of AZ80/pure aluminum by placing a pure aluminum plate on a sheet of AZ80 Mg alloy followed by FSP. The formation of Mg-Al solid solution and Mg-Al intermetallic phases in the composite region was observed after FSP. From the corrosion studies, the composite layer exhibited better performance compared with AZ80 Mg alloy and the performance was observed as poor compared with pure aluminum. Later on Liu et al., [4] adopted another way of secondary phase introduction to embedding multi-walled carbon nanotubes (MWCNTs) into 1016 alloy. Holes of same depth with different diameters were produced on the surface and filled with different % of MWCNTs and FSP was carried out.

Thangarasu et al., [5] developed AA1050/TiC composite by FSP and strong bonding between the composite layer and the aluminum substrate was observed. The distribution of TiC particles was observed as homogeneous, and the composite exhibited 45% higher hardness compared with unprocessed alloy. Khorrami et al., [6, 7] studied the effect of pre-strain on the recrystallization behavior during FSP. This phenomena was interesting to observe as the stored strain certainly influence the microstructure at the atomic level while subjected to secondary processing. Initially, aluminum sheets were processed by constrained groove pressing (CGP) and followed by FSP. From the microstructural studies, a significant grain growth was noticed in the stir zone which can be attributed to the higher strain stored in the material which unleashed the grain growth effect during the recrystallization during FSP. Later, the authors studied the effect of adding dispersing particles by FSP into the CGPed sheets on microstructure evolution and mechanical properties [Khorrami et al., [8].

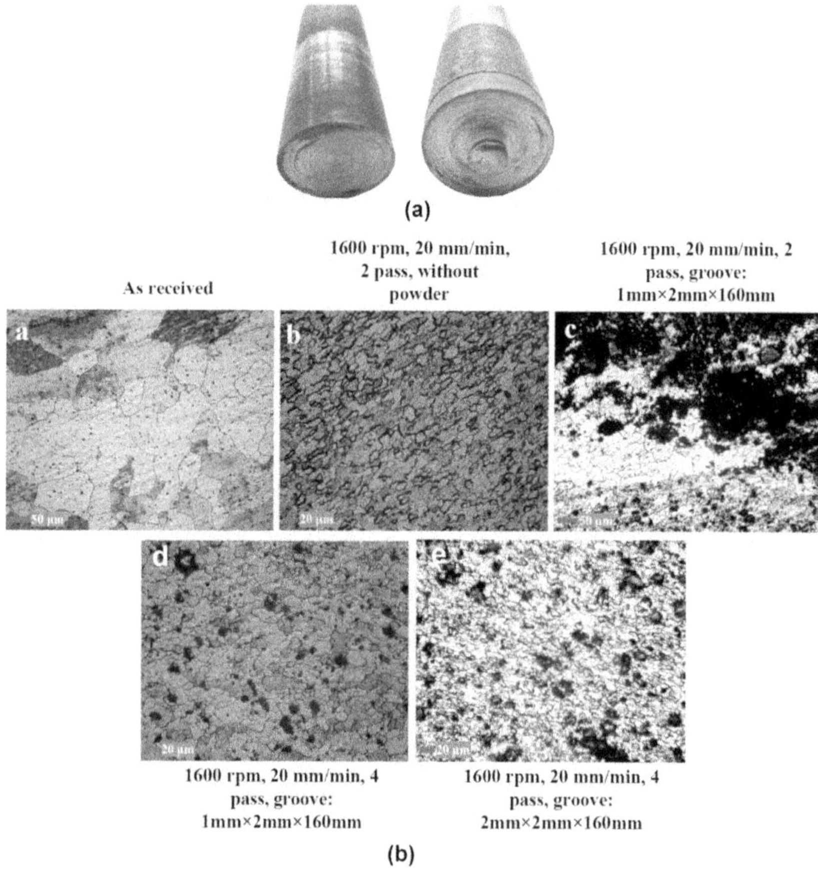

FIGURE 10.2 (a) Photograph of FSP tools used to develop composites, and (b) optical microscope images of unprocessed and FSPed samples. (*Source:* Reprinted with permission Zeidabadi et al., [9] .© 2017 Elsevier.)

1050 Aluminum sheets were processed by CGP for 1, 2 and 3 cycles to introduce a stored strain value of 1.16, 2.32, and 3.48, respectively. Then, around 1.5 V% of nano-SiC particles were added by FSP. From the results, it was observed that the higher stored strain values resulted in the formation of banded structure (combination of fine and coarse grains) after FSP. By adding SiC particles, the banded structure was refined. In the stir zone, regions with a fine distribution of nano-SiC showed excellent grain refinement which is attributed to the pinning effect achieved by the present of nano-SiC. On the contrary, in the stir zone, where the presence of nan-SiC was insufficient, grain growth was observed as represented with large grain sizes. From the works of Khorrami et al., [6–8] it can be understood that the stored strain in the sheets leads to higher nucleation of new grains and subsequent rapid grain growth during FSP. By introducing nanoparticles into the sheet, after the higher nucleation of new grains due to stored strain, the grain growth can be suppressed by utilizing the advantage of the pinning effect brought from the presence of nanoparticles.

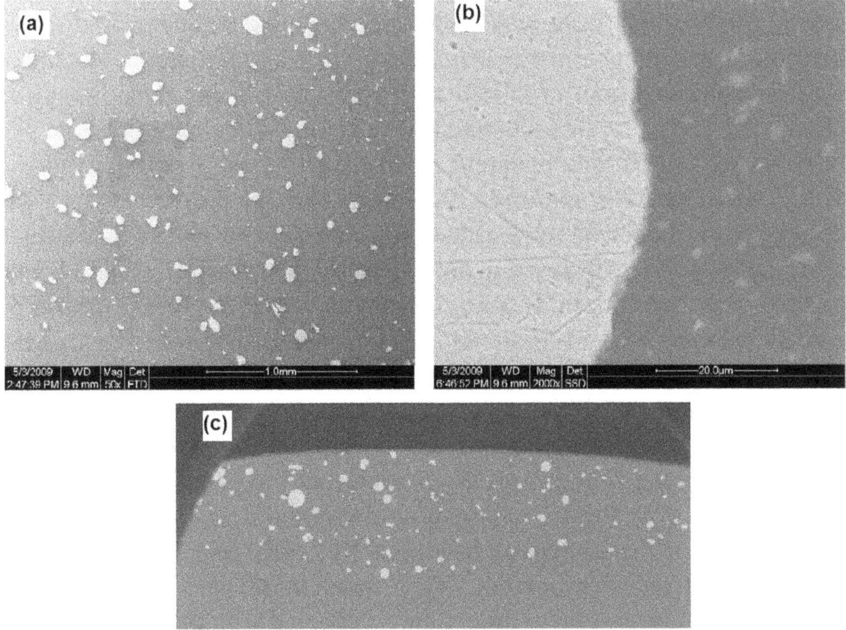

FIGURE 10.3 Microstructure of Al-Ni composite observed by usingSEM: (a) distribution of Ni particles in the composite, (b) strong bonding at the interface of Al matrix and Ni particle and (c) cross-section of stir zone. (*Source:* Reprinted with permission Yadav and Bauri [11]. © 2011 Elsevier.)

Zeidabadi et al., [9] introduced Nb particles into A11050 alloy by using FSP. Here developing intermetallic phases in the Al matrix was achieved by FSP as shown in Figure 10.2. The reinforcement of Nb was done by providing two different groove dimensions (1×2×160 mm and 1×2×160 mm) by doing two and four passes of FSP. During FSP, due to the exothermic reaction between the matrix (Al) and the dispersing phase (Nb) an intermetallic phase (Al_3Nb) was observed as formed in the stir zone. As the number of FSP passes are increased to four, dispersed particles were observed as uniform, and therefore hardness and tensile strength were observed as increased. Yadav and Bauri [10] used FSP to develop in-situ composites of Al-TiB_2, and uniform distribution of TiB_2 was observed after two FSP passes. A cast sheet of aluminum and K_2TiF_6 and KBF_4 precursors was obtained from liquid processing route and subjected to FSP.Due to FSP, the agglomerates were observed as broken and uniform distribution was achieved. The defects resulted during casting of the composite were also eliminated after FSP. Very fine grains were achieved of them around 60 % grains were with high angle boundaries. Interestingly, the strength of the composite was increased up to 1.5 times and the ductility was also enhanced from 11 to 19% after FSP. Later, Yadav et al., [11] also introduced Ti powder into aluminum to develop particulate composite by FSP and studied particle twinning, microstructure, and thermal stability. The Ti particles were uniformly distributed, and the presence of Ti was observed as in elemental form without forming any new phases. Compressive type twins were observed in the Ti particles and fracture of individual particles was also noticed. Higher yield strength up to 3.5 times compared with the base material and more stability to withstand against grain growth was also observed for the composite due to the presence of Ti. Similarly, Yadav and Bauri [12] added Ni particles into commercial pure aluminum (1050) by FSP and formation of any intermetallics was not observed. A strong interface between the Ni particles and the matrix was clearly demonstrated from the microstructural observations as shown from the SEM images of the composite in Figure 10.3. The composite exhibited excellent mechanical properties. Yield strength of the composite was measured as increased up to three times compared with the base material without much sacrificing the ductility.

Bhat et al., [13] introduced Fe powder into pure aluminum by FSP, and fine grains were observed in the stir zone along with the presence of fine Fe, aluminides and high dislocations. The creep resistance was observed as poor for the composite in spite of the presence of Fe particles. Mahmoud et al., [14] developed hybrid composites by introducing SiC and Al_2O_3 into AA1050-H24 alloy by FSP to investigate wear properties. The distributed particles were observed as uniform in the nugget zone after FSP. With SiC addition, hardness was measured as increased up to 100%. By adding, Al_2O_3, the hardness was observed as decreased in the hybrid composite.

From the wear studies, lower coefficient of friction and increased wear resistance was observed with 20% Al_2O_3 and 80% SiC. By varying the applied load from 2 N to 10 N, the hybrid composites showed different wear characteristics which suggest the load dependent performance of the composite.

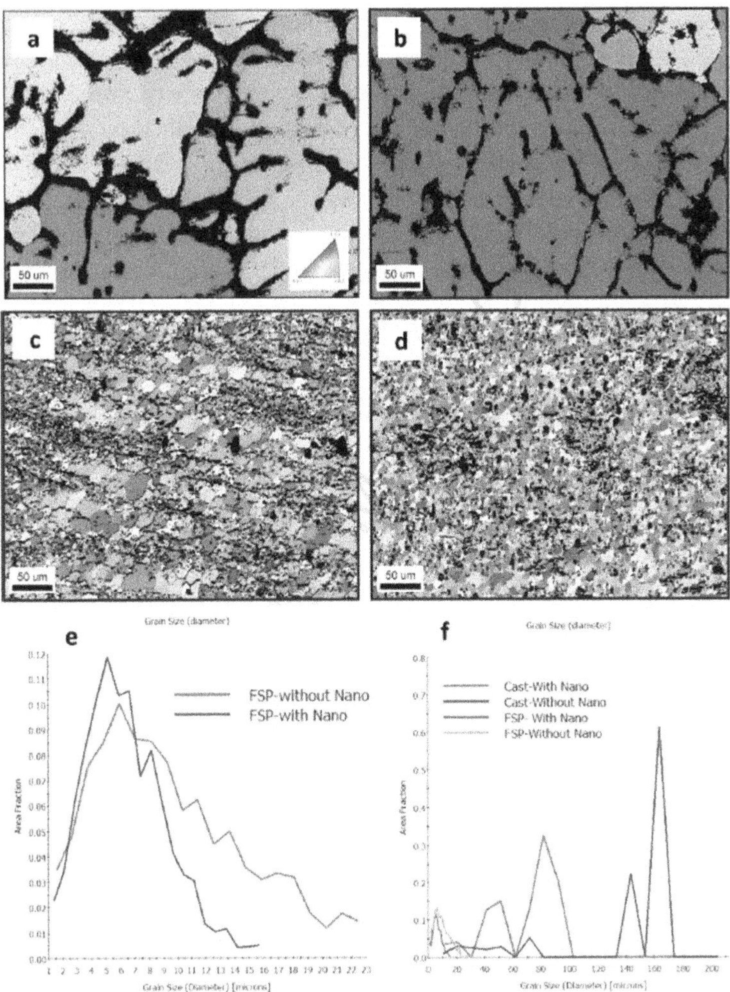

FIGURE 10.4 (See color insert.) Orientation image maps with respect to FSP normal direction: (a) as cast without reinforcement, (b) as cast with reinforcement, (c) FSPed without reinforcement, (d) FSPed nanocomposite, (e) and (f) comparison of grain size distribution. (*Source:* Reprinted with permission Hoziefa et al. [18]. © 2016 Elsevier.)

10.1.2 2xxx Series

Next to 1xxx series, the presence of copper as the main constituting alloying element gives 2xxx alloys which have also been widely used as matrix materials to develop surface composites by FSP. Cavaliere et al., [15] developed 2618-Al_2O_3 composite and mechanical behavior were investigated. From the stress amplitude–fatigue life studies, composites exhibited a classical behavior and increased fatigue life with decreased cyclic stress amplitude. A stable crack growth at microscopic level was observed in the composite in the failed regions which was characterized by brittle failure. Zhao et al., [16] developed nano-ZrB_2/2024Al composites by in situ mechanism using 2024Al– K_2ZrF_6–KBF_4 system. Then the composites were subjected to FSP to modify the microstructure. After FSP, the composite exhibited refined grain size and the agglomerated nano-ZrB_2 particles were observed as distributed uniformly throughout the matrix. The superelasticity of the FSPed composites was observed as increased due to the modified microstructure after FSP. The superplastic elongation of 292.5% was measured for the FSPed composite compared with cast condition (67.2%). Zahmatkesh and Enayati [17] produced Al-Al_2O_3 surface layer on Al2024 substrate by air plasma spraying. Then the surface was processed by FSP to distribute Al_2O_3 particles into the substrate. After FSP, Al_2O_3 particles were observed as uniformly distributed into the substrate up to 600 μm. The surface composite layer exhibited excellent bonding strength with the substrate and also better wear properties.

Hoziefa et al., [18] developed metal matrix composite by introducing 1 wt.% of Al_2O_3 into AA2024 alloy in the semi-solid state by mechanical stirring. The composite was then subjected to a single FSP pass. For comparison, FSP was also carried out for alloy without Al_2O_3. Due to the addition of Al_2O_3, a grain refinement was achieved in the composite before FSP as compared in Figure 10.4. After FSP, the level of grain refinement was further increased and achieved up to 2.7 μm. Mechanical properties were observed as increased for the composite due to the presence of the dispersed phase. FSP uniformly distributed Al_2O_3 in the AA2024 matrix and further increased the ultimate tensile strength and yield strength up to 71% and 30%, respectively. In addition, the authors suggested that by adopting FSP defects such as porosity and shrinkage cavity usually produced during casting can be eliminated.

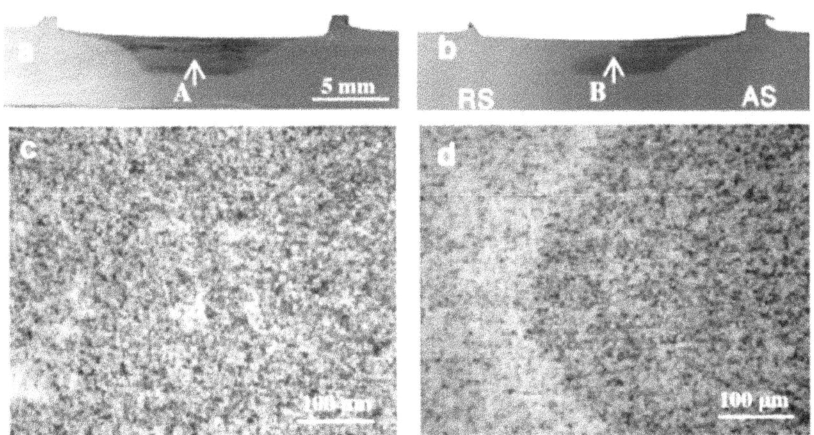

FIGURE 10.5 Macroscope images at the cross section of FSPed samples: (a) same tool rotational direction between passes, (b) opposite direction of the tool rotation between the passes, (c) optical microscope image of the "A" region, and (d) optical microscope image of B region from the processed zones (*Source:* Reprinted with permission Dolatkhah et al., [19]. © 2012 Elsevier.)

FIGURE 10.6 Cross section macroscope and microscope images of FSPed composite produced by carrying multi FSP passes by changing the orientation after each pass. (*Source:* Reprinted with permission Dolatkhah et al., [19]. © 2012 Elsevier.)

10.1.3 5xxx Series

Among all aluminum alloys, 5xxx series alloys with magnesium as the main constituting element were extensively used as matrix materials for developing composites by FSP. Dolatkhah et al., [19] developed Al5052/SiC composites by FSP to investigate the effect of shifting the tool rotation direction between passes on the distribution of the dispersing phase. It is interesting to see a higher level of homogeneity in the particle dispersion after four passes. By changing the rotational direction between the passes, distribution of the particles was observed as uniform in the advancing side as well as in the retreating side as shown in Figure 10.5. Therefore, the level of homogeneity of particles dispersion is increased as shown in Figure 10.6 and which further introduced better wear resistance. Ranjit Bauri et al., [20] introduced Ni powder into Al 5083 alloy by friction stir processing at different tool rotational speeds (1000, 1200, 1500 and 1800 rpm) and travel speeds (0.1, 0.2, 0.3, 0.4 and 0.5 mm/s). A combination of optimized parameters were suggested (1200 rpm with 0.4 mm/s) to successfully develop the composite without any defect. Grin refinement was achieved from a starting size of 25 μm to 3 μm after FSP. A good bonding between the Ni particles and the matrix was noticed. Providing geometrical features on the pin and shoulder was observed as the best strategy to uniformly disperse the powder. Increased hardness from 81 to 91 Hv and increased tensile strength from 303 to 339 MPa with increased % elongation (from 25% to 31%) achieved in the composite.

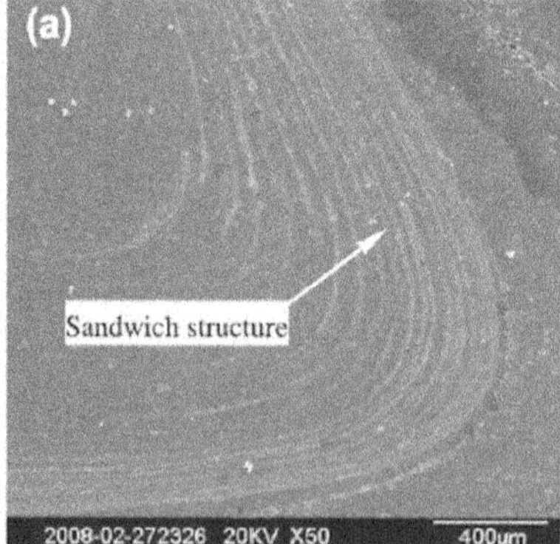

FIGURE 10.7 SEM images of composite produced from 5A06 alloy sheet and $Al_{84.2}Ni_{10}La_{2.1}$ amorphous phase through FSP: (a) low magnification, and (b) high magnification. (*Source:* Reprinted with permission Liu et al., [21]. © 2013 Elsevier.)

Liu et al., [21] introduced $Al_{84.2}Ni_{10}La_{2.1}$ amorphous phase in 5A06 phase by FSP. A groove was produced on the surface of 5A06 alloy sheet and was filled with $Al_{84.2}Ni_{10}La_{2.1}$ amorphous thin strip. After FSP, the amorphous phase was observed as initiated crystallization. A sandwich structure of different chemical composition was observed in the stir zone as shown in Figure 10.7. Very fine grains of 0.3-1 μm were observed in the composite, and interestingly, ultrafine-grained fine bulk structures of a grain size of 90-400 nm were also observed in the stir zone from transmission electron microscope (TEM) observations.

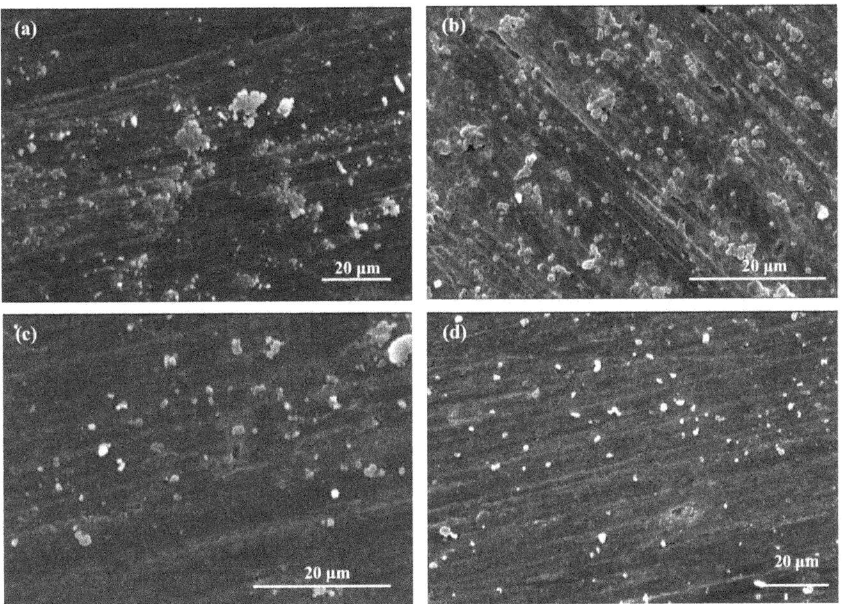

FIGURE 10.8 SEM images showing the distribution of $ZrSiO_4$ particles: (a) 1st pass, (b) 2nd pass, (c) 3rd pass, and (d) 4th pass. (*Source:* Reprinted with permission from Rahsepar and Jarahimoghadam [26]. © 2016 Elsevier.)

Due to these microstructural changes, higher hardness and tensile strength were measured for the composite.From the corrosion test conducted in salt solution, higher corrosion resistance was observed for as received 5A06 alloy compared with the composite. Yuvraj et al., [22] produced $Al5083/B_4C$ composite by one pass and three pass FSP to investigate the effect of incorporating B_4C on mechanical and tribological properties. The hardness, tensile strength, and wear rate of the composite were measured as 124.8 Hv, 360 MPa and 0.00327 mg/m, respectively. These values are higher when compared with the base material values 82 Hv, 310 MPa and 0.0057 mg/m, respectively. The distribution of dispersing phase was observed as uniform after three passes compared with a single pass. B_4C particles were used of two different sizes one at nano-size and other at micro-size. Compared with micro-particles, the effect to increase the mechanical and wear properties were observed as higher when nanoparticles were dispersed. Amra et al., [23] introduced nano-CeO_2 and SiC particles into Al5083 alloy in various amounts individually as well as combined through FSP and developed nanocomposites. All of the FSPed composites exhibited higher hardness and tensile strength compared with the unprocessed condition. Among them, composite with 100%SiC has shown higher tensile strength. Interestingly, the addition of SiC decreased the pitting corrosion resistance, and the addition of CeO_2 increased the corrosion resistance. The main reason behind the decreased corrosion resistance in the composite with SiC was explained as the formation of micro-galvanic cells. Composite with 75%CeO_2 + 25%SiC combination of dispersing phase was observed as optimum in obtaining better mechanical properties and corrosion properties.

Sahandi Zangabad et al., [24] developed hybrid composites by dispersing TiO_2 nano-particles into Al-Mg alloy (AA5052) by doing FSP up to 6 passes. Increased tensile properties (yield strength – 90% and ultimate tensile strength 31%) and fatigue behavior was observed for the composites as the number of FSP passes were increased. Yuvaraj et al., [25] also developed surface nano-hybrid composites by through FSP by using Boron carbide (B4C) and Titanium carbide (TiC) as dispersing phases and Al5083 alloy as matrix material. Higher wear resistance was observed for the hybrid composite due to the presence of TiC. M. Rahsepar and H. Jarahimoghadam [26] developed composites of Al5052- $ZrSiO_4$ by FSP and the distribution of the dispersed phase after four passes was investigated. With the increase in the FSP passes, the distribution of the particles from agglomerated state to well-dispersed state was observed as shown in Figure 10.8. With better distribution of the dispersing particles, good ductility and corrosion resistance was observed for the composites after multiple FSP passes.

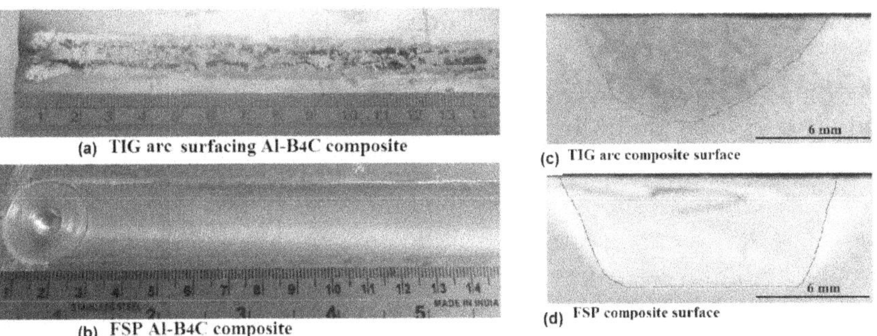

FIGURE 10.9 (See color insert.) Comparison of the surface composites produced by two different methods: (a) photograph showing the surface composite produced by TIG welding, (b) surface composite by FSP, (c) cross-section comparison of composite produced by TIG, and (d) FSP. (*Source:* Reprinted with permission from Yuvraj et al., [29]. 2017 Elsevier.)

Mohammadzadeh Jamaliana et al., [27] adopted a novel approach during friction stir welding of AA5086-H34 aluminum alloy sheets. Nano-SiC particles were added in the weld zone to develop a weld zone of AA5086-H34/SiC composite. By altering the process parameters such as tool rotational speed and travel speed, grain size was observed as altered, but the distribution of SiC in the stir zone was nonuniform for single pass FSP. However, by increasing the number of passes to three, higher level of grain refinement and uniformity in the distribution of SiC was observed from the microstructural observations. Single-pass weld joint has given poor mechanical performance due to the agglomeration of the SiC particles. After three, four and six FSP passes, interestingly the mechanical properties of the composite were observed as improved. As the number of FSP passes are increased, the interface bonding between the matrix and SiC was observed as increased and therefore, the mode of failure in the composites with multi-pass FSP was noticed as a ductile fracture. Whereas in single pass FSP weld joint, brittle mode of fracture was observed, this was attributed to the agglomeration of SiC particles.

Huang et al., [28] developed a surface coating of SiC reinforced AA5056 composite on Al2024 substrate. Then, FSP was carried out to investigate the microstructure and mechanical properties. From the observations, the porosity in the surface coating was observed as decreased due to FSP. The distribution of SiC particles was also observed as uniform in the FSPed region. Yuvraj et al., [29] developed surface composites of

Al5083 by introducing B_4C particles through TIG welding and FSP. The distribution of B_4C particles in the composites produced by both methods was compared. Figure 10.9 compares the macroscopic view of the surface and the cross section of the respective composites produced by TIG and FSP. The authors reported a higher thickness in the composite produced by FSP compared with the composite produced by TIG welding. This is true that FSP lead to uniform distribution of secondary phase and also by introducing grain refinement effect, structure-sensitive properties such as mechanical, chemical, surface, wear and bio-properties are certainly influenced to a great extent as reported by several other authors (Sharifitabar et al., [30], Soleymani [31], and Liu et al., [32]).

10.1.4 6xxx Series

Among all other aluminum alloys, 6XXX series alloys have also been widely used as the matrix material to develop composites by FSP.Yang et al., [33] developed AA6061/Al_2O_3 composite by holes filling method. Three passes were produced on the surface in three parallel paths. By increasing the axial force and number of passes from one to three, homogeneous distribution of Al_2O_3 particles in the matrix was observed. Interestingly, nanosize of Al_2O_3 has found to have no distinct effect on the macrohardness. The authors explained that the softening of base material resulted from overaging was the reason behind unaffected hardness of the composite. Anvari et al., [34], developed AA6061 alloy/Cr_2O_3 composites and interestingly in situ formation of other phases was observed after 6 passes. Phases such as Al_2O_3, Cr_2O_3, Cr, $Al_{13}Cr_2$ and $Al_{11}Cr_2$ were found in the matrix. As per the author's explanation, the temperature that was measured during FSP in the stir zone (347°C) led to develop these phases. Increased hardness (147 Hv) and tensile strength (518 MPa) was observed for the composites due to the Orowan mechanism and the presence of a high density of dislocations. Additionally, Anvari et al. [35] investigated the wear properties of the composite and found that the wear mechanism for the composite as delamination wear unlike in the case of the base material and FSPed AA6061 alloy without nanoparticle dispersion exhibited adhesive and abrasive wear due to the lack of hard phase.

Aoh et al., [36] introduced copper coated SiC into Al6061 matrix and formation of a cohesion layer between the dispersed particles, and the substrate was clearly observed. Due to this interfacial layer, a higher level of diffusion was also observed between the dispersed particles and the matrix which led to improve the mechanical properties of the composite. Further, the authors adopted T5 heat treatment technique to restore the strength of the composite. Interestingly, the tensile strength of the composite with Cu coated SiC was observed as higher compared with the base and the composite with Cu less SiC particles. These findings strongly suggest that the pre-treatment of the dispersing phase to yield maximum benefit. Devaraju et al., [37] developed Al6061 based hybrid composites by FSP method using several combinations of SiC, Al_2O_3, and graphene (Gr) phases. By separately adding Gr to SiC and Al_2O_3, hybrid composites ofAl/Gr-SiC and Al/Gr-Al_2O_3 were produced, and increased hardness was observed in both the composites due to the addition of the secondary phase particles. The wear properties of the hybrid composites were compared with Al/Al_2O_3+SiC composite and found promising wear behavior for the hybrid composites due to the presence of Gr which acted as a solid lubricant.

Thangarasu et al., [38] dispersed TiC particles into AA6082 alloy by FSP, and uniform distribution of SiC particles was achieved. The interface between the SiC particles and the matrix was observed as free from any intermetallics. With an increased amount of TiC particles, the stiffness of the matrix was increased, and from the fracture studies, the formation of the voids was observed as decreased with the increased TiC amount. The wear mechanisms were also influenced by the TiC particles amount. The wear mode was observed as changed from adhesion to abrasion for the composite. Naresh and Kumar [39] developed a composite of Al6061-Al_2O_3 by FSP and increased hardness was measured for the composites. Zhao et al., [40] introduced B_4C into Al6061 alloy by FSP and observed a uniform distribution of B_4C after four passes. Hardness was also observed as higher with lower variations within the FSPed zone. From the wear studies, lower coefficient of friction values were observed for the composite which is an indication for improved tribological behavior.

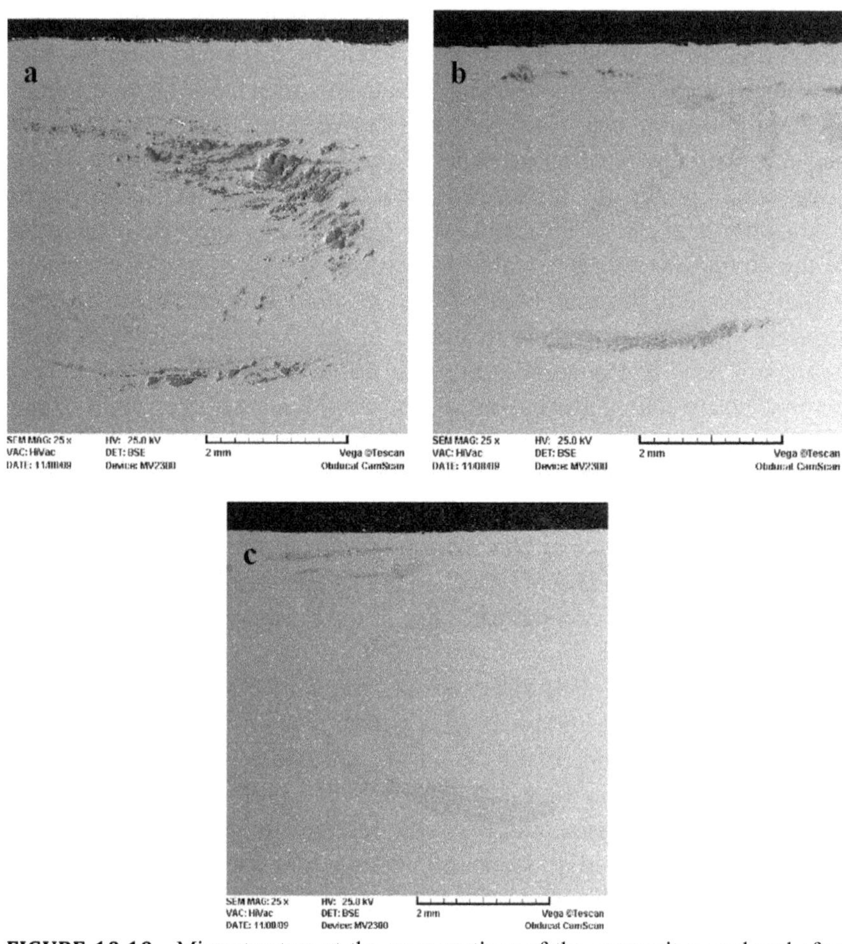

FIGURE 10.10 Microstructure at the cross sections of the composites produced after number of FSP passes: (a) one, (b) three, and (c) four passes. (*Source:* Reprinted with permission from Shafiei-Zarghani et al., [44]. 2011 Elsevier.)

Du et al., [41] introduced CNT into Al6061 alloy and investigated the level of dispersion of CNT into the substrate with respect to number of FSP passes. After three passes the distribution of CNT was observed as uniform. Interestingly, the hardness was decreased after three FSP passes in the samples without CNT. However, due to the presence of CNT, hardness was measured as increased in the composite with increased number of passes. Improved mechanical properties as reflected from the increased tensile strength and ductility in CNT embedded FSP samples suggest the promising role of CNT in enhancing the resistance to failure of the alloy.Kurt et al., [42] investigated the effect of process parameters and the dispersing phase on the wear behavior of FSPed Al6061 alloy. Better wear properties were observed for 50% TiC+ 50% Al_2O_3 dispersed hybrid composite. Similarly, Yang et al., [43] produced nano-Al_2O_3 dispersed AA6061 composites by FSP. As number of FSP passes was increased to three, uniform dispersion of the particles was observed. Other interesting observation that the authors demonstration in the composite is the decreased hardness. The reason was explained by considering aging effect due to the FSP process which suppressed the effect of addition of nano-particles on increasing the hardness. Shafiei-Zarghani et al., [44] also introduced Al_2O_3 into Al6082 alloy by FSP and developed surface composite to investigate the addition of Al_2O_3 on the wear properties of Al6082. Figure 10.10 shows the cross-sectional microstructure of the composites produced after one, three and four passes, respectively.

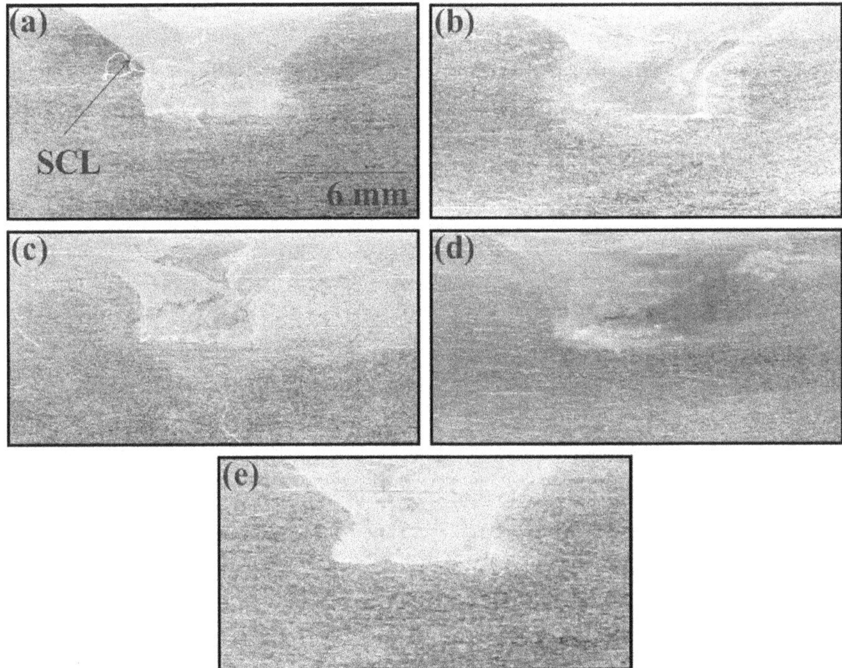

FIGURE 10.11 Microstrucire of the composites with varying dispersing phases: (a) 100% TiC, (b) 75% TiC + 25% B_4C, (c) 50% TiC + 50% B_4C, (d) 25% TiC + 75% B_4C, and (e) 100% B_4C. (*Source:* Reprinted with permission from Rejil et al., [45]. 2012 Elsevier.)

Rejil et al., [45] produced Al-based hybrid composites by dispersing TiC + B_4C into AA6360 alloy using FSP. The produced surface composite layer (SCL) was observed as uniform with the dispersion of the phases, and no defect was observed at the interface of the SCL and the matrix material. The particles were dispersed without any agglomeration in the stir zone as shown in Figure 10.11. Better wear resistance was observed for the hybrid composite compared with the base material.

10.1.5 Other Al Alloys

Aluminum alloys (7xxx series) with zinc (Zn) as the main constituting alloying element were also used as matrix materials to produce surface composites. Alidokht et al., [46, 47] developed hybrid composite by dispersing SiC and MoS_2 by FSP a significant grain refinement was observed from a starting size of 100 μm to 4 μm after the first pass. At higher tool rotation speed (1600 rpm), uniform distribution of the dispersed particle was observed. Additionally, FSP helped to break the silicon particles and aluminum dendrites. The hybrid composite exhibited higher hardness compared with the base material and lower compared with single phase dispersed composites (A356/SiC). Due to the presence of SiC and MoS2, wear properties of the hybrid composite were also improved compared with the base material. As the load was increased from 10 N to 40 N during wear studies, the material removal mechanism was changed from delamination wear mechanism to abrasion mechanism. Hussain et al., [48] developed Al7075-TiN composite by FSP to study the effect of adding high hard and brittle TiN phase at the surface of the Al alloy. The authors used three different pin profile, i.e., (i) square, (ii) threaded, and (iii) triangular. The microstructure of the composite was observed as influenced by the pin profile. Formation of secondary phases between the dispersed particles and the matrix material was observed. Due to the presence of the brittle phase, decreased ductility was noticed for the composites. Eskandari et al., [49] used two different dispersing phases of two different particle sizes. The authors used FSP to disperse micro-and nano-sized TiB_2 and Al_2O_3 particles into AA8026 alloy. Mono-composites and hybrid composites were produced by doing multiple FSP passes by overlapping each pass. With increased tool rotational speed, lower travel speeds and number of passes resulted uniform distribution of the dispersing phase into the matrix. Higher hardness and further better wear properties were recorded for the hybrid composites compared with nano-composites .

10.2 COPPER

Producing copper (Cu)-based surface composites by FSP was also reported by several authors. Cartigueyen and Mahadevan [50] developed Cu based surface composites by dispersing SiC particles through FSP. SiC particles of size ≈ 12 μm were introduced into the surface of pure Cu by filling the holes produced on the surface of the Cu sheet with SiC particles followed by carrying FSP. From the observations, it was noticed that the temperature at the advancing side is increased with increased rotational speed and decreased at higher travel speed and tilt angle. Hardness was observed as increased for the composites due to the dispersed SiC particles. The distribution was observed as free from agglomeration, and a well bonding was also observed at the interface of the SiC particles and the matrix. Jafari et al., [51] introduced CNT into the surface of pure Cu by FSP and improved mechanical and wear were observed due to the addition of nano-featured dispersing phase (CNT). Due to high input heat generated by using FSP tool with larger pin dimensions, damage to CNT was significantly observed. As the number of passes increased to three, the dame of CNT was observed as higher. Increased hardness and wear resistance was also observed for the composite due to the presence of CNT.

Sabbaghian et al., [52] introduced TiC particles into pure Cu sheets by adopting 3° spindle tilt angle. From the XRD studies, no intermetallic compound was found at the interface of TiC, and the matrix and no indication for interfacial reaction was observed.Fine-grained structure resulted from FSP and incorporated TiC increased the hardness of the composite. From the wear studies, change in coefficient of friction was observed as lower and due to the lower contact of the matrix in the composite, better wear properties were observed. Barmouz et al., [53] developed Cu based surface composites by using pure Cu as matrix and SiC as dispersing phase through FSP route. Due to the continuous dynamic recrystallization, fine grains were appeared after FSP. A small variation in the grain size was observed by changing the tool travel speed from 40, 80 to 200 mm/min. Presence of SiC introduced pinning effect and hence, the grain growth was also suppressed. From the microhardness measurements, it was observed that the adjacent sides of stir zone were softened as reflected in the lowered hardness due to the annealing effect. However, within the stir zone due to the grain refinement and the presence of SiC higher hardness was observed in the composites. A weak bonding between the SiC particles and the matrix decreased tensile strength and % elongation. Due to the reduction of the direct load on the material and increased hardness in the composites, better wear properties were observed.

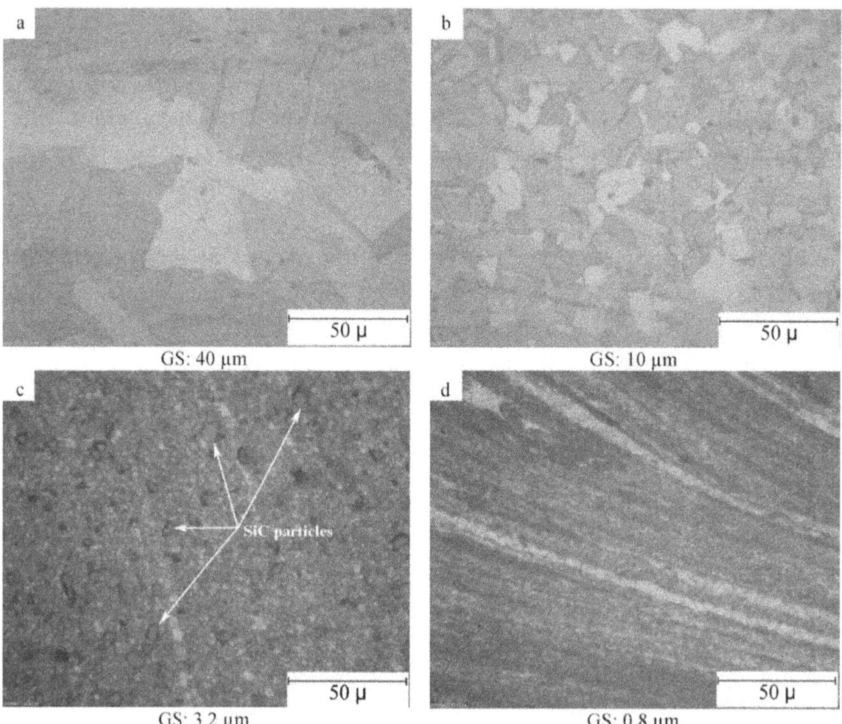

FIGURE 10.12 Optical microscope images: (a) unprocessed pure Cu, (b) FSPed Cu without SiC addition, (c) composite of Cu-microSiC, and (d) composite of Cu-nanoSiC. (*Source:* Reprinted with permission from Barmouz et al., [54]. © 2011 Elsevier.)

Further, Barmouz et al., [54] produced Cu-SiC composite by dispersing SiC particles of size 5 μm and 30 nm into pure Cu using FSP. The authors compared the effect of addition of SiC with FSPed Cu without SiC. After FSP, the grain size was decreased compared with the base material. However, addition of SiC has a significant role on arresting the grain growth and resulted higher level of grain refinement. The level of grain refinement was observed as more in the composite produced by using nano-SiC. Figure 10.12 compares the microstructure of base material, FSPed Cu and the composites produced by using micro-SiC and nano-SiC. The hardness of the composites was observed as increased with the higher volume fraction of the dispersing phase. All composites exhibited more hardness compared with the unprocessed and FSPed Cu due to the presence of SiC. The effect was observed as higher for composite with nano-SiC. Similarly, Akramifard et al., [55] also produced Cu-SiC composite by introducing SiC particles into holes drilled on the surface of the Cu plate followed by FSP. Uniform distribution of SiC particles was observed due to the more of particle filling (hole filling). Very fine grains were observed in the stir zone with no formation of intermetallics. The Cu-SiC composite has shown better wear properties, and mechanical behavior compared with the unprocessed Cu.

Sathiskumar et al., [56, 57] introduced B_4C particles into pure Cu plate and developed surface MMCs by FSP. The interfacial bonding between the B_4C particles and the matrix was observed as excellent in the composites. Furthermore, the effect of process parameters on the formation of the composites was also investigated. The area of the surface composite was observed as increased with decrease of traverse speed and decreased with the increase of traverse speed. The distribution of the B_4C particles was observed as better with lower travel speeds. From the hardness measurements, more hardness was measured in the composites which were produced at higher traverse speeds. Sabbaghian et al., [58] produced Cu-TiC composite by FSP and mechanical and wear properties were investigated. Formation of intermetallic was not observed in the stir zone due to the lower processing temperature results during FSP. Stir zone exhibited very fine grains and uniform distribution of TiC particles. Higher hardness was measured for the composites which were attributed to the presence of TiC and the smaller grain size. From the wear studies, lower coefficient of friction was observed for the composite, and better wear properties were recorded compared with the unprocessed Cu.

10.3 MAGNESIUM

Pure magnesium and several magnesium alloys were processed to develop surface composites by FSP. Most of the work was done using AZ series (aluminum and zinc) magnesium alloys as matrix material. A few other alloys were also processed along with pure Mg and different properties such as hardness, tensile, corrosion and bioproperties were observed as greatly influenced in the composites.

10.3.1 Pure Mg

Ratna Sunil et a., [59], introduced nano-hydroxyapatite (nHA) of crystallite size 32 nm into pure Mg with an aim to develop biodegradable composites for medical implant applications. A grain refinement from a starting size of around 2000 μm to 5 μm was observed in the Mg-nHA composites. nHA powder was introduced into the matrix by groove filling method. In parallel, FSPed Mg was also produced without adding nHA powder in order to compare and assess the role of addition of nHA on the material properties [60]. Uniform distribution was observed at the surface, and nonuniform distribution was found in the thickness direction. Figure 10.13 shows the photographs of the FSPed regions and corresponding microstructures obtained at the cross-section.

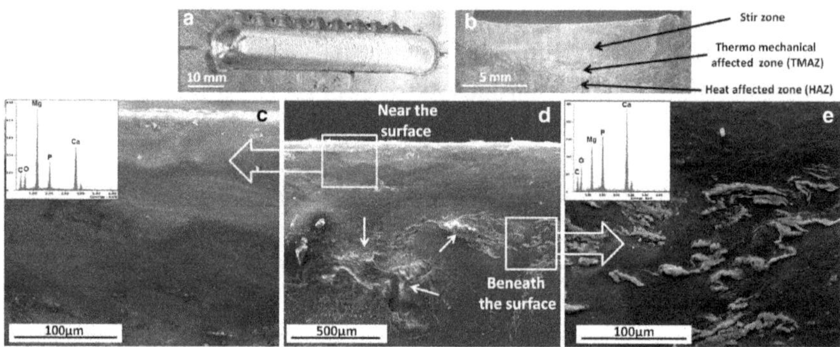

FIGURE 10.13 (a) Photograph of Mg-HA composite showing the surface,(b) cross-section, (c) uniform distribution of HA powder near the surface, (d) low magnified image, and (e) agglomeration of HA at the sub-surface. (*Source:* Reprinted with permission from Ratna Sunil et al., [59]. © 2014 Elsevier.)

Due to the grain refinement, surface energies of FSP Mg and Mg-nHA composite was measured as increased. From the in vitro biomineralization tests, excellent bioactivity was observed for the Mg-nHA composite. Due to the higher biomineralization, reduced corrosion rate was observed for the composite compared with the pure Mg due to the quick formation of a passive layer and decreased intensity of galvanic couple resulted from the grain size reduction. Improved cell adhesion and proliferation was noticed from the cell culture studies assessed by using rat skeletal muscle (L6) cells. Mertens et al. [61] produced Mg/carbon composites by sandwich method. The introduced carbon played a role in achieving grain refinement due to pinning effect of the carbon. Furthermore, Navazani and Dehghani [62] developed Mg/ZrO_2 composite by FSP, and better mechanical performance was observed because of the fine grain structure and the presence of the ZrO_2 particles.

10.3.2 AZ31 Mg Alloy

AZ series alloys are the widely used group of Mg alloys to develop composites using FSP. Among them, AZ31 Mg alloy occupies first place. Morisada et al. [63] developed a composite of AZ31 Mg alloy- multi-walled carbon nanotubes (MWCNTs) by FSP at different tool travel speeds (25, 50, and 100 mm/min). Better dispersion of MWCNTs in AZ31 Mg alloy was observed at 25 mm/min. Due to the presence of MWCNTs, higher level of grain refinement was observed in the composite due to pinning effect similar to what reported by Mertens et al. [64] in developing carbon fiber reinforced AZ31B Mg alloys by FSPusing sandwich method. Later, Azizieh et al. [65] introduced nano-Al_2O_3 into AZ31Mg alloy by FSP using three different tools consisting of a columnar circular pin, columnar circular pin with threads and flutes keeping the other dimensions are the same. The Al_2O_3 powder was selected in three different sizes (approximately 35, 350 and 1000 nm) to fabricate the composites. Onion ring type pattern, due to the material flow from the hot zone beneath the tool shoulder to the relatively cooler zones in the thickness direction [63], which is a general finding in the stir zone, observed after processing with 1000 and 1200 rpm rotational speeds.

Huang et al. [66] introduced a new kind of tool which facilitates supplying secondary phase during processing through the FSP tool in producing AZ31-SiC composite. A specially designed tool known as direct friction stir processing (DFSP) tool was used to directly introduce SiC powder. A hole is provided within the tool which is filled with secondary phase powder and processed with a tilt angle. Higher level of grain refinement, more stir zone width were achieved in AZ31-SiC composite produced with DFSPed tool compared with usual FSP tool. More amount of secondary phase was introduced into the matrix using DFSP tool compared with FSP tool as shown in Figure 10.14. Using DFSP tool, initial machining processes such as producing grooves or holes to fill the secondary phase particles on the surface of the workpiece can be eliminated. However, the thickness of the composite layer that is produced by DFSP tool is relatively less compared with that of layer produced by FSP tool.

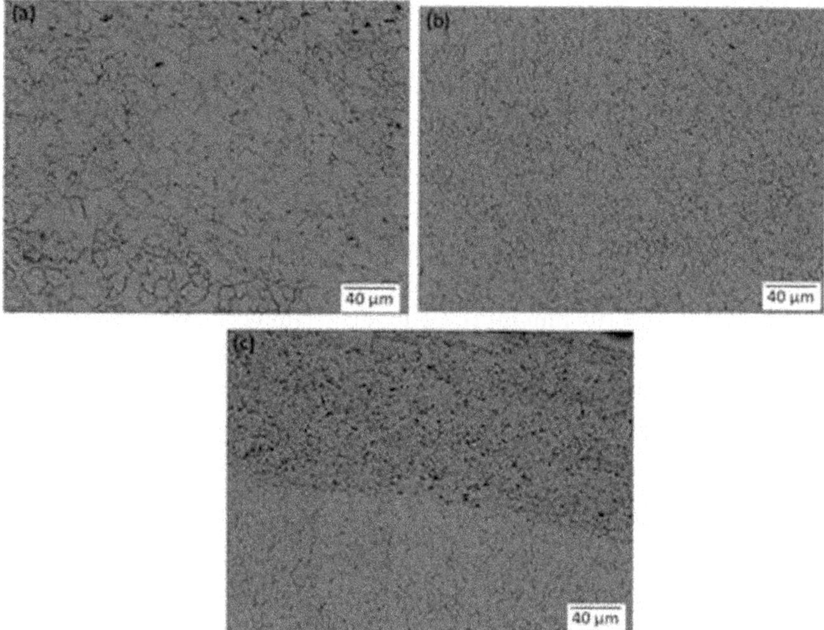

FIGURE 10.14 Optical microscope images: (a) unprocessed, (b) processed with FSP tool, and (c) DFSP tool. (*Source:* Reprinted with permission from Huang et al., [66]. © 2014 Elsevier.)

Recently, nanohydroxyapatite (nHA) powder was dispersed into AZ31 Mg alloy by FSP to develop biodegradable composites targeted for degradable implant applications [67]. Due to the fine grains and incorporated nHA, higher bioactivity, better corrosion resistance, and enhanced cell activities are observed. Incorporating bioactive ceramic powder into the surface of AZ31 Mg alloy really improved the cell kinetics at the surface which further improves the healing rate and decreases the implant rejection during the recovery of the fractured bone. Further, Jiang et al. [68] dispersed nano-SiO_2into AZ31 Mg alloy by FSP and two times increase in the hardness was observed in the composite. Balakrishnan et al. [69] fabricated AZ31/TiC composites by FSP and uniform distribution was observed without any formation of intermetallics due to reactions between TiC and AZ31 matrix.

10.3.3 AZ61 Mg Alloy

Lee et al. [70] produced AZ61–SiO_2 MMCs by dispersing 5–10 vol. % nano-SiO_2. After FSP, clusters of size 0.1 to 3 μm of nano-SiO_2 were observed in the composite, and the clustering was observed as decreased after higher number of passes. Formation of MgO and Mg_2Si particles due to the reaction ofSiO_2 and Mg at the higher temperature also confirmed by XRD analysis. However, the level of formation of the secondary phase was lower compared with the liquid state methods. Higher hardness was observed up to two times for the composite compared with unprocessed AZ61 alloy. High strain rate superplasticity (HSRSP) was observed for the composites from the tensile experiments conducted at 350°C at two different strain rates 1×10^{-2} s^{-1} and 1×10^{-2} s^{-1}.

10.3.4 AZ91 Mg Alloy

AZ91Mg alloy is the next alloy of AZ series widely used to develop composites by FSP. SiC and $Al_2O_3$are the most common dispersing phases used to develop composites. Asadi et al. [71] produced AZ91/SiC surface nanocomposite by FSP up to 8 passes and a grain refinement from 150 μm to 600 nm. Due to the generation of a higher amount of heat at higher tool rotational speeds, grain growth was observed. But the increased tool traverse speed resulted in lower grain size and higher hardness. By changing the tool rotation direction, better uniformity in the particle dispersion was noticed. This kind of behavior is similar to what reported by Yoones Erfan et al. [72] in developing nano-SiC dispersed AZ31 composite. The occurrence of tunneling defect was evident at the higher tool travel speeds. By comparing the works of Asadi et al. [71], and Yoones Erfan et al. [72] it can be understood that the tool travel speed must be optimized.

Later on, Asadi et al. [73] produced AZ91/SiC composites and a grain refinement was achieved from 150 μm to 7.17 μm, and hardness improvement from 63 to 96 Hv was measured. Furthermore, the grain refinement level was observed as higher when the SiC dispersing phase was in nanosize [74]. In parallel, Al_2O_3 particles were also used to develop AZ91 composite and evolution of fine grains, increased mechanical properties, and improved wear resistance was observed for the composites produced by using SiC compared with the composites produced by Al_2O_3 particles [75]. Recently, Mostafa Dadaei et al., [76] also developed composites using SiC and Al_2O_3 particles and observed similar effect of SiC addition compared with Al_2O_3 in enhancing the properties of the composites made of AZ91. Mertens et al. [64] dispersed carbon fibers in AZ91D alloy by adopting sandwich method. The carbon fibers were observed as nonuniformly distributed with porosity and defects in the stir zone when tool rotational speeds and travel speeds were increased. Presence of more amount of secondary phase ($Mg_{17}Al_{12}$) was found as the reason that may require sufficient force at different combinations of tool rotational speeds and travel speeds to break and dissolve in the secondary phase. From the work of Mertens et al. [64], it can be understood that the processing parameters play very important role and are which needed to be optimized even to process the same series of alloys having slight difference in chemical composition.

Chen et al. [77] developedcomposite by dispersing SiC particles into thixoformed AZ91 magnesium alloy by FSP. The added SiC particles were observed as uniformly distributed in the matrix with slight agglomeration. Multi-passes were suggested to reduce the agglomeration. From the results, increased hardness and tribological properties were found for the composite. Khayyamin et al. [78] developed AZ91/SiO_2 MMCs by FSP at different processing conditions. At a tool rotational speed of 1250 rpm, three different tool travel speeds (20, 40, 63 mm/min) were adopted. Decreased grain size and increased hardness was measured when the tool travel speeds were increased as also observed by Asadi et al. [73] while developing AZ91 composites with SiC powder. These studies demonstrate that the higher tool travel speeds reduce the localized heat concentration and controls the grain growth. Faraji et al. [80] produced AZ91-Al_2O_3 composites by FSP using two different tools which contain square and circular pin profiles. The effect of the tool travel speed on the width of the thermomechanical affected zone (TMAZ) was found to be significant.

The microstructural observations clearly demonstrated that at the lower travel speeds a thick TMAZ was measured. A uniform microstructure was observed in the composite processed with square pin profile. Grain refinement from a starting size of 130 μm to around 5–6 μm was measured. Due to the grain refinement and introduced secondary phase increased hardness was observed in the composites. Furthermore, Faraji et al. [80] studied the role of the Al_2O_3 particle size on the evolution of fine microstructure, agglomeration of added secondary phase particles and mechanical proprieties by adding Al_2O_3 with three different particle sizes (3000, 300 and 30 nm) while producing the composites by FSP. From the microstructural observations, it was evident that the effect of nanoparticles (30 nm) on refining the grain size as higher compared with the other particles (3000 and 300 nm). From the results reported by Faraji et al. [80], the type of pin found to be played a crucial role in microstructure modification and for the increase of the hardness. Triangle pin profile was produced fine grains with improved mechanical properties due to shape and sharp edges on the triangle profile that induced optimum material flow during FSP compared with square pin profile [79]. This fact was also confirmed in the other works of Faraji et al. [80] that the triangular profile as optimum in producing higher grain refinement compared with other profiles such as circular, triangular and square profiles.

10.3.5 ZM21 Mg Alloy

Madhusudhan Reddy et al. [81] adopted holes filling and closing method to introduced SiC and B_4C powders into the surface of ZM21 Mg alloy by FSP to develop surface composite. Grain refinement was noticed from a starting size of 40 μm to 20 μm in the composite. The embedded particles were also found as well dispersed that promoted to reduce the grain growth. The level of grain refinement achieved was observed as lower compared with the other FSPed Mg alloys. The addition of SiC and B_4C by FSP increased the surface hardness in the composite. During wear studies, combination of adhesive and abrasive wear was found for ZM21 alloy. But the mode of wear was changed from high abrasive wear to medium adhesive wear in the composites due to the presence of dispersing phase. Decreased average friction coefficient (0.025) was measured for the composite compared base ZE21 alloy (0.2).

TABLE 10.2 Brief Summary of the Work Done to Develop Mg-Based Composites By FSP in Chronological Order

Material system	Number of FSP passes (Max.)	Method of secondary phase introduction	Grain refinement (µm)	Significant findings in the composite	Reference
AZ31/ MWCNTs	1	Groove filling	Up to 0.5	• Increased the hardness. • Higher uniformity in distribution of MWCNTs at lower tool travel speed	Morisada et al., [63].
AZ61/ amorphous SiO_2	4	Groove filling and covering	Up to 0.8	• Uniform distribution of SiO_2 with four passes • Formation of MgO and Mg_2Si phasesduring FSP	Lee et al., [70]
Thixoformed AZ91/ SiC	6	Groove filling	120 to 3.1	• Uniform distribution of SiC and increased strength • Excellent tribological properties	Chen et al., [77]
AZ91/SiC	1	Groove filling and covering	150 to 7.1	• Grain refinement with increased tool rotational and travel speed. • Increased microhardness in the composites	Asadi et al., [73]
AZ91/SiC AZ91/Al_2O_3	8	Groove filling and covering	150 to 1.8	• Smaller grain size, higher hardness, increased strength, elongation and wear resistance SiC compared with Al_2O_3	Asadi et al., [74]
AZ91/SiC	2	Groove filling and covering	150 to 5	• Homogenous microstructure with nano-sized SiC particles • Grain growth with higher tool rotational speed. • Decreased grain size with second FSP pass	Asadi et al. [75]
AZ31/Al_2O_3	4	Groove filling and covering	70 to 2.9-4.4	• Enhanced particles distribution with higher tool rotation speed. • Threaded profile resulted better material flow	Azizieh et al., [65]
AZ91/Al_2O_3	1	Groove filling	130 to 6	• Square pin profile found to beoptimum compared with circular • Thick TMAZ at lower travel speeds • Al_2O_3 particles reduced the grain size in the composite	Faraji et al., [79]

AZ91/Al_2O_3	3	Groove filling	150 to 2.8	• Higher grain refinement and increased hardness due to nano-Al_2O_3 particles. • Uniform and smaller grain size with increased number of FSP passes	Faraji et al., [80]
i. AZ31/ carbon fibers ii. AZ91/ carbon fibers	1	Sandwich of C fabric between two metal sheets	Up to 10.0±3.6 (AZ31) and 6.2 ± 1.9 (AZ91)	• The optimized processing parameters were completely different for AZ31 and AZ91 alloys • Precipitation hardening of FSPed AZ91 increased the hardness further	Mertens et al. [64]
ZM21/SiC and B4C	1	Hole filling and covering	40 to 20	• Uniform distribution of SiC and B_4C particles • Better properties in the composites with B_4C compared with SiC	Madhusudhan Reddy et al., [81]
AZ31/SiC	1	Direct supply through the FSP tool	16.57 to 1.24	• Direct friction stir processing (DFSP) tool has been introduced • Smaller grin size and increased hardness in the composites produced by DFSP tool compared with FSP tool.	Huang et al., [66]
AZ31/nano-SiC	1	Groove filling	Up to 1	• Equi-axed ultrafine grains with sizes of less than 1 μm. • Increased hardness up to two times than that of the original alloy.	Jiang et al., [68]
AZ31/ nano-hydroxyapatite	1	Groove filling	1500 to 3.5	• Higher bioactivity and lower degradation in simulated body fluid • Negligible toxicity toward rat skeletal muscle (L6) cells and improved cell response	Ratna Sunil et al., [59]
Mg/nano-hydroxyapatite	1	Groove filling	56 to 2	• Grain boundary pinning effect led to higher level of grain refinement • Improved bioactivity, degradation, and cell adhesion	Ratna Sunil et al., [67]
AZ31/TiC	1	Groove filling and covering	Not reported	• Uniformly distribution of TiC without forming clusters and interfacial phases	Balakrishnan et al., [69]

Table 10.2 briefly summarizes the significant developments happened in producing magnesium-based composites by FSP. Several other groups of Mg alloys were developed for various applications in the material field. However, only a few alloy systems were selected as matrix materials to develop composites by FSP. The available information on developing Mg-based composites by FSP projects this technique as promising to develop surface MMCs.

10.4 TITANIUM

Titanium is the most attractive material in the modern engineering applications. Titanium-based composites were also produced by FSP as we can find in the literature. Farnoush et al., [83] introduced nano-calcium phosphate particles into Ti-6Al-4V alloy to develop Ti-CaP composites targeted for biomedical applications.

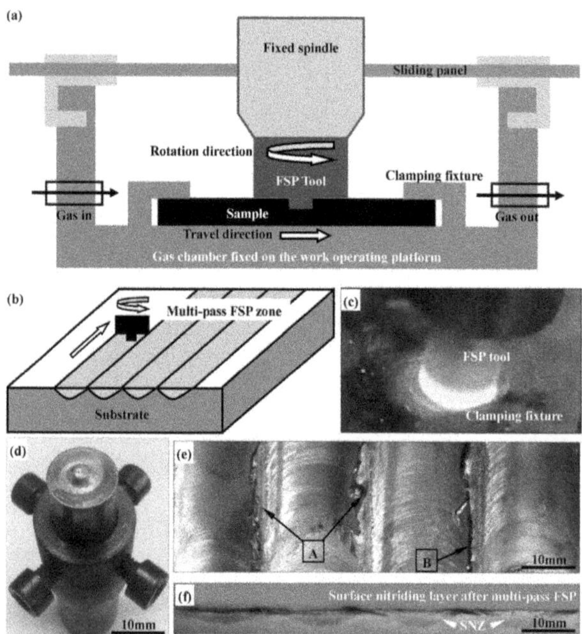

FIGURE 10.15 (See color insert.) (a) Schematic diagram of special set-up developed to provide gas supply during FSP, (b) Procedure adopted for multipass FSP, (c) photograph showing FSP of Ti-6Al-4V, (d) photograph of FSP tool shoulder after one pass, (e) macrograph at the surface and microstructure at the cross section after surface nitriding by FSP. (*Source:* Reprinted with permission from Li et al., [84]. © 2013 Elsevier.)

It is true that titanium is an excellent material and the best choice in the medical field to be used as promising candidate for load-bearing implants and medical devices. However, Ti suffers from lower bioactivity. In physiological environment, Ti exhibits bioinert nature. On the other hand, CaP mineral phases are best known for their high bioactive nature widely used in several biomedical applications. The authors proposed to get the combined benefit of Ti and CaP for biomedical applications. Better corrosion and performance and tribo-corrosion properties were observed for the Ti-CaP composite produced by FSP. Li et al., [84] done FSP in the presence of Nitrogen atmosphereto do surface nitriding of Ti-6Al-4V. Special equipment was developed as shown in Figure 10.15 (a) to provide nitrogen gas during FSP. A thick surface layer with high quality was produced. Typical photographs obtained during FSP and macroscopic and microscopic images at the surface and the cross-section are also shown in 10.15. The authors successfully demonstrated the formation of nitrides in the surface up to 300 μm. Similar studies were also carried out by Shamsipur et al., [85] to develop in-situ TiN particles in the surface. FSP of commercial pure Ti was carried out in the presence of nitrogen atmosphere, and FSP was carried out up to three passes. The hardness of the composite layer was measured as six times higher compared with the unprocessed Ti. Furthermore, Li et al., [86], also used Ti-6Al-4V alloy as base material and introduced Al particles into grooves produced on the surface. Then FSP was carried out to develop in-situ composite layer. Formation of Ti_3Al particles in the FSPed region increased the surface hardness and the wear properties considerably.

10.5 STEELS

Steels occupy an important space in the materials technology which are widely used in several industrial applications. However, information on developing steel base composites by FSP is limited in the literature. Salekrostam et al., [87] produced 316L-SiC composite by FSP at different tool rotational and traverse speeds. Better properties and microstructure were observed for the composite processed at 1000 rpm with 14 mm/min traverse speed. Due to grain refinement, Ghasemi-Kahrizsangi and Kashani-Bozorg [88] introduced TiC nano-particles into mild steel plate with an aim to develop steel/TiC composite. With the increased number of FSP passes, the agglomerated TiC particles were observed as broken into tiny particles and after four FSP passes, the distribution was observed as uniform. The addition of TiC nano-particles has a significant role on increasing the hardness, and the surface layer has shown very fine grain size. Compared with FSP material without addition of TiC, the composite has exhibited excellent mechanical and wear properties.

The overall work carried out in producing surface composite is mainly focused on aluminum based and magnesium-based composites. Selection of tool material and design are the important factors which play crucial role in processing other hard metals with high melting point. Further research work is in a great need in other metals such as steels, titanium and other hard materials.

KEYWORDS

- **aluminum**
- **copper**
- **magnesium**
- **steels**
- **titanium**

REFERENCES

1. Mishra, R. S., Ma, Z. Y., & Charit, I., (2003). Friction stir processing: a novel technique for fabrication of surface composite, *Materials Science and Engineering, A341,* 307/310.
2. Kurt, A., Uygur, I., & Cete, E., (2011). Surface modification of aluminum by friction stir processing. Journal of Materials Processing Technology 211, 313-317. doi:10.1016/j.jmatprotec.2010.09.020.
3. Fencheng Liu, Qiang Liu, Chunping Huang, Kun Yang, Chenggang Yang, & Liming Ke, (2013). Microstructure and Corrosion Resistance of AZ80/Al Composite Plate Fabricated by Friction Stir Processing, *Materials Science Forum, 747–748,* 313–319.
4. Liu, Q., Ke, L., Liu, F., Huang, C., & Xing, L., (2013). Microstructure and mechanical property of multi-walled carbon nanotubes reinforced aluminum matrix composites fabricated by friction stir processing. *Materials and Design 45,* 343–348. doi:10.1016/j.matdes.2012.08.036.

5. Thangarasu, A., Murugan, N., Dinaharan, I., & Vijay, S. J. (2012). Microstructure and microhardness of AA1050/TiC surface composite fabricated using friction stir processing, Sadhana, *37*(5), 579–586.
6. Khorrami, M. S., Kazeminezhad, M., & Kokabi, A. H. (2012). *Mater. Sci. Eng., A,543,* 243–248.
7. Khorrami, M. S., Kazeminezhad, M., & Kokabi, A. H. (2012). *Mater. Des., 40,* 364–372.
8. Sarkari Khorrami, M., Kazeminezhad, M., & Kokabi, A. H., (2015). Influence of Stored Strain on Fabricating of Al/SiC Nanocomposite by Friction Stir Processing, *Metallurgical and Materials Transactions, A.,* DOI: 10.1007/s11661-015-2776-9.
9. Seyyed Reza Hosseini Zeidabadi, & Habib Daneshmanesh, (2014). Fabrication, and Characterization of in-situ Al/Nb Metal/Intermetallic Surface Composite by Friction Stir Processing, *Materials Science & Engineering, A.,* http://dx.doi.org/10.1016/j.msea.2017.03.014.
10. Devinder Yadav & Ranjit Bauri, (2015). Friction Stir Processing of Al-TiB2 In Situ Composite: Effect on Particle Distribution, Microstructure, and Properties, Journal of Materials Engineering and Performance, DOI: 10.1007/s11665-015-1404-6.
11. Devinder Yadav, Ranjit Bauri, Alexander Kauffmann, & Jens Freudenberger, (2016). Al-Ti Particulate Composite: Processing and Studies on Particle Twinning, Microstructure, and Thermal Stability, *Metallurgical and Materials Transactions, A.,* DOI: 10.1007/s11661-016-3597-1.
12. Devinder Yadav, & Ranjit Bauri, (2011). Processing, microstructure and mechanical properties of nickel particles embedded aluminum matrix composite, *Materials Science and Engineering A 528,* 1326–1333.
13. Udaya Bhat, K., Rajendra Udupa, K., Prakrathi, S., & Prashant Huilgol, (2016). Microstructure and Impression Creep Behavior of Al-Based Surface Composite Produced by Friction Stir Processing, *Trans Indian Inst Met 69*(2), 623–627.
14. Essam Mahmoud, R. I., Makoto Takahash, Toshiya Shibayanagi, & Kenji Ikeuchi, (2010). Wear characteristics of surface-hybrid-MMCs layer fabricated on aluminum plate friction stir processing, *Wear, 268,* 1111–1121.
15. Cavaliere, P., (2005). Mechanical properties of Friction Stir Processed 2618/Al2O3/20p metal matrix composite, *Composites: Part A 36,* 1657–1665.
16. Yutao Zhao, Xizhou Kai, Gang Chen, Weili Lin, & Chunmei Wang, (2016). Effects of friction stir processing on the microstructure and superplasticity of insitunano-ZrB2/2024Al composite, Progress in Natural Science: *Materials International, 26,* 69–77.
17. Zahmatkesh, B., & Enayati, M. H., (2010). A novel approach for development of surface nanocomposite by friction stir processing. *Materials Science and Engineering: A 527,* 6734–6740.
18. Hoziefa, W., Toschi, S., Ahmed, M. M. Z., Al. Morri, Mahdy, A. A., El-Sayed Seleman, M. M., El-Mahallawi, I., Ceschini, L., & Atlam, A., (2016). *Influence of Friction Stir Processing on the Microstructure and Mechanical Properties of a Compocast AA2024-Al_2O_3 Nanocomposite,* doi: 10.1016/j.matdes.2016.05.114.
19. Dolatkhah, A., Golbabaei, P., Givi, M. K. B., & Molaiekiya, F., (2012). Investigating effects of process parameters on microstructural and mechanical properties of

Al5052/SiC metal matrix composite fabricated via friction stir processing. *Materials and Design 37,* 458–464. doi:10.1016/j.matdes.2011.09.035.

20. Ranjit Bauri, Janaki Ram, G. D., Devinder Yadav, & Shyam Kumar, C. N., (2015). Effect of process parameters and tool geometry on fabrication of Ni particles reinforced 5083 Al composite by friction stir processing, Materials Today: Proceedings *2* 3203– 3211.
21. Peng Liu, Qing-Yu Shi, & Yuan-Bin Zhang, (2013). Microstructural evaluation and corrosion properties of aluminum matrix surface composite adding Al-based amorphous fabricated by friction stir processing, *Composites: Part B 52,* 137–143.
22. Narayana Yuvaraj, Sivanandam Aravindan, & Vipin, (2015). Fabrication of Al5083/ B_4C surface composite by friction stir processing and its tribological characterization, http://dx.doi.org/10.1016/j.jmrt.2015.02.006.
23. Amra, M., Khalil Ranjbar, & Dehmolaei, R., (2015). Mechanical Properties and Corrosion Behavior of CeO_2 and SiC Incorporated Al5083 Alloy Surface Composites, *Journal of Materials Engineering and Performance, 24,* 3169–3179.
24. Sahandi Zangabad, P., Khodabakhshi, F., Simchi, A., & Kokabi, A. H., (2016). Fatigue fracture of friction-stir processed Al–Al3Ti–MgO hybrid nanocomposites, *International Journal of Fatigue 87,* 266–278.
25. Yuvaraj, N., Aravindan, S., & Vipin, (2017). Wear Characteristics of Al5083 Surface Hybrid Nano-composites by Friction Stir Processing, *Trans Indian Inst Met, 70* (4), 1111–1129.
26. Mansour Rahsepar, & Hamed Jarahimoghadam (2016). *The influence of multi-pass friction stir processing on the corrosion behavior and mechanical properties of zircon-reinforced Al metal matrix composites, 671,* 214–220.
27. Hasan Mohammadzadeh Jamalian, Hadi Ramezani, Hasan Ghobadi, Mohammad Ansari, Saeed Yari, & Mohammad Kazem Besharati Givi, (2016). Processing-structure-property correlation in nano-SiC-reinforced friction stir welded aluminum joints, *Journal of Manufacturing Processes 21,* 180–189.
28. Chunjie Huang, Wenya Li, Zhihan Zhang, Marie-Pierre Planche, Hanlin Liao, & Ghislain Montavon, (2016). Effect of Tool Rotation Speed on Microstructure and Microhardness of Friction-Stir-Processed Cold-Sprayed SiCp/ Al5056 Composite Coating, *Journal of Thermal Spray Technology, 25* (7), 1357–1364.
29. Yuvaraj, N., Aravindan, S., & Vipin, (2017). Comparison studies on mechanical and wear behavior of fabricated aluminum surface nanocomposites by fusion and solid state processing, *Surface & Coatings Technology 309,* 309–319.
30. Sharifitabar, M., Sarani, A., Khorshahian, S., &Shafiee Afarani, M., (2017). Fabrication of 5052Al/Al_2O_3 nanoceramic particle reinforced composite via friction stir processing route, *Materials and Design 32,* 4164–4172.
31. Soleymani, S., Abdollah-Zadeh, A., & Alidokht, S. A., (2012). Microstructural and tribological properties of Al5083 based surface hybrid composite produced by friction stir processing, *Wear 278–279,* 41– 47.
32. Qiang Liu, Liming Ke, Fencheng Liu, Chunping Huang, &Li Xing, (2013). Microstructure and mechanical property of multi-walled carbon nanotubes reinforced aluminum matrix composites fabricated by friction stir processing. Materials & Design.45, 343-348.

33. Min Yang, Chengying Xu, Chuansong Wu, Kuo-chi Lin, & Yuh Chao, J., (2010). Linan An Fabrication of AA6061/Al_2O_3 nano ceramic particle reinforced composite coating by using friction stir processing, *J Mater Sci45,* 4431–4438.
34. Anvari, S. R., Karimzadeh, F., & Enayati, M. H., (2013). A novel route for development of Al–Cr–O surface nano-composite by friction stir processing, *Journal of Alloys and Compounds 562,* 48–55.
35. Anvari, S. R., Karimzadeh, F., & Enayati, M. H. (2013). Wear characteristics of Al–Cr–O surface nano-composite layer fabricated on Al6061 plate by friction stir processing, *Wear 304,* 144–151.
36. Jong-Ning Aoh, Chih-Wei Huang & Wei-JuCheng (2014). Fabrication of Al6061-AMC by adding copper-coated SiC reinforcement by Friction Stir Processing (FSP), *Materials Science Forum, 783–786,* 1721–1728.
37. Devaraju, A., Kumar, A., & Kotiveerachari, B., (2013). Influence of addition of Grp/Al2O3p with SiCp on wear properties of aluminum alloy 6061-T6 hybrid composites via friction stir processing, *Trans. Nonferrous Met. Soc. China 23,*1275−1280.
38. Thangarasu, Murugan, N., Dinaharan, I., & Vijay, S. J., (2014). Synthesis and characterization of titanium carbide particulate reinforced AA6082 aluminum alloy composites via friction stir processing, *Archives of Civil and Mechanical Engineering. 15*(2), 324–334.
39. Naresh, P., & Adepu Kumar, (2015). Effect of Nano Reinforcement on Fabrication of Al/Al_2O_3 Surface Composite by Friction Stir Processing, *Materials Science Forum, 830–831,* 467–471.
40. Yong Zhao, Xiaolu Huang, Qiming Li, Jian Huang, & Keng Yan, (2015). Effect of friction stir processing with B4C particles on the microstructure and mechanical properties of 6061 aluminum alloy, *Int. J. Adv. Manuf. Technol.,* DOI 10.1007/s00170-014-6748-9.
41. Zhenglin Du, Ming-Jen Tan, Jun-Feng Guo, & Jun Wei, (2016). Friction stir processing of Al–CNT composites, Proc I Mech E Part L: *J Materials: Design and Applications,230*(3) 825–833.
42. Halil Ibrahim Kurt, Murat Oduncuoglu, &Ramazan Asmatulu, (2016). Wear Behavior of Aluminum Matrix Hybrid Composites Fabricated through Friction Stir Welding Process, *Journal of Iron and Steel Research International, 23* (10): 1119-1126.
43. Min Yang, Chengying Xu, Chuansong Wu, Kuo-Chi Lin, & Yuh Chao, J., (2010). Linan An, Fabrication of AA6061/Al2O3 nano ceramic particle reinforced composite coating by using friction stir processing, *J Mater Sci 45,* 4431–4438. DOI 10.1007/s10853-010-4525-1.
44. Shafiei-Zarghani, A., Kashani-Bozorg, S. F., & Zarei-Hanzaki, A., (2011). Wear assessment of Al/Al_2O_3 nano-composite surface layer produced using friction stir processing, *Wear 270,* 403–412.
45. Maxwell Rejil, C., Dinaharan, I., Vijay, S. J., & Murugan, N., (2012). Microstructure and sliding wear behavior of AA6360/(TiC + B4C) hybrid surface composite layer synthesized by friction stir processing on aluminum substrate, Materials Science and Engineering A *552,* 336– 344.
46. Alidokht, A., Abdollah-Zadeh, A., Soleymani, S., & Assadi, H., (2011). Microstructure and tribological performance of an aluminum alloy based hybrid composite produced by friction stir processing, *Materials and Design, 32,* 2727–2733.

47. Alidokht, S. A., Abdollah-Zadeh, A., & Assadi, H., (2013). Effect of applied load on the dry sliding wear behavior and the subsurface deformation on hybrid metal matrix composite, *Wear 305,* 291–298.
48. Hussain, G., Hashemi, R., Hashemi, H., Khalid, & Al-Ghamdi, A., (2015). An experimental study on multi-pass friction stir processing of Al/TiN composite: some microstructural, mechanical, and wear characteristics, *Int J Adv Manuf Technol* DOI 10.1007/s00170-015-7504-5.
49. Eskandari, H., Taheri, R. & Khodabakhshi, F. (2016). Friction-stir processing of an AA8026-TiB_2-Al_2O_3 hybrid nanocomposite: Microstructural developments and mechanical properties, *Materials Science and Engineering A 660,* 84–96.
50. Cartigueyen, S. & Mahadevan, K. (2015). Effects of heat generation on microstructure and hardness of Cu/SiCp surface composite processed by friction stir processing, *Materials Science Forum, 830–831*, 472–475.
51. Jalal Jafari, Mohammad Kazem Besharati Givi, & Mohsen Barmouz, (2014). Mechanical and microstructural characterization of Cu/CNT nanocomposite layers fabricated via friction stir processing, Int J Adv Manuf Technol, DOI 10.1007/s00170-014-6663-0.
52. Sabbaghian, M., Shamanian, M., Akramifard, H. R., & Esmailzadeh, M., (2014). Effect of friction stir processing on the microstructure and mechanical properties of Cu–TiC composite, *Ceramics International, 40*(8B), 12969–12976.
53. Mohsen Barmouz, Mohammad Kazem Besharati Givi, & Javad Seyfi, microstructure, microhardness, wear, and tensile behavior, Materials Characterization, *62*(1), (2011). *On the role of processing parameters in producing Cu/SiC metal matrix composites via friction stir processing: Investigating,* 108–117.
54. Barmouz, M., Asadia, P., Besharati Givi, M. K., Taherishargh, M., (2011). Investigation of mechanical properties of Cu/SiC composite fabricated by FSP: Effect of SiC particles' size and volume fraction, *Materials Science and Engineering A 528,* 1740–1749.
55. Akramifard, H. R., Shamanian, M., Sabbaghian, M., & Esmailzadeh, M., (2014). Microstructure and mechanical properties of Cu/SiC metal matrix composite fabricated via friction stir processing, *Materials and Design 54,* 838–844.
56. Sathiskumar, R., Murugan, N., Dinaharan, I., & Vijay, S. J., (2013). Effect of Traverse Speed on Microstructure and Microhardness of Cu/B4C Surface Composite Produced by Friction Stir Processing, *Trans Indian Inst Met 66*(4), 333–337.
57. Sathiskumar, R., N.Murugan, I.Dinaharan, & Vijay, S. J., (2014). Fabrication and characterization of Cu/B_4C surface dispersion strengthened composite using friction stir processing, *Archives of Metallurgy and Materials, 51,* 83—87.
58. Sabbaghian, M., Shamanian, M., Akramifard, H. R., & Esmailzadeh, M., (2014). Effect of friction stir processing on the microstructure and mechanical properties of Cu–TiC composite, Ceramic International, 40(8), 12969–12976.
59. Ratna Sunil, B., Sampath Kumar, T. S., Chakkingal, U., Nandakumar, V., & Doble, M., (2014). Friction stir processing of magnesium–nanohydroxyapatite composites with controlled in vitro degradation behavior. *Mater. Sci. Eng. C 39,* 315–324.
60. Ratna Sunil, B., Sampath Kumar, T. S., & Chakkingal, U., (2012). Bioactive Grain Refined Magnesium by Friction Stir Processing, *Mater. Sci. Forum 710,* 264–269.

61. Mertens, A., Simar, A., Adrien, J., Maire, E., Montrieux, M. H., Delannay, F., et al., (2015). *Mater. Charact. 107,* 125–133.
62. Navazani, M., & Dehghani, K., (2016). *J. Mater. Process. Technol. 229,* 439–449.
63. Morisada, Y., Fujii, H., Nagaoka, T., & Fukusumi, M., (2006). *Mater. Sci. Eng. A 419,* 344–348.
64. Mertens, A., Simar, A., Montrieux, H. M., Halleux, J., Delannay, F., Lecomte-Beckers, J., Poole, W. J., & Kainer, K. U. (Eds.), *Proceedings of the 9th International Conference on Magnesium Alloys and Their Applications*, Vancouver, Canada, pp. 845–850. http://hdl.handle.net/2268/120134.
65. Azizieh, M., Kokabi, A. H., & Abachi, P., (2011). *Mater. Des. 32,* 2034– 2041.
66. Huang, Y., Wang, T., Guo, W., Wan, L., &Lv, S., (2014). *Mater. Des. 59,* 274–278.
67. Ratna Sunil, B., Sampath Kumar, T. S., Chakkingal, U., Nandakumar, V., & Doble, M., (2014). *J.Mater. Sci. Mater. Med. 25,* 975–988.
68. Jiang, Y., Yang, X., Miura, H., & Sakai, T., (2013). *Rev. Adv. Mater. Sci. 33,* 29–32.
69. Balakrishnan, M., Dinaharan, I., &Palanivel, R., (2011). Sivaprakasam, R. J. Magnes. Alloys *3,*76–78.
70. Lee, C. J., Huang, J. C., & Hsieh, P. I., (2006). Scr. Mater. *54,* 1415–142.
71. Asadi, P., Besharati Givi, M. K., Faraji, G., (2010). *Mater. Manuf. Process. 25,* 1219–1226.
72. Erfan, Y., & Kashani-Bozorg, S. F., (2011). *Int. J. Nanosci. 10,*1073.
73. Asadi, P., Faraji, G., Besharati Givi, M. K., (2014). *Int. J. Adv. Manuf. Technol. 51,*247–260.
74. Asadi, P., Faraji, G., Masoumi, A., Besharati Givi, M. K., (2011). *Metall. Mater. Trans. A 42A* 2820–2832.
75. Asadi, P., Besharati Givi, M. K., Rastgoo, A., Akbari, M., Zakeri, V., Rasouli, S., (2012). *Int. J. Adv. Manuf. Technol. 63,* 1095–1107.
76. Dadaei, M., Omidvar, H., Bagheri, B., Jahaji, M., Abbasi, M., (2014). *Int. J. Mater. Res. 105* (4) (2014). 369–374.
77. Chen, T., Zhu, Z., Ma, Y., Li, Y., Hao, Y., & Wuhan, J. (2010). *Univ. Technol. Mater. Sci. Ed.* 223–227, doi:10.1007/s11595-010-2223-0.
78. Khayyamin, D., Mostafapour, A., & Keshmiri, R., (2013). *Mater. Sci. Eng. A 559* 217–221.
79. Faraji, G., Dastani, O., & Akbari Mousavi, S. A. A., (2011). *Proc. Inst. Mech. Eng. B: J. Eng. Manufacture 225,* 1331–1345.
80. Faraji, G., Dastani, O., & Akbari Mousavi, S. A. A., (2011). *J. Mater. Eng. Perform. 20,* 1583–1590.
81. Madhusudhan Reddy, G., Sambasiva Rao, A., & Srinivasa Rao, K., (2013). *Trans. Indian Inst. Met. 66* (1), 13–24.
82. Ratna Sunil, B., G. Pradeep Kumar Reddy, Hemendra Patel, & Ravikumar Dumpala, (2016). Magnesium based surface metal matrix composites by friction stir processing, *Journal of Magnesium and Alloys 4,* 52–61.
83. Hamidreza Farnoush, Ashkan Abdi Bastami, Ali Sadeghi, Jamshid Aghazadeh Mohandesi, & Fathollah Moztarzadeh, (2013). Tribological and corrosion behavior of friction stir processed Ti-CaP nanocomposites in simulated body fluid solution, *Journal of the Mechanical Behavior of Biomedical Materials, 20,* 90–97.

84. Bo Li, Yifu Shen, & Weiye Hu, (2013). Surface nitriding on Ti–6Al–4V alloy via friction stir processing method under nitrogen atmosphere, *Applied Surface Science 274,* 356– 364.
85. Ali Shamsipur, Seyed Farshid Kashani-Bozorg, &Abbas Zarei-Hanzaki, (2013). Production of in-situ hard Ti/TiN composite surface layers on CP-Ti using reactive friction stir processing under nitrogen environment, *Surface and Coatings Technology 218,* 62–70.
86. Bo Li, Yifu Shen, Luo Lei, & Weiye Hu, (2014). Fabrication and Evaluation of Ti3Alp/Ti–6Al–4V Surface Layer via Additive Friction-Stir Processing, *Materials and Manufacturing Processes,29,* 412–417.
87. Salekrostam, R., Besharati Givi, M. K., Asadi, P., & Bahemmat, P., (2010). Influence of Friction Stir Processing Parameters on the Fabrication of SiC/316L Surface Composite, *Defect and Diffusion Forum 297–301,* 221–226. doi:10.4028/www.scientific.net/DDF.297-301.221.
88. Ahmad Ghasemi-Kahrizsangi, & Seyed Farshid Kashani-Bozorg (2012). Microstructure and mechanical properties of steel/TiC nano-composite surface layer produced by friction stir processing, *Surface and Coatings Technology 209,* 15–22.

CHAPTER 11

Properties of the Composites Produced by FSP

Surface composites produced by FSP certainly exhibit different properties compared with that of base materials. The important factors which influence the properties of the composites produced by FSP are explained below:

I. Grain Refinement Due to FSP

In polycrystalline materials, grain boundaries are the regions of mismatch or defective regions with an insufficient number of atoms. Grain boundaries are classified as two-dimensional surface defects. With the decrease of the grain size, the fraction of grain boundary is increased to a great extent. If the level of defects is higher in a polycrystalline material, structure-sensitive properties such as yield strength, electronics properties, electrical properties, corrosion, and bioproperties are severely affected. During FSP, recrystallization is happened which is usually at a temperature of 0.7–0.8 times of that of the base material melting point which is called as dynamic recrystallization. The main reason for grain refinement in FSP is dynamic recrystallization.

II. Addition or In-Situ Formation of Secondary Phases

Along with the effect of smaller grain size achieved after FSP, the addition of secondary phase particles introduces dispersion strengthening in the FSPed composites. Furthermore, the nature of the usual secondary phases is high hard with brittleness. Compared with conventional liquid state methods, FSP offers to distribute secondary phase uniformly. In some cases, the secondary phase can be produced during FSP due to the reaction of the added dispersing phase with the matrix which is further distributed to develop the composites. By adding an appropriate amount of secondary phase as a single phase or mixture of multiple phases (hybrid) mechanical properties are altered.

III. Supersaturated Grains

In alloys, solid solutions are a category and most widely useful for several applications. If the solubility of the alloying element exceeds its maximum limit at the room temperature, then such alloys can be called as supersaturated solid solutions. Usually, with a higher amount of alloying elements beyond equilibrium limit, secondary phases (intermetallics) are formed which can be seen easily at the grain boundaries. For example, Figure 11.1(a) shows the presence of $Mg_{17}Al_{12}$ at the grain boundaries as shown in Figure 11.1. The solid solution grains are represented by "α" and the intermetallic phase at the grain boundaries is represented with "β." It is possible to increase the solubility of the alloying element either by mechanical route or by heat treatment. Figure 11.1 (b) shows the decreased amount of $Mg_{17}Al_{12}$ phase after FSP. The fine grains have become rich in Al, and the amount of secondary phase has been reduced to a great extent due to FSP. Similarly, Figure 11.1 (c) shows the microstructure of heat treated AZ91 at 410°C for 24 h. The amount of $Mg_{17}Al_{12}$ was observed as decreased as also observed from the XRD analysis as shown in Figure 11.2 and therefore, it is suggested that the grains became supersaturated. Usually, supersaturated grains introduce lattice distortion and bring higher surface energy which further influences the material properties.

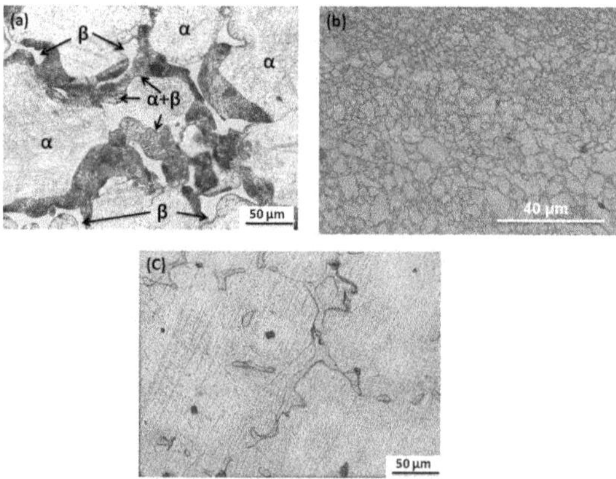

FIGURE 11.1 Optical microscope images of AZ91 Mg alloy: (a) unprocessed, (b) FSPed, and (c) heat-treated at 410°C.

IV. Texture Effect

Next to the grain size, the orientation of the grain also plays an important role on the material properties. In crystallographic planes, high-density planes are associated with high surface energy. The surface energy of a polycrystalline material with preferred orientation is different compared with its starting condition. Orientation of the grains can be changed by mechanical processing such as metal forming works including extrusion, rolling, drawing, etc. In FSP, the material which is stirred under the shoulder exhibit orientation in the most possible plastically deformed orientation or in other words, close-packed orientation. Figure 11.2 shows a typical example of the orientation change of AZ91 Mg alloy after FSP. The intensities of certain peaks were decreased, and for certain peaks, the intensity was observed as increased. For example, (002) which is a close-packed plane in hcp crystal structure intensity was increased after FSP. This orientation change influences mechanical and corrosion properties.

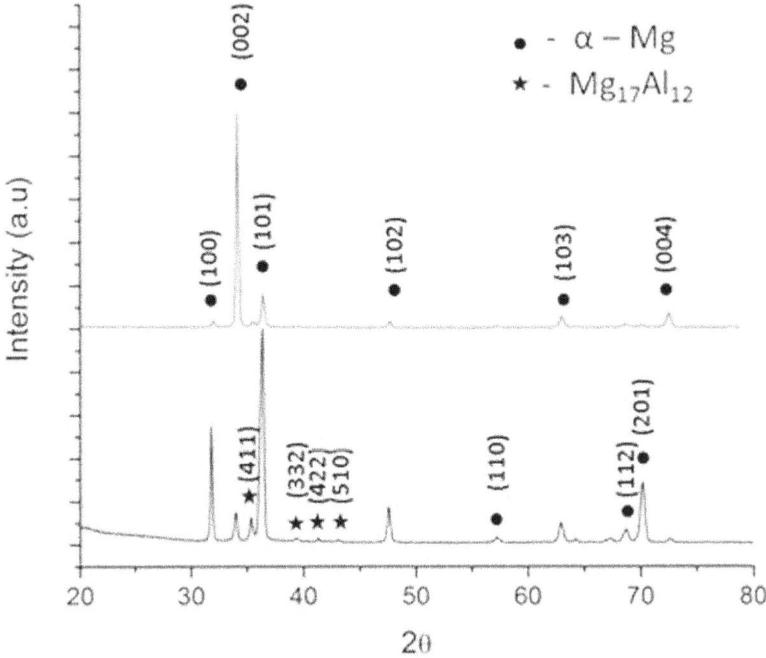

FIGURE 11.2 XRD patterns of AZ91 Mg alloy before and after FSP.

V. Modified Secondary Phase of the Base Material

The amount of secondary phase is also altered due to FSP. Usually, the intermetallic phases present in any alloy exhibit a hard and brittle nature. When the amount of these phases is decreased due to FSP, and also the distribution in the form of very fine particles resulted from FSP alter the bulk properties. The composite becomes hybrid if the parent secondary phase in the base material is distributed as fine particles due to FSP, as additional secondary phase powder is added externally. The total amount of the secondary phase is decreased after FSP compared with the unprocessed condition.

11.1 MECHANICAL PROPERTIES

The important group of material properties influenced by FSP is mechanical properties. Due to the aforementioned important factors in FSP, mechanical properties such as hardness, tensile strength, ductility, bending and fracture resistance are greatly affected by altering the grain size and adding secondary phase. One important mechanism to increase the strength of FSPed materials is grain size reduction. As per the Hall-Petch equation, as given below, the strength of a material can be increased by decreasing the grain size.

$$\sigma_o = \sigma_i + k/\sqrt{D} \qquad (1)$$

where, σ_i is frictional stress, k is a factor and D is the grain size. Therefore, the strength of a material is increased if the grain size (D) is decreased.

Another important mechanism by which the mechanical properties are influenced is "dispersion strengthening." The addition of secondary phase during FSP is completely different compared with adding secondary phase powder in the liquid state. In FSP, the added secondary phase is embedded within the material while recrystallization is undergoing and therefore, the particles arrest the grain growth and also results smaller grains by introducing pinning effect. The presence of these dispersed phases enhances the mechanical properties.

Multi-pass FSP is one more important factor that introduces different levels of microstructure modification and results different effects in altering the mechanical properties. Figure 11.3 demonstrates the effect of multi-pass FSP in developing composites on altering the mechanical properties [1].

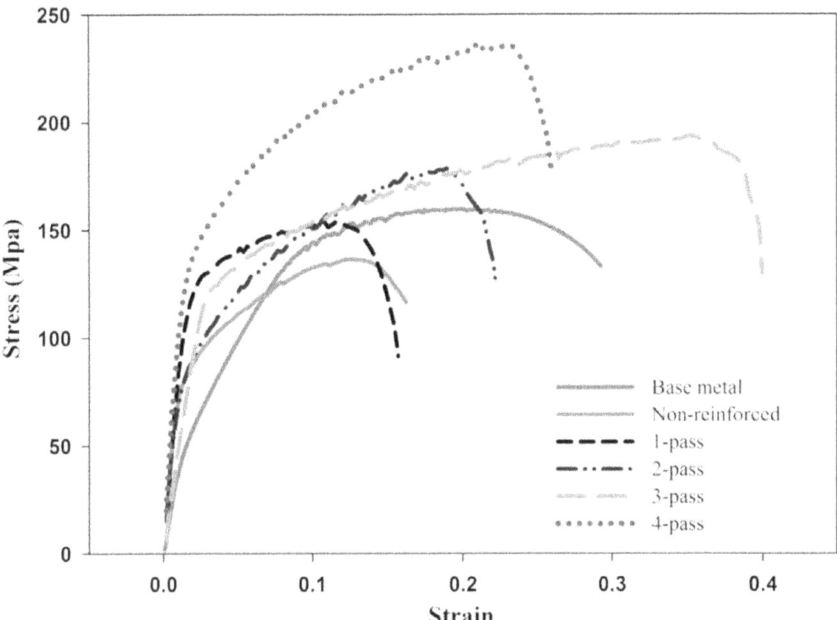

FIGURE 11.3 Stress-strain curves of unprocessed and processed AA5052 with $ZrSiO_4$. (*Source*: Reprinted with permission from Rahsepar and Jarahimoghadam [1]. © 2016 Elsevier.)

As explained by Rahsepar and Jarahimoghadam [1] (shown in Figure 11.3), by increasing number of FSP passes, % elongation can be increased in AA5052-$ZrSiO_4$. With the addition of secondary phase after three and four FSP passes, the yield strength, ultimate tensile strength, and % elongation were observed as increased which is really interesting. This is similar to what reported by Narimani et al., [2]. In their work, increased hardness was measured for the FSPed AA6063 alloy from a starting value of 65 Hv. By adding TiB_2 nanopowder, hardness was measured as increased up to 140 HV which is more than three times compared with the unprocessed condition. Also, from the tensile tests, composites exhibited higher UTS compared with FSPed samples and as a received sample as shown in Figure 11.4.

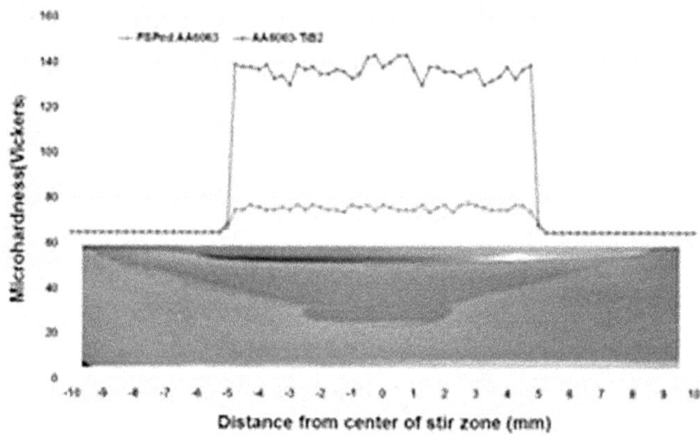

(a)

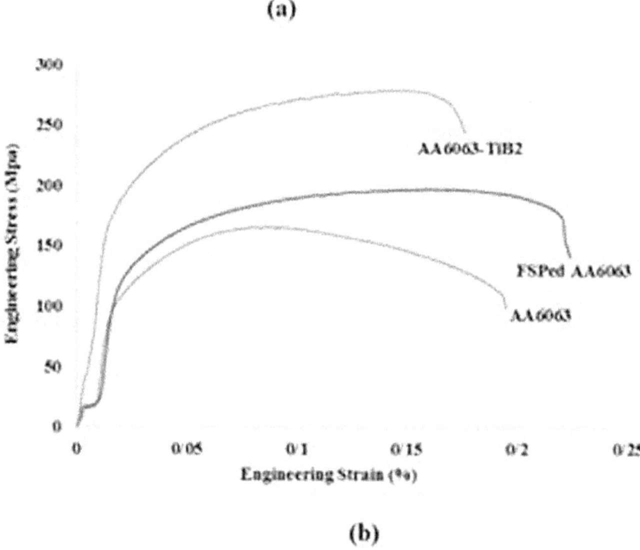

(b)

FIGURE 11.4 Mechanical properties of AA6063 unprocessed, FSPed and FSPed composite: (a) hardness profile, and (b) stress-strain curves. (*Source:* Reprinted with permission from Narimani et al., [2]. © 2016 Elsevier.)

11.2 TRIBOLOGICAL PROPERTIES

Another important group of properties which are highly influenced are the tribological properties. In particular, applications which involve, relative movement of surfaces or contact of surfaces demands high wear resistance.

Mass loss due to mechanical contact at the interface of counter surfaces is the main observation in the material degradation by wear. By providing high hard phases at the surface, better wear resistance can be introduced. Instead of making the entire surface as hard, introducing hard phases in the amounts of required fractions is the best strategy in surface engineering. The composites developed by FSP offer better wear properties compared with the composites produced by liquid state methods. By introducing hard particles into the surface by FSP alter the wear mechanisms. In most of the works done by FSP, suggest the reduced contact area while performing the experiments due to the incorporated secondary phase particles.

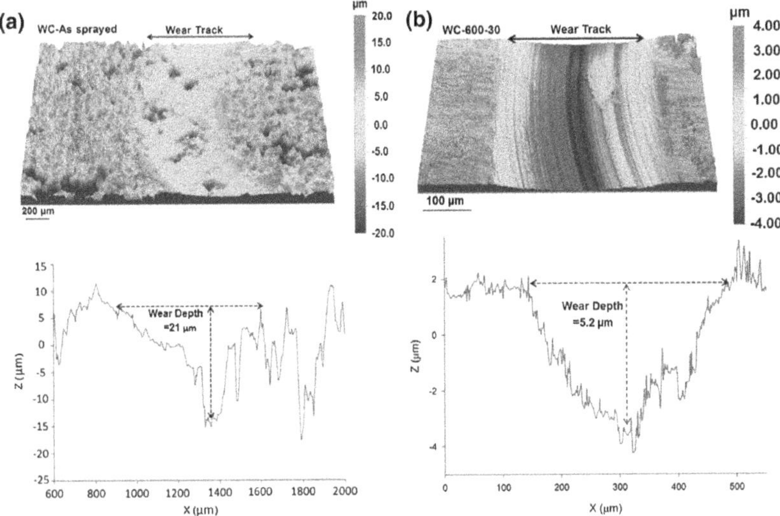

FIGURE 11.5 (See color insert.) Wear topography of the surfaces: (a) WC surface coating, and (b) WC coating followed by FSP. (*Source:* Reprinted with permission from Rahbar-Kelishami et al., [3]. © 2015 Elsevier.)

Rahbar-Kelishami et al., [3] developed WC-12%Co surface coating on 52100 steel and follow up FSP was carried out. Then, wear studies were done on the samples before FSP and after FSP. Figure 11.5 compares both the surfaces after wear studies as shown in 3D topography. Maximum wear depth in the sample without FSP was measured as 21 μm. After FSP, the wear track depth was measured as 5.2 μm as shown in Figure 11.5 (b). The authors also suggested the formation of new intermetallics during FSP which further enhanced the wear resistance. Anvari et al., [4] applied Cr_2O_3 powder on AA6061 plate by atmosphere plasma spray method, and FSP was carried out for six passes. Without reinforcements also, FSP was able to decrease the material loss by reducing the hardening Particles during FSP. The wear mechanism itself has been changed from adhesion and abrasion to de-lamination in the nano-composites compared with unprocessed AA6061 alloy and FSPed alloy without reinforcement phase. Alidokht et al., [5] developed hybrid composites by introducing SiC and MoS_2 into A356 alloy and wear properties were investigated. In parallel, they developed monocomposites by dispersing SiC and MoS_2 into A356 alloy. Compared with monocomposites, hybrid composite exhibited excellent wear properties. Additionally, as applied load is increased in the composites, the wear behavior is also better for the hybrid composites compared with the unprocessed and monocomposites. The thickness of the subsurface effected by wear was measured as lower for the hybrid composite.

11.3 CORROSION PROPERTIES

With the modified microstructure and the presence of different phases after FSP, the corrosion properties are also greatly influenced. The grain boundary is the region of the pseudo-crystal structure with less number of required atoms per unit cell in any polycrystalline material. This region is usually called as mismatch region as schematically shown in Figure 11.6. The atoms at the grain boundaries possess a lower number of bonds compared with the atoms at the grain interior. Therefore, chemical affinity at the grain boundaries is higher for the atoms present at the grain boundaries compared with the atoms present at the interior. Therefore, the surface energy is increased with decreasing the grain size in polycrystalline materials. Then, in principle, the corrosion rate is increased with grain refinement. However, along with the grain size other microstructural features also contribute to the corrosion management and the overall effect may decrease the corrosion rate. Grain refinement and the corrosion rate is also material dependent. In some material systems, a higher amount of grain boundary brings the advantage of quick passivation and followed by corrosion protection. In a few materials, due to the higher amount of grain boundaries, corrosion attack is initiated at the grain boundaries, and corrosion rate is increased. This is the reason why several contradicting results were reported in the literature indicating increased corrosion rate as well as decreased corrosion rate with grain refinement. For example, Wang et al. [6] processed AZ31 Mg alloy through hot rolling and a significant effect of grain refinement was observed in altering the fracture resistance and also the corrosion resistance examined in Hank's solution. Similarly, Hoog et al. [7], produced grain refined pure Mg by equal channel angular pressing (ECAP) and also by surface mechanical attrition treatment (SMAT). From the corrosion experiments, controlled corrosion behavior was observed for grain refined Mg produced by ECAP. While assessing the corrosion of fine-grained AZ31 Mg alloy, Alvarez-Lopez et al. [8] also reported the best corrosion resistance in for smaller grain size.

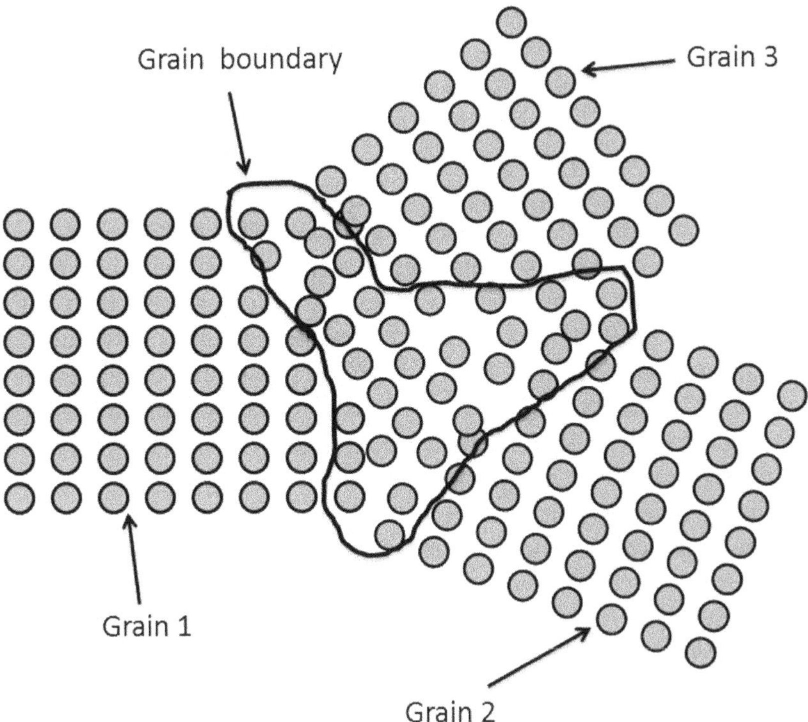

FIGURE 11.6 Schematic representation of grain boundaries and grains in a polycrystalline material.

Mostaed et al. [9] developed grain refined ZK60 Mg alloy by ECAP a grain size of 600 nm was achieved, and due to the smaller grain size, uniform corrosion and lower pitting was noticed in phosphate buffer solution (PBS). Gu et al. [10] processed AZ31 Mg alloy by ECAP and the effect of grain size on the corrosion behavior was observed. Corrosion resistance of ECAPed alloy was decreased after 1-2 passes and increased after 3rd and 4th pass Hank's solution. This study itself indicates the diversified corrosion behavior with respect to grain size. Ratna Sunil et al., [11] developed grain refined AZ31 Mg alloy by groove pressing, and increased corrosion resistance was observed due to grain refinement tested in simulated body fluid (SBF). Same alloy processed with ECAP and obtained smaller grain size relatively compared with that of grain refinement obtained in groove pressing. However, the corrosion behavior for ECAPed AZ31 Mg alloy was observed as decreased for first 2 passes. This kind of behavior was also reported by Hamu et al., [12] in ECAPed AZ31 Mg alloy. Along with grain size, other factors such as high dislocation density and presence of more number of twins influence the corrosion behavior. Hoog et al. [7], achieved better grain refinement (2.5 μm) by adopting SMAT and compared with ECAPed Mg. Interestingly, ECAPed pure Mg has shown better corrosion resistance compared with SMAT Mg. It is true that the corrosion behavior is affected with the presence of stresses. In SMAT higher peening stress is accumulated at the surface compared with ECAP sample. Therefore, it can be understood that the lattice imperfections such as dislocations, stacking faults, twins, and presence of secondary phase particles influence the corrosion behavior along with the grain size effect.

In the composites produced by FSP, along with grain size reduction effect, the presence of secondary phase influences the corrosion behavior. If the electrochemical behavior of the secondary phase completely different compared with the matrix, the galvanic couple is generated between the dispersed phase and the surrounding matrix material and galvanic corrosion is elevated. Also due to the presence of these secondary phase particles, the formation of pits is another common observation in the composites produced by FSP. The medium which is used to test the corrosion behavior also considered as an influencing factor.

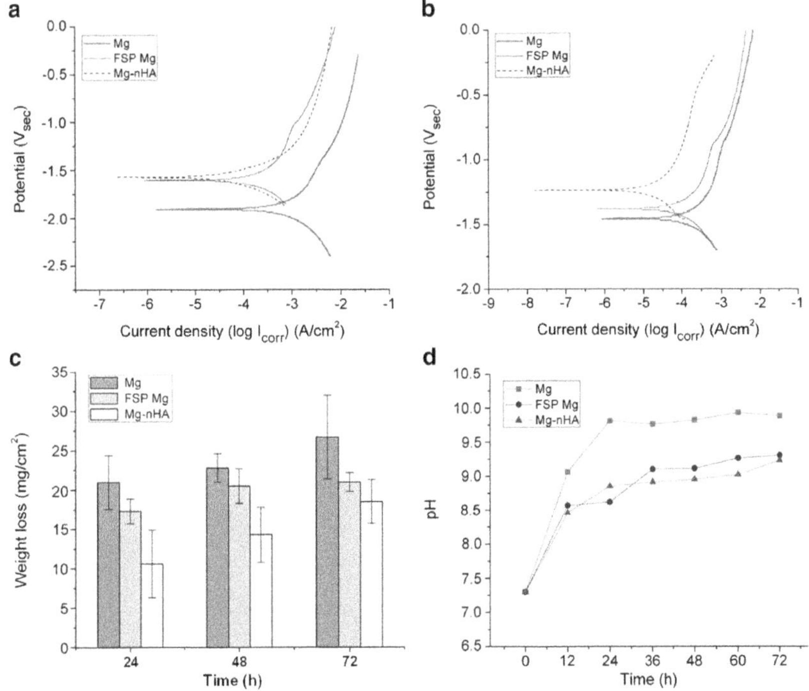

FIGURE 11.7 Corrosion behavior of pure Mg, FSPed Mg and Mg-nHA composite: (a) potentiodynamic polarization curves before immersion and (b) after immersion studies, (c) weight loss measurements from the immersion studies, and (d) pH change during the immersion study carried out for 72 h. (*Source:* Reprinted with permission from Ratna Sunil et al., [13]. © 2014 Elsevier.)

In the works of Ratna Sunil et al., [13, 14], grain refinement was achieved in pure Mg and AZ31 Mg alloy by using FSP. Also, nano-hydroxyapatite (nHA) powder was used to disperse into the surface with an aim to develop degradable Mg-based composites. Figure 11.7 shows the corrosion behavior of pure Mg, FSPed Mg (without reinforcement), Mg-nHA composite developed by FSP. After FSP, slight increased corrosion resistance was noticed for FSPed Mg. By introducing nHA into Mg by FSP enhanced deposition of calcium phosphate mineral phases from the SBF on the surface of the composite and these mineral depositions further increased the corrosion resistance. From the weight loss measurements also, better degradation properties were observed for the composite as the immersion time was increased to 72 h which was also reflected from the pH measurements carried out at different intervals of time.

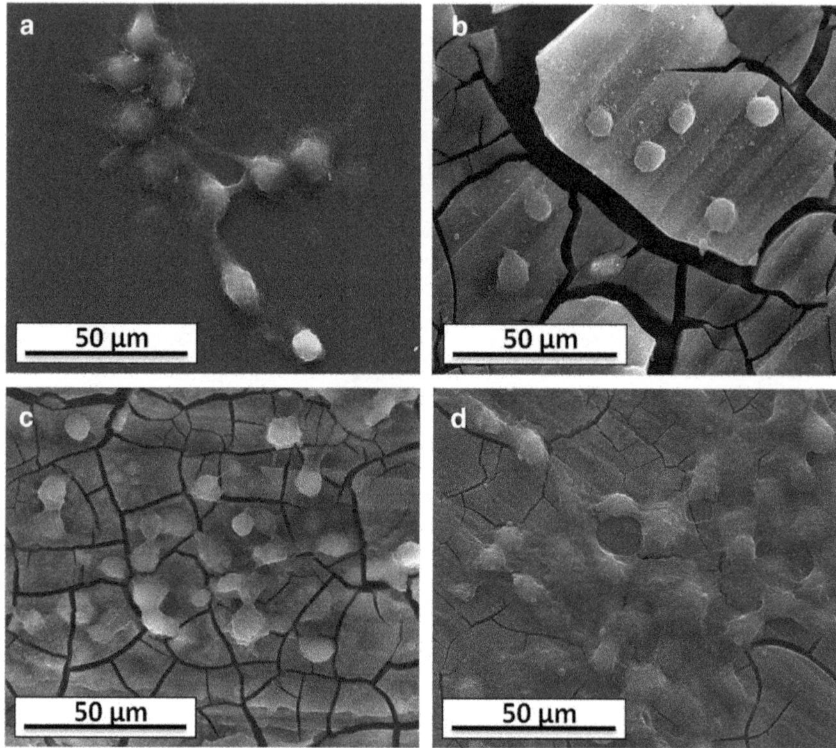

FIGURE 11.8 SEM images of rat skeletal muscle (L6) cells incubated for 24 h on (a) standard tissue culture plate, (b) pure Mg, (c) FSPed Mg and (d) Mg-nHA composite developed by FSP. (*Source:* Reprinted with permission from Ratna Sunil et al., [13]. © 2014 Elsevier.)

11.4 OTHER PROPERTIES

Composites developed by FSP enhance several other properties along with mechanical, tribological and corrosion properties. Ratna Sunil et al., [13] demonstrated increased surface energy for the FSPed and FSPed Mg-nHA and AZ31-nHA composites. Due to the grain refinement after FSP, increased surface energy was calculated from water contact angle measurements. FSPed AZ31 has 31.7 ± 2.9 mJ/m^2 surface energy which is higher compared with 16.2 ± 1.4 mJ/m^2. For the composite, the surface energy was measured as 33.3 ± 1.7 mJ/cm^2,which is higher than FSPed AZ31. Due to the presence of nano-hydroxyapatite, grain growth during FSP was decreased, and higher grain refinement was measured for the composite compared with FSPed AZ31. This leads to increase the surface energy. Similarly, higher surface energy was measured for FSPed AZ31 Mg alloy and AZ31-nHA composite [14]. The increased surface energy promoted biomineralization and further bioactivity. Figure 11.8 shows the SEM images of rat skeletal muscle (L6) cells. Compared with unprocessed Mg and FSPed Mg, a higher level of cell activities were recorded on the surface of the composite incubated for 24 h. Higher cell activities indicate increased bioactivity due to smaller grain size and the presence of nHA introduced by FSP.

Sing and Pal [15] investigated the damping properties of aluminum-based composites produced by FSP. Initially, SiC particles were encapsulated with MgO and nanocrystalline $MgAl_2O_4$ spinel via sol-gel technique and the composites were produced through stir casting. Then the as-cast material was friction stir processed and mechanical, and damping studies were carried out. FSP resulted equiaxed grains with very fine grain size distribution. From the damping studies, as cast composite exhibited higher level of storage modulus (65%) compared with as cast aluminum without reinforcement. Whereas the composites which were processed by FSP have exhibited 61% more storage modulus (61%) compared with the composites produced from stir casting method. Ni et al., [16] developed a composite by dispersing NiTi particles into Commercial 6061Al-T651 alloy. Usually, interfacial reaction is a problem with NiTi–Al composites. But through FSP, no interfacial reaction was observed. The shape memory effect of NiTi particles as observed from reversible transformation from martensite to austenite was well established before comparing FSPed aluminum composite. The NiTi particles also retained the thermal expansion effect as observed from the thermal expansion curves. Compared with the Al alloy without reinforcement, the composites exhibited better damping properties due to the presence of NiTi particles.

The major fraction of research work carried out in developing surface composites by FSP demonstrate enhanced mechanical, tribological and corrosion properties along with little information on thermal behavior, bio-properties, and damping properties. Extensively, more studies on other properties may widen the scope of FSP in developing surface composites for the manufacturing industry which yet to be explored.

KEYWORDS

- **corrosion properties**
- **FSP**
- **mechanical properties**
- **NiTi-Al composites**
- **SiC particles**
- **tribological properties**

REFERENCES

1. Mansour Rahsepar, & Hamed Jarahimoghadam (2016). The influence of multi-pass friction stir processing on the corrosion behavior and mechanical properties of zircon-reinforced Al metal matrix composites, Mater. *Sci. Eng. A, 671*, 214–220.
2. Mohammad Narimani, Behnam Lotfi, & Zohreh Sadeghian, (2016). Investigating the microstructure and mechanical properties of Al-TiB_2 composite fabricated by Friction Stir Processing (FSP), *Materials Science and Engineering A 673*, 436–442.
3. Rahbar-Kelishami, A., Abdollah-Zadeh, A., Hadavi, M. M., Banerji, A., Alpas, A., & Gerlich, A. P., Effects of friction stir processing on wear properties of WC–12%Co sprayed on 52100 steel, *Materials and Design 86* (2015). 98–104.
4. Anvari, S. R., Karimzadeh, F., & Enayati, M. H. (2013). Wear characteristics of Al–Cr–O surface nano-composite layer fabricated on Al6061 plate by friction stir processing, *Wear 304*, 144–151.
5. Alidokht, S. A., Abdollah-Zadeh, A., & Assadi, H., (2013). Effect of applied load on the dry sliding wear behavior and the subsurface deformation on hybrid metal matrix composite, *Wear 305*, 291–298.
6. Wang, H., Estrin, Y., Fu, H., Song, G. L., & Zúberová, Z., (2007). 'The effect of pre-processing and grain structure on the bio-corrosion and fatigue resistance of magnesium alloy AZ31'. *Adv Eng Mater,9*, 967–972.
7. Hoog, C., Birbilis, N., & Estrin, Y., (2008). 'Corrosion of pure Mg as a function of grain size and processing route.' *Adv Eng Mater*, *10*, 579–582.
8. Alvarez-Lopez, M., Pereda, M. D., Valle, J. A., Fernandez-Lorenzo, M., Garcia-Alonso, M. C., Ruano, O. A., Escudero, M. L., (2010). 'Corrosion behavior of AZ31 magnesium alloy with different grain sizes in simulated biological fluids'. *Acta Biomater,6*, 1763–1771.
9. Mostaed Ehsan, Hashempour Mazdak, Fabrizi Alberto, Dellasega David, Bestetti Massimiliano, Bonollo Franco, & Vedani Maurizio, (2015). Microstructure, texture evolution, mechanical properties and corrosion behavior of ECAP processed ZK60 magnesium alloy for biodegradable applications, *Journal of the Mechanical Behavior of Biomedical Materials*, http://dx. doi.org/10.1016/j.jmbbm.2014.05.024.

10. Gu, X. N., Li, N., Zheng, Y. F., Kang, F., Wang, J. T., & Liquan, R., (2011). 'In vitro study on equal channel angular pressing AZ31 magnesium alloy with and without back pressure'. *Mat Sci Eng, B., 176*, 1802–1806.
11. Ratna Sunil, B., Arun Anil Kumar, Sampath Kumar, T. S., & Uday Chakkingal, (2013). 'Role of biomineralization on the degradation of fine-grained AZ31 magnesium alloy processed by groove pressing'. *Mater Sci Eng, C., 33,*1607–1615.
12. Hamu, G. B., Eliezer, D., & Wagner, L., (2009). 'The relation between severe plastic deformation microstructure and corrosion behavior of AZ31 magnesium alloy,' *J Alloy Compd, 468*, 222–229.
13. Ratna Sunil, B., Sampath Kumar, T. S., Uday Chakkingal, Nandakumar, V., & Mukesh Doble, (2014). 'Friction stir processing of magnesium – nanohydroxyapatite composites with controlled in vitro degradation behavior', *Mater Sci Eng, C., 39,* 315–324.
14. Ratna Sunil, B., Sampath Kumar, T. S., Chakkingal, U., Mukesh Doble, N. V. (2014). Nano-hydroxyapatite reinforced AZ31 magnesium alloy by friction stir processing: a solid-state processing for biodegradable metal matrix composites, *J. Mater. Sci. Mater. Med., 25*, 975–988.
15. Subhash Singh, & Kaushik Pal, (2017). Influence of surface morphology and UFG on damping and mechanical properties of composite reinforced with spinel $MgAl_2O_4$-SiC core-shell microcomposites, *Materials Characterization 123,* 244–255.
16. Ni, D. R., Wang, J. J., & Ma, Z. Y., (2015). Shape Memory Effect, Thermal Expansion and Damping Property of Friction Stir Processed NiTip/Al Composite, *Journal of Materials Science and Technology,* http://dx.doi.org/doi: 10.1016/j.jmst.2015.12.013.

CHAPTER 12

Challenges and Perspectives

Grain refinement and the addition of secondary phase particles are the two major factors along with some minor influencing factors such as modification of phase distribution, supersaturation, and texture which alter the material properties. Developing surface composites by FSP has been demonstrated as an excellent route to alter the following properties.

I. Elastic Modulus

Elastic modulus (E) is a constant which can be modified by altering the chemical composition. Addition of secondary phase (ceramic particles or intermetallic particles) alters the slope of the stress-strain curve from which modified elastic modulus can be calculated.

II. Hardness

Hardness of a material is an important property and engineers necessarily consider hardness of a material as prime information to select a material for applications where mechanical behavior of the structure is crucial. The resistance to plastic deformation or permanent deformation that is offered by the material is measured in terms of the applied load and the impression that is placed on the material. By developing fine grains, hardness can be increased. Addition of hard and brittle particle into the matrix by FSP also increases the hardness.

III. Tensile Strength

Among all the mechanical properties, information on tensile behavior of a material is crucial when the structure is targeted for load-bearing applications. There are several strengthening mechanisms by which the tensile strength of a material is increased. In FSP, the modified microstructure and the dispersion of the secondary phase particles enhance the tensile properties. However, from the reported information, most of the studies demonstrated that the increase in strength decreased the percentage of elongation. Hence, the selection of an optimum combination of strength and ductility is important based on the application.

IV. Wear Resistance

When two surfaces have a relative motion at their interfaces, material loss by mechanical means results wear of a material. Usually, wear resistance is demanded for several applications where surface degradation by material loss is common. To counterfeit this, protecting hard and wear-resisting surface coatings are developed. Then the wear characteristics of such surface depend on the quality of the surface coating. Instead of providing surface coatings, making the surface up to certain thickness hard may yield relatively better results. Introducing hard phases into the surface in the form of dispersing particles is a good strategy. Additionally, grain refinement has well demonstrated as a better strategy to introduce high wear resistance.The combined effect of grain refinement and embedding hard secondary phase particle in FSP brings advantage compared with other routes.

V. Fatigue

Fatigue is another important mechanical property which also can be influenced by grain refinement and the addition of very fine, i.e., nanosized dispersing phases. Particularly by adding shape memory particles, fatigue properties can be enhanced for composite produce by FSP.

VI. Electrical and Thermal Conductivity

The electrical and thermal conductivity of the composite can be altered by selecting appropriate secondary phase. Depending on the electrical and thermal properties of the dispersed phase,the resulting composite after FSP exhibit lower or higher thermal/electrical conductivity.

VII. Magnetic Properties

Magnetite particles can be added in FSP, and localized magnetic properties can be introduced in a nonmagnetic material. This helps to develop tailored surfaces which exhibit magnetic behavior at certain locations while the rest of the structure shows nonmagnetic behavior.

VIII. Damping Properties

It is also interesting to observe from the literature that the damping properties of a material can be altered by adding suitable secondary phase particles into matrix by FSP. Usually, the materials which help to increase damping behavior such as shape memory alloys and piezoelectric materials can be dispersed in the form of fine particles, and damping behavior can be introduced for a structure.

IX. Corrosion Properties

The factors which affect the corrosion behavior are completely play different roles for different materials. For example, not in all the cases, grain refinement increases the corrosion resistance. In developing surface composites by FSP, contradictory reports on increased corrosion rate and decreased corrosion rate was observed for different material systems. The electrochemical behavior of matrix and the dispersing phase when they together present in a medium determines the bulk corrosion behavior of the composite. Hence, it can be understood that the corrosion behavior of FSPed composites are said to be completely material dependent and mostly drawing a common conclusions for all material systems is difficult.

12.1 DEFECTS IN SURFACE COMPOSITES PRODUCED BY FSP

The defects which are found in FSP are similar to that of its basic process FSW. The selection of process parameters plays an important role on developing defect-free surface composite. The material flow during FSW and FSP without reinforcing particles is slight different compared with the material flow in the presence of dispersing phase. When a secondary phase is added to the matrix material, thermal conductivity and the local material flow in the stir zone are affected by the added secondary phase. Therefore, occurrence of defect is more during developing surface composites by FSP. Defects which are commonly observed in FSP such as wormhole, lack of penetration, lack of fusion, surface lack of fill, scalloping, root flow defect, surfacing galling and nugget collapse are also can be seen in developing surface composites by FSP [1]. Figure 12.1 demonstrated different defects observed in the composites produce by FSP as explained by W. J. Arbegast [1]. These defects are formed due to lack of appropriate combination of process parameters. Excess amount of heat or insufficient amount of heat results combination of these defects in the composites developed by FSP.

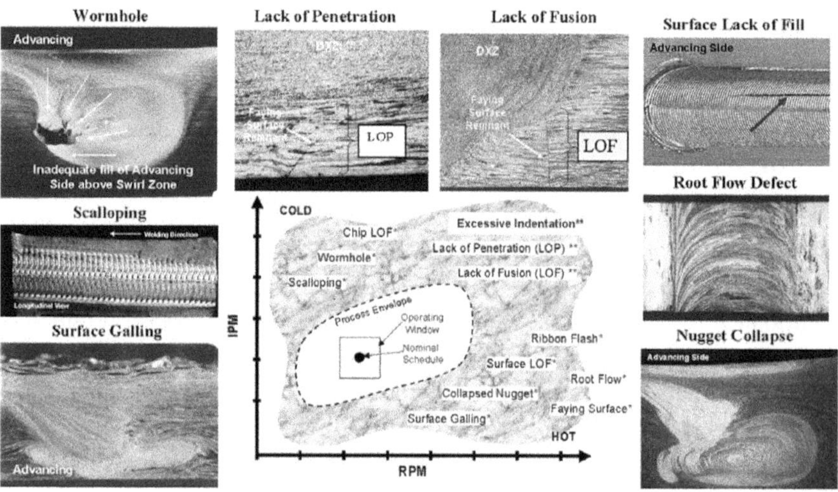

FIGURE 12.1 Different defects appeared in surface composites produced by FSP. (*Source:* Reprinted with permission from Arbegast [1]. © 2008 Elsevier.)

If the produced heat is higher, excessive material flow is resulted and that further leads to flash formation, surface galling and nugget collapse defects. Lack of fill, wormhole and other defects which are influenced by the consolidation of material are resulted due to poor material flow if the produced heat is insufficient. The material flow rates and transfer rate of material are different at the advancing side and retreating side. Due to this imbalance in the material flow, usually, defect are occurred at the advancing side as the material flow leads to develop an abrupt change in the microstructure compared with retreating side [2]. Tool penetrating depth is another crucial factor that needs attention in developing surface composites. When tool penetration depth is sufficient, the shoulder touches the workpiece surface and generates heat due to friction. Insufficient penetration depth may result longitudinal cracks and tunneling defects.

12.2 CHALLENGES

Although FSP has proven as an excellent tool to develop composite materials within the solid-state, still the techniques suffer from a few limitations and challenges as given below.

- For every material system, optimizing the process parameters is required, and for every tool design again the process parameters are different.
- The area of the surface that is being produced in FSP is smaller compared with other routes. In order to increase the area of the surface composite, number of adjacently FSP passes needed to be carried out.
- The amount of dispersing phase is uncontrolled. It is difficult to produce the composites with specific composition. Usually, the powder is filled in grooves or hole or may be supplied on the surface prior to FSP. The total amount of the dispersing phase that is incorporated into the matrix is not measurable.
- The distribution of the secondary phase is influenced by number of passes. The dispersing phase is distributed in agglomerated condition after the first FSP pass. In order to distribute the phase uniformly throughout the stir zone, multi-pass FSP may be required which increases the production cost.
- Variation in the distribution of the secondary phase within the stir zone. Due to the difference in the material flow, the added secondary phase is distributed at different levels in the thickness direction from the surface.
- FSP also results processed regions of varying thickness moving from the nugget to the heat affected zone.
- Workpiece surface geometry also influences the feasibility of the process. Flat surfaces can be easily processed through FSP to develop surface composites compared with surfaces of concave or convex geometry.
- Using automation and robotics can be adopted in FSP, but the traditional robots which are designed for different application are relatively less stiff to use for FSP. Therefore, high-end robots and automation increases the capital cost for the manufacturing industries.

- Limitation with the thickness of the workpiece is another challenge in FSP. Material flow is difficult for the workpieces with higher thickness (more than 6 mm).Processing thin sheets (less than 1 mm) is also difficult. The limitation in the thickness of the workpiece is due to the capability of an FSP tool and its pin design. As the length of the pin is limited between certain values from minimum to maximum, the resulting surface composite also depends on the tool dimensions.
- FSP tool wear is another issue needs additional study as the material loss at the tool shoulder alter the tool design and at the end, influences the success of the process. The dispersing phase in developing surface composites by FSP is usually very hard and brittle. These particles lead to tool wear at the shoulder due to which, pin profile and dimensions are completely altered.

12.3 FUTURE DEVELOPMENTS

Information on developing composites by using new alloys as matrix material is still insufficient. For example, more work can be expected in the near future in developing composites of steels or titanium alloys as matrix material. Increasing the amount of secondary phase which is dispersed in the matrix by developing tailored process designs may be one of the potential areas needs further research [3]. With the development of new tool materials, processing of new alloys will be foreseen in future developments. By adopting automation methods in controlling the process parameters, higher production rate of the surface composites can be achieved without defects. Developing optimum combination of dispersing phases to produce hybrid composites is another potential area that requires elaborate studies. Tool design is crucial in FSP, particularly, while dispersing the secondary phase particles into the surface of the matrix, novel tool designs yield higher level of dispersion and distribution [4, 5]. Developing special surface coatings on tool shoulder and pin to decrease the tool wear is another important research area requires attention [6,7]. From the overall observation made for the past two decades, developing surface composites by FSP has sufficiently demonstrated as a potential tool in surface engineering to tailor the surfaces with enhanced mechanical, corrosion, tribological and bio-properties within the solid-state. Future developments in this area certainly will offer viable solutions to several industrial issues.

KEYWORDS

- **challenges**
- **corrosion properties**
- **elastic modulus**
- **fatigue**
- **hardness**
- **wear resistance**

REFERENCES

1. Arbegast, W. J., (2008). A flow-partitioned deformation zone model for defect formation during friction stir welding. *Scripta Materialia 58*, 372–376. doi:10.1016/j.scriptamat.2007.10.031.
2. Nandan, R., Debroy, T., &Bhadeshia, H., (2008). Recent advances in friction-stir welding-Process, weldment structure, and properties. *Progress in Materials Science 53*, 980–1023. doi:10.1016/j.pmatsci.2008.05.001.
3. Ratna Sunil, B. (2016). Different strategies of secondary phase incorporation into metallic sheets by friction stir processing in developing surface composites, *International Journal of Mechanical and Materials Engineering 11,* 12., https://doi.org/10.1186/s40712-016-0066-y.
4. Thompson, B., &Babu, S. S., (2010). Tool degradation characterization in the friction stir welding of hard metals. *Welding Journal 89,* 256–261.
5. Miyazawa, T., Iwamoto, Y., Maruko, T., &Fujii, H., (2011). Development of Ir based tool for friction stir welding of high-temperature materials. *Science and Technology of Welding and Joining 16, 188*–192. doi:10.1179/1362171810y.0000000025.
6. Buffa, G., Fratini, L., Micari, F., &Settineri, L. (2012). On the choice of tool material in friction stir welding of titanium alloys. In: *Transaction of North American Manufacturing Research Conference of SME June 4–8, 2012 Notre Dame*, Indiana, USA, pp. 785–794.
7. Sato, Y., Miyake, M., Kokawa, H., Omori, T., Ishida, K., Imano, S., Park, S., & Hirano S., (2011). Development of a cobalt-based alloy FSW tool for high-softening-temperature materials, In: Mishra, R. S., Mahoney, M. W., Sato, Y., Hovanski, Y., &Verma, R. (Eds.), *Friction Stir Welding and Processing VI.* TMS, San Diego, California, USA, pp. 3–9.

PART III

Friction Surfacing

CHAPTER 13

Introduction to Friction Surfacing

Modifying the surface properties by adding additional material of different chemical composition at the surface by different techniques such as cladding, weld deposition, thermal spraying, plasma spraying, etc. is another important field in the surface engineering. In all the processes, the quality of metallurgical bonding between the coating layer and the substrate is crucial. The objectives of developing such engineered surfaces by depositing different materials at the surface of the structures are given below.

- To develop high wear resistance surfaces.
- To deposit material on the worn or damaged surfaces to compensate the material loss.
- To seal a crack or pore by material deposition.
- To provide high corrosion resistance at the surface.
- To enhance the biofunctionalization properties.
- To produce high hydrophobic surface.
- To alter the wettability of the surface.
- To alter the thermal and electric conductivity at the surface.

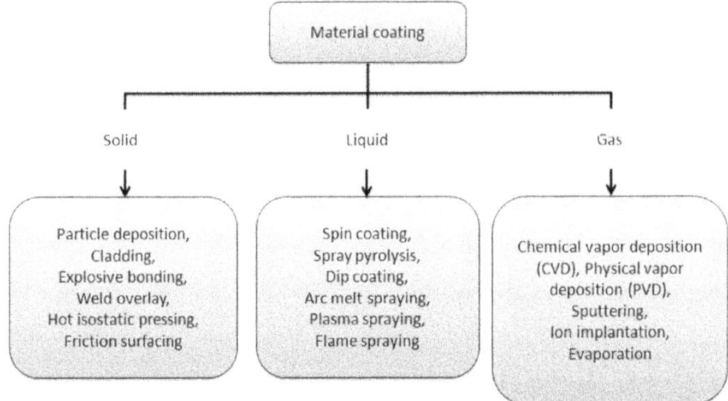

FIGURE 13.1 Schematic diagram shows the three basic routes of surface coating techniques.

All these methods used to achieve one or more of the above property enhancement can be grouped as Solid, liquid and gas state methods. In the solid-state methods; no melting is seen during the process, and in liquid state methods, the deposited material is melted during the process. Coating material is transferred to the substrate as free atoms or ions in gas state methods. Figure 13.1 shows the classification of surface coating techniques based on the state of the coating material.

13.1 CLADDING PROCESSES

These are a group of processes in which a desirable material is deposited in the form of a strip to introduce resistance to corrosion, abrasion, and oxidation and to provide protection to the base material of the structure. Hot roll bonding, cold roll bonding, explosive bonding, brazing and weld overlaying are a few examples of cladding processes. Figure 13.2 schematically shows different cladding processes.

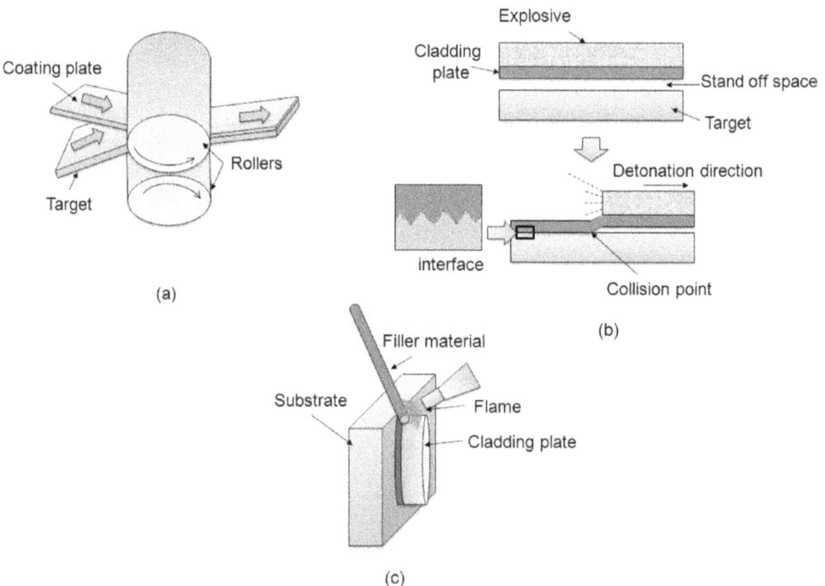

FIGURE 13.2 Schematic illustration of different cladding processes: (a) roll bonding, (b) explosive bonding, and (c) brazing.

In *hot rolling process,* the cleaned plates are stacked and subjected to a combination of suitable heat and pressure to bring them to plastic state and then the cladding plate is bonded at the surface. The thickness of the cladding is usually represented with% of the total thickness. In the most of commercial application, while using low alloy steels or stainless steels, the cladding thickness obtained by hot rolling process is ranging from 10 to 20%. Whereas in *cold rolling process,* the cleaning of the plates is done either by chemical or mechanical means and the plates are subjected to a high pressure to decrease the thickness (50 to 80% in a single pass) followed by sintering to complete the bonding at the interface. In *Explosive bonding*, the cladding is completed in a very-short-duration with high-energy impulse due to the explosion. The two surfaces of metals together subjected to cleaning by flushing out the surface oxide films or any contaminants during the explosion and a perfect metallic bond is developed. A sudden collision does not result at the interface of the two surfaces but rather the joint is formed progressively due to plastic deformation caused by the pressure generated during explosion. Angle bonding and parallel-plate bonding are the two basic geometric configurations used in explosive bonding. Angle bonding configuration is used for bonding sheets or tubes, where the bandwidth is limited to 20 times of the flyer plate thickness. Parallel-plate geometry is used to clad larger flat areas and concentric cylinders [1–3]. In *brazing*, the cladding and the backing materials are arranged as a layered sandwich with a brazing material placed between the surfaces to be bonded. Then the sandwich is heated in a furnace in a protective atmosphere or under vacuum continuous to a certain temperature to melt the brazing material. The liquid brazing material forms intermetallic zones with the substrate, and the cladding plate results a strong bond between the plates [4, 5].

13.2 WELDING OVERLAY PROCESSES

Developing hard surfaces is the prime objective in welding based surface engineering processes. Weld cladding or weld overlay cladding are the other names used in the industry. The material that is intended to be deposited on the substrate is used in the form of electrode. Several fusion based welding processes such as manual metal arc welding, gas welding, submerged arc welding and tungsten inert gas welding can be adopted.

The prime objective is not to permanently join two surfaces but to utilize the heat generated in the welding process to melt the electrode material and to transfer to the targeted surface. Figure 13.3 shows the schematic illustration of the welding process used to deposit surface coating.

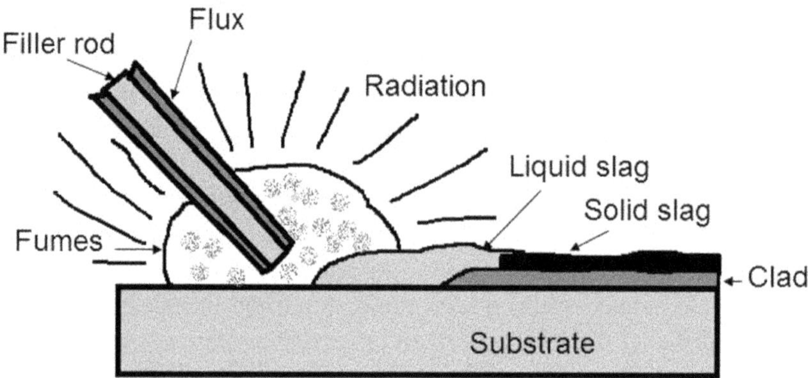

FIGURE 13.3 Schematic illustration showing weld overlay process in developing surface coatings.

Weld overlay processes are useful to develop thick surface coatings up to 100 mm thickness. These processes can be easily adopted as no special equipment is required and the existing welding equipment in the welding industry is sufficient. The basic information about the processes and different mechanisms behind melting and transfer of material are widely known in the manufacturing industry, and therefore no special training is required and can be readily adopted. Another variant in weld overlay process is powder spray and fuse method. In this method, the depositing material is supplied in the form of powder into the flame along with fluxing. The powder is heated to melting point, and a layer of molten metal is coated on the substrate. Usually, oxyacetylene flames are used in this method. The deposition of material is done in different patterns such as dot, stringer beads, discontinuous beads, contiguous beads, Zig-Zag beads, continuous layers, etc. based on the requirement. However, formation of slag layers, higher amount of dilution, formation of unwanted secondary phases, solidification associated thermal stresses, requirement of skilled operator are a few limitations of weld overlay processes. Wide industrial applications can be found in which weld overlay processes are used to repair or to build the surfaces including crusher jaws and cones, ends of material handling buckets, shovels, evacuator bucket teeth, geological drilling tools, etc.

13.3 SURFACE HARDENING BY HEAT TREATMENT

Particularly for steels, several surfaces or case hardening methods were developed in which increasing the hardness and wear resistance by martensite transformation at the surface up to certain depth is the main objective. Martensite is a nonequilibrium phase resulted when diffusion-less phase transformation is happened in steels during cooling. The heat treatment process which results matersite phase after cooling is called as hardening that involves heating the component to certain temperature (above the upper critical temperature in the iron-iron carbide phase diagram) and cooling to room temperature such a way the phase transformation from pearlite to ferrite + cementite is interrupted. The carbon usually must diffuse out during the phase transformation is entrapped in the lattice, and instead of body-centered cubic (bcc), body-centered tetragonal (bct) crystal structure is resulted which is known as martensite. The mechanical properties of hardened steel are superior in hardness and wear resistance compared with non-heat treated steels. Martensite transformation depends on the carbon content of the steel. Low carbon steels are difficult to get martensite transformation. High carbon steel can be easily undergone martensite transformation. However, follow up tempering process is also required to control the brittleness that is induced due to the hardening. By transforming the surface of a component to martensite up to a certain thickness, the corresponding hardness and wear resistance can be increased without affecting the core.

There are two different approaches depending on the amount of carbon in the steels to achieve martensite transformation as shown in Figure 13.4. If the components are with low in carbon content, the processes involve two stages. In the first stage, carbon is introduced into the surface, and in the second stage, martensite transformation is achieved. On the other hand, if the component is rich in carbon, directly heat-treated for martensite transformation. Carburizing, nitriding, and carbo-nitriding are the processes known as thermochemical diffusion processes alter the chemical composition at the surface. In carburizing, carbon is introduced by using carbon enriching agents in the form of solid, liquid or gases. In nitriding, nitrogen is introduced to develop nitrides at the surface of the component. In carbonitriding, the combined effect of carburizing and nitriding is achieved.

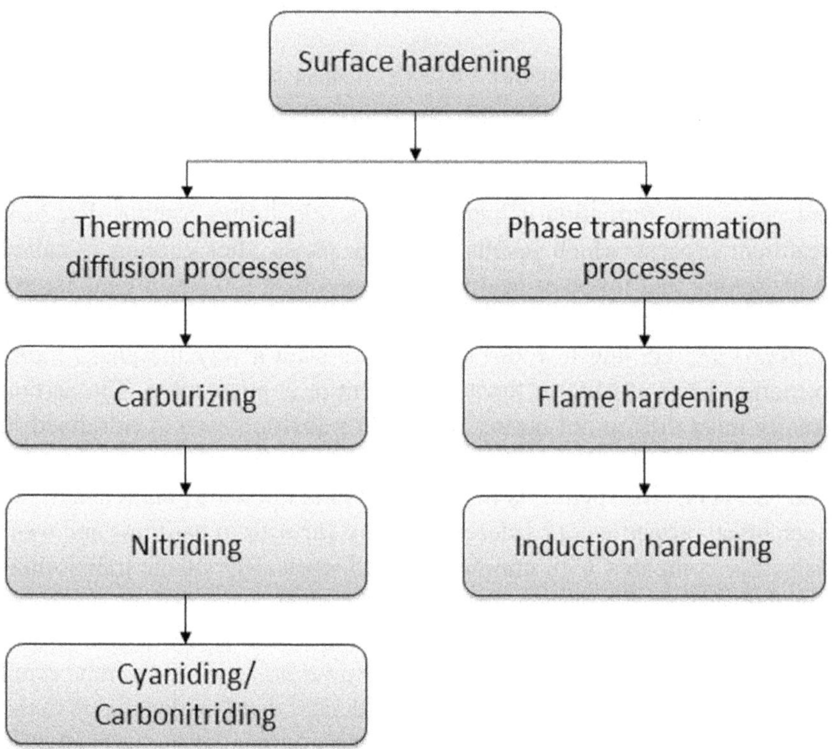

FIGURE 13.4 Schematic illustration of surface hardening techniques.

Induction hardening and flame hardening are the other categories of heat treatments which do not involve any change in chemical composition. The surface is heated to a temperature above the upper critical temperature with the help of induction or by applying localized flame. Once the surface is reached to austenitic zone, then the cooling is done by spraying water or in some alloys simply by air cooling. Depending on the amount of heat generated at the surface and the depth of the surface undergone an austenitic transformation, a thick martensite transformed surface is resulted.

Laser beam surface hardening (LBSH) and electron beam surface hardening (EBSH) are another category of processes in which high radiant energy is used to melt the surface at selective regions. These processes offer several technical and economical benefits compared with the other conventional heat treatment processes. The energy density is high, and the heating and cooling steps are completed very quickly. Martensite transformation is occurred due to the self-quenching process as the substrate act as a sink. Gears/gear teeth, camshafts, gear housing, cylinder liners, axles, exhaust valves, and valves guide are a few components usually surface hardened by LBSH. In order to process the components in EBSH, vacuum chambers are required to avoid the interference of the electron beam with the gas molecules. Workpiece thickness should be higher in order to process with EBSH. Auto/agricultural machine parts, machine tool components, ball bearing racings, piston rings, gears, crankshafts, and camshafts are the components heat treated by EBSH.

13.4 THERMAL SPRAYING

Thermal spraying occupies a special place in surface engineering. A group of processes have been developed in which material is heated rapidly in a hot gaseous medium and simultaneously projected at a high velocity on to a specially prepared surface (target) where it builds up the desired coating. The following steps are involved in developing a surface coating in thermal spraying (if the coating material is assumed to be used in the form of powder as shown in Figure 13.5).

- Spraying particles are supplied through a hopper and are transported through a carrying gas.
- A flame is produced due to the supply of fuel gas.
- The particles travel through the flame and impact on the surface.
- Thermal energy transfer happened from particle to substrate.
- The molten, semi-molten or solid particles are deposited on the substrate.

Due to the adhesion, inter-diffusion and mechanical interlocking, a sound coating is developed. These kinds of coatings are different from the spray atomization and behave similar to that of overlay coatings.

Coating material can be used in the form of powder, wire or rod. However, depending on the type of coating material, the material feed mechanism is altered. The level of dilution is minimum or completely eliminated, and therefore, the substrate is unaffected. Wide variety of materials (metals, non-metals, plastics, and ceramics) can be coated. Usual thickness that can be obtained from thermal spraying ranges from 0.1 to 1 mm. Thermal barrier coatings, abrasive resistance coatings, restoration of dimensions and surface coatings in medical applications are a few examples Where thermal spraying play vital role. Flame spraying, ceramic rod spraying, electric arc spraying, plasma spraying, detonation gun spraying and high-velocity oxy-fuel (HVOF) spraying are a few best known thermal spraying processes.

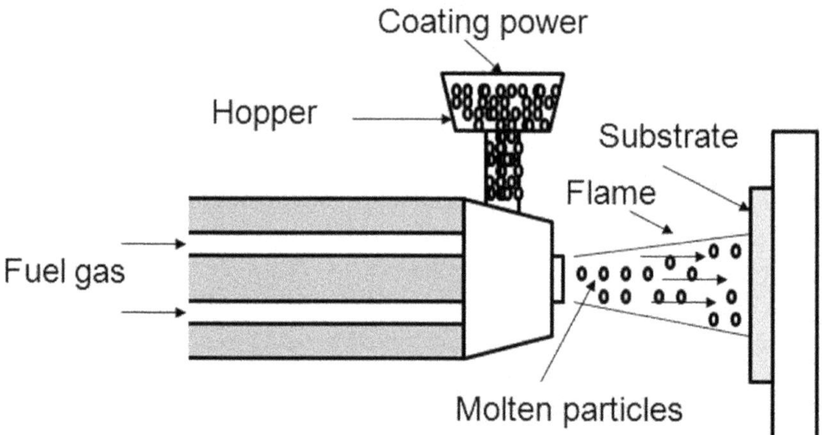

FIGURE 13.5 Schematic illustration of a typical thermal spraying process.

Friction surfacing is one of friction-assisted techniques developed on the principle of material transfer to the substrate by plastically deforming with the aid of heat generated due to friction by using a consumable rod [6]. Unlike friction welding, friction stir welding and friction stir processing; the intended material that is coated in the form of a consumable rod called mechtrode is used in friction surfacing. Compared with weld cladding process and thermal spray process, friction surfacing gives a moderate thickness of surface layer from 0.5 to 6 mm in one pass [7].

KEYWORDS

- **brazing**
- **cladding processes**
- **explosive bonding**
- **heat treatment**
- **mechtrode**
- **rolling process**
- **surface hardening**
- **thermal spraying**
- **welding overlay processes**

REFERENCES

1. Banker, G. & Reinke, E. G., (1993). Explosion Welding, Brazing, and Soldering, Vol. 6, *ASM Handbook*, ASM International.
2. Linse, V. D., (1993). Procedure Development and Process Considerations for Explosion Welding, Brazing, and Soldering, Vol 6, *ASM Handbook*, ASM International.
3. Patterson, R. A., (1993). Fundamentals of Explosion Welding, Welding, Brazing, and Soldering, ASM Handbook, ASM International.
4. Jha, S., Karavolis, M., Dunn, K., & Forster, J., (1993). Brazing with Clad Brazing Materials, Welding Brazing, and Soldering, Vol 6, ASM Handbook, ASM International.
5. Karavolis, M., Jha, S., Forster, J., & Meeking, K., (1993). Application of Clad Brazing Materials Welding, Brazing, and Soldering, Vol 6, *ASM Handbook*, ASM International.
6. Nicholas, E. D., & Thomas, W. M., (1986). *Weld. Journal. 5,* 17–27.
7. Bedford, G. M., (1990).Friction surfacing for wear applications. *Metals. Mater.6*(11), 702–705.

CHAPTER 14

Friction Surfacing of Metals

14.1 PROCESS DEVELOPMENT

Initial reports on friction surfacing can be found in the middle of the twentieth century. In 1941, this technique was demonstrated and patented by Klopstock and Neeleands [1]. Literature shows that the process was later tried in the welding industry as a surface modification technique particularly in USSR in the second half of the twentieth century [2]. Among the important developments in the evolution of friction surfacing, works of Dunkerton and Thomas (1984) are worthy to mention as they proposed using friction surfacing to seal the localized crack at the surface [2]. Later, W. M.Thomasin 1985 [3] and Nicholas and Thomas in 1986 [4] demonstrated friction surfacing as a promising variant of friction-assisted processes as a potential tool to address several issues in the surface engineering. Bedford and Richards in 1985 [5] studied the interface of the surface produced after friction surfacing and demonstrated the absence of dilution. Additionally, lower amount of residual stresses as the melting and solidification were completely eliminated in friction surfacing process was also observed.

In the late 80s, the quantity of work done on friction surfacing was limited until the interest on advanced manufacturing methods is grown in the late 90s [6–8]. Furthermore, Nicholas, in 1993 and 2003 [9, 10] demonstrated the potential of friction surfacing in developing surface coating with improved wear and corrosion properties. Later on, several research groups adopted friction surfacing and developed various combinations of surface coatings on different substrates. However, significant research findings were demonstrated in the past twenty years. Several material systems were processed by friction surfacing to coat wide variety of alloys including mild steels, tool steels, stainless steels, copper and its alloys, nickel alloys, aluminum alloys, magnesium alloys, and titanium.

Based on the development history, friction surfacing is not entirely new to the welding industry but the interest on adopting this technique to address several issues in the manufacturing industry is gradually increasing, and as an alternative to some of the existing surface coating methods, friction surfacing is now emerging as a potential technique.

Friction surfacing can be categorized as solid-state surface coating technique in which the material is transferred from a plasticized consumable rod to the substrate as schematically explained in Figure 14.1 [2]. As shown in Figure 14.1(a), a rotating rod is pressed against the substrate by applying a suitable amount of axial load. Due to the friction between the end surface of the consumable rod and the surface, heat is generated, and a viscoplastic boundary layer is formed at the tip of the rod as shown in Figure 14.1(b). At a certain combination of temperature and pressure, due to the interdiffusion process at the interface of the consumable rod and the substrate, a strong metallic bond is established between the plasticized material and the substrate. The heat that is dissipated through the substrate from the depositing material lead to consolidate the layer at the substrate and then a viscoplastic shearing interface is formed between the deposited layer and the consumable rod. With continuous heat dissipation through the deposited layer and the substrate, the viscoplastic shearing interface is shifted to the next regions of the consumable rod as shown in Figure 14.1(c).

As the tool travels over the substrate, the viscoelastic material is then transferred from the rotating consumable rod to the substrate as a continuous surface coating as shown in Figure 13 (d). From the beginning of the formation of a diffusion layer to the end of the material transfer from the consumable rod to the substrate, the entire process does not result melting of the material, i.e., the entire process is completed within the solid-state itself. Usually, the surface produced by friction surfacing exhibits fine microstructure and strong metallurgical bonding with the substrate. The coating region is free from any defects such as cracks, pores, and contamination [6].

Friction surfacing helps to coat similar or dissimilar materials on different substrates. Particularly, developing dissimilar surface coatings is difficult. Obtaining strong bonding between the coating and the substrate is technically challenging. Different combinations of materials were already investigated and reported. Developing steel surfaces using friction surfacing is now accepted in the industry for several commercial applications [11]. Using different alloy steels to produce high wear and corrosion resistant surface was well reported by several authors [12–17]. Chandrasekaran et al. [18–20] observed that the coating of aluminum on steel substrates is difficult but compared with Inconel due to the role of thermal stability of both the material at high temperatures. Most of the studies were carried out on different combinations of steels. Limited studies were done on nonferrous metals.

FIGURE 14.1 Photographs showing different stages in friction surfacing of an AA6082-T6 aluminum alloy over AA2024-T3 substrate: (a) rotating consumable rod approaches the target, (b) formation of plasticized layer due to the frictional heat generated, (c) initial deformation of the plasticized material, and (d) material transfer from the consumable rod to the targeted substrate. (*Source:* Reprinted with permission from Gandra et al. [2]. © 2014 Elsevier.)

Prasad Rao et al., in 2013 [21] presented a detailed feasibility study on friction surfacing on nonferrous substrates. Coatings were done by friction surfacing using different combinations of materials. Low carbon steel, commercially pure aluminum, titanium and copper rods were selected as coating materials and nonferrous materials such as aluminum, magnesium, copper, and Inconel 800 were selected as substrates. Figure 14. 2 shows typical photographs and corresponding microstructure at the interface of copper-steel and aluminum-steel dissimilar surface coatings. As observed by Prasad Rao et al., [21] a continuous coating was successfully obtained when steel consumable rods were used to coat on copper, aluminum and Inconel 800 substrates due to the wide difference in the mechanical and thermal properties of the coating material and the substrates. Whereas in the remaining combinations, several issues were recorded during the experiments. For example, aluminum and magnesium could not withstand their strength at high temperature generated due to the rotation of the consumable rod. In the case of titanium substrates, low strength at high temperature and also oxidation during the process are the important issues noticed during the experiments. Copper and titanium coatings were not possible on any kind of substrates, and aluminum coating was successfully done over aluminum substrate alone. These findings explain the role of material type and the selection of appropriate process parameters to successfully develop nonferrous surface coatings.

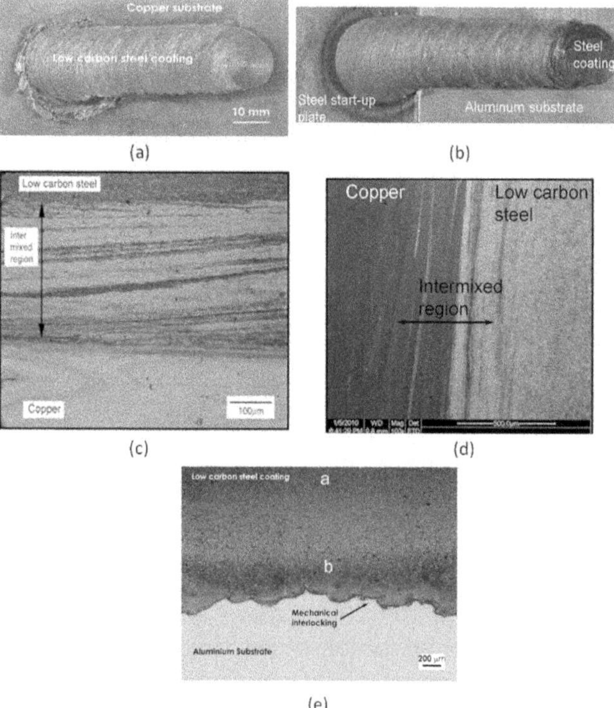

FIGURE 14.2 Friction surfacing of dissimilar materials: (a) low carbon steel over copper substrate, (b) low carbon steel over aluminum substrate, (c) interface of copper and low carbon steel low magnification, (d) corresponding interface high magnification, and (e) interface of aluminum and low carbon steel. (*Source:* Reprinted with permission from Rao et al., [21]. © 2012 Springer.)

Another interesting development in using friction-surfacing process is to produce metal matrix composite coatings on substrates. This variant in friction surfacing was first successfully demonstrated by Madhusudhan Reddy et al. in 2009 [22] by producing a surface composite coating of SiC reinforced AA2124 aluminum alloy over A356 aluminum alloy. In order to develop such composite coatings to improve the surface performance, hard ceramic powder particles or refractory materials are coated on the substrates by different surface modification routes such as laser melt treatment [23], electron beam irradiation [24], cast sintering [25], plasma spraying [26, 27], etc. All these processes deposit composite material on a surface as liquid phase at high processing temperature. It is true that the reinforcement material in the coating may react with the substrate at the interface and may form unwanted secondary phases due to the higher processing temperature. Here comes the advantage of using solid-state processes such as friction surfacing to deposit composite coatings. From the works of Madhusudhan Reddy et al. [22, 28], the composite layer was observed as defect free with improved particle distribution in the matrix. A strong metallurgical bonding between the coating and the substrate was also clearly observed. X-Ray diffraction analysis showed no peaks corresponding to Al_4C_3 that is an indication to stability of SiC without any dissociation into the matrix during the process. Improved wear resistance and also resistance towards pitting corrosion were observed for the coating. Similarly, titanium substrate was coated with AA2124-SiC composite layer using friction surfacing, and improved wear properties were observed for the composite surface compared with titanium substrate [22].

Another significant approach in developing surface composite coatings by using a consumable rod filled with dispersing phases was demonstrated by Gandra et al., [29]. In which, holes are provided in a consumable rod and are filled with secondary phase powder. During friction surfacing, the secondary phase powder that is filled in the consumable rod are dispersed and transferred along with the viscoelastic material to the substrate. Figure 14.3 shows the surface coatings of AA 6082-T6 over AA 2024-T3 alloy with and without SiC particles. Figure 14.3 (a) shows the photograph of the consumable rod cross-section and the macrostructure of the coating interface. Figure 14.3 (b) shows the photograph of consumable rod with a hole filled with SiC powder and the corresponding coating interface. Microstructural observations carried out at the cross-section of the coating reveal the distribution of SiC as per the material flow (Figure 14.3

(c)). Additionally, Gandra et al., [29] developed multilayer coatings by repeating the friction surfacing on the same substrate layer-by-layer up to three iterations. This is similar to what reported by Batchelo et al., [30] in developing multilayered coatings using friction surfacing. Figure 14.4 shows the macroscopic view of the three layers at the cross section and the corresponding microstructure of AA 6082-T6 over AA 2024-T3 alloy. The interface between the individual coating layers and the substrate was metallurgically sound without any defect. As expected in friction surfacing, refinement was achieved up to 30% compared with the starting gain size of the consumable rod. This work clearly demonstrates the potential of developing multilayered surface coatings using friction surfacing technique.

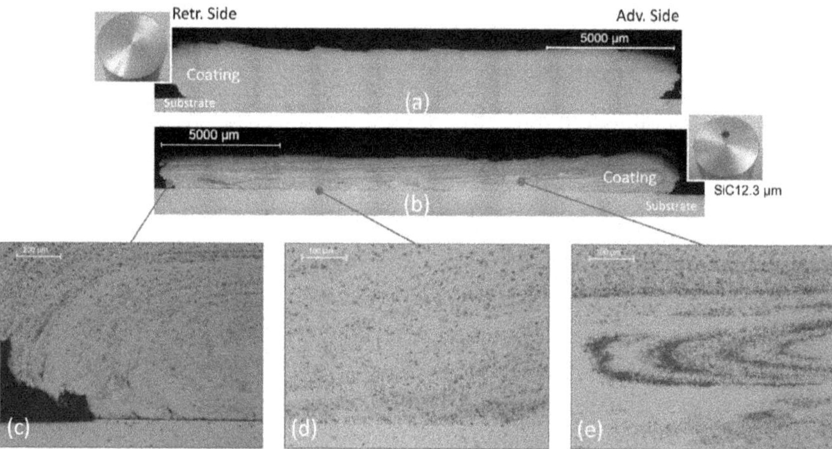

FIGURE 14.3 Cross section macrograph of SiC reinforced coating and consumable rod configuration. (a) FS coating of AA 6082-T6 over AA 2024-T3; (b) 12.3 lm SiC reinforced composite coating; (c) poorly bonded edge; (d) fully bonded interface; and (e) SiC particles arranged according to material flow patterns. (*Source:* Reprinted with permission from Gandra et al., [29]. © 2013 Elsevier.)

Furthermore, multilayer deposition by friction surfacing can be utilized to alter the composition in the thickness direction by using different consumable rods as presented in the same work of Gandra et al., [29]. Figure 14.5 shows the functionally graded composite coatings developed by friction surfacing. Consumable rods of having different concentration of SiC particles filled in different number of holes provided in the consumable rods were used to deposit different layers. The resulted surface coating consisting of graded composition in the thickness direction can be seen in macroscopic observations. From the microstructural observations done at different layers in the thickness direction, it can be seen that the variation in the concentration and distribution of SiC (Figure 14.5 (a), (b) and (c)) is increased as the number of holes in the consumable rods were increased from 1 to 3. The wear behavior of these coatings was also promising from the wear studies due to the presence of SiC and the microstructural features. These findings establish the potential of friction surfacing process to develop multi-layered coatings with variation in the composition within the solid-state itself and actually opened interesting new research areas in connection with additive manufacturing and functionally graded materials.

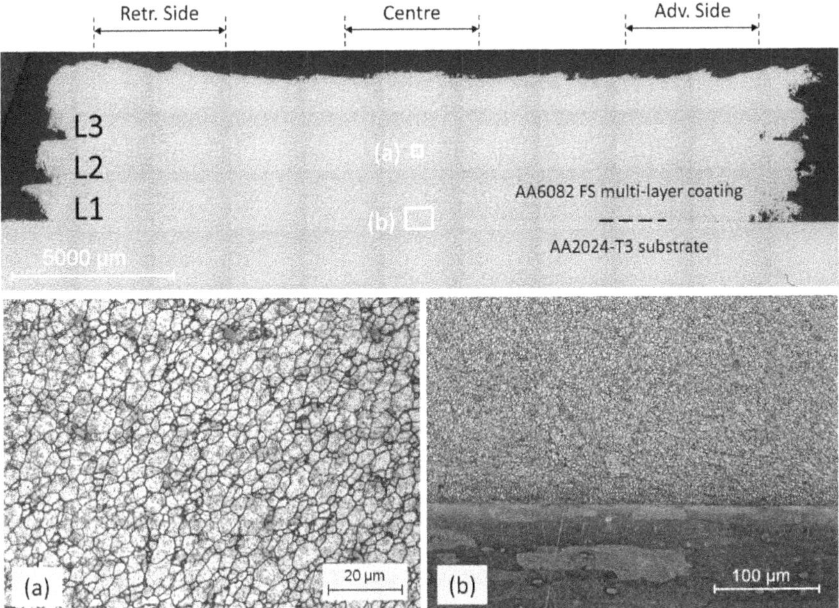

FIGURE 14.4 Cross section macrograph of multi-layer coating. (a) Coating microstructure, and (b) bonding interface to the substrate. (*Source:* Reprinted with permission from Gandra et al., [29]. © 2013 Elsevier.)

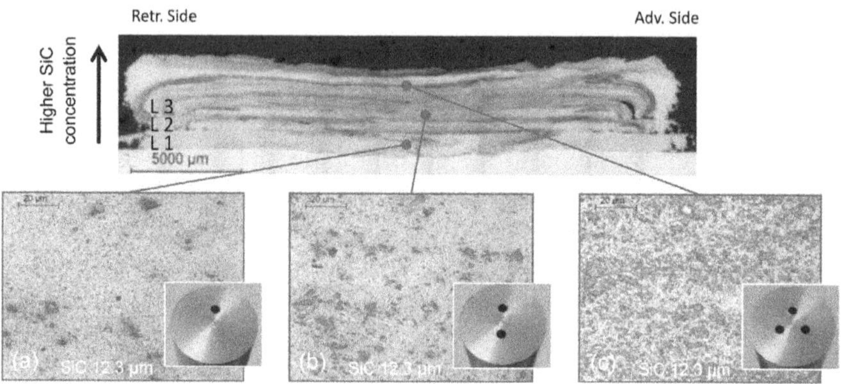

FIGURE 14.5 FGM multi-layer coating presenting an increasing SiC fraction area gradient along thickness direction. (*Source:* Reprinted with permission from Gandra et al., [29]. © 2013 Elsevier.)

Producing underwater surface coatings is another development in friction surfacing research as demonstrated by Li and Shinoda [31]. Underwater welding and localized repairing of the surfaces is necessary to increase the life span of the offshore structures. In such applications, conventional fusion based welding processes are not effective due to the surrounding environment and different thermal events during solidification. Several studies on underwater friction welding demonstrate the advantage of adopting solid-state welding processes for offshore applications [31–35] used AISI 440Chardenable martensitic stainless steel with 1%C and 17% Cr as consumable rod and a low carbon steel (SM50C) as substrate and demonstrated underwater friction surfacing by using tap water as the medium. It was observed that the offset of the coating centerline with reference to the sliding axis of the rotating consumable rod was almost eliminated in underwater friction surfacing unlike in friction surfacing carried out in air. More uniform shape was observed for the deposited coating in the case of underwater friction surfacing. The deposition efficiency which is usually consumable rod rotation speed dependant was observed as less insignificant in the underwater compared with air. Additionally, the deposited surface layers exhibited uniform fine microstructure and hardness distribution compared with deposits developed in air. From the overall findings, it can be understood that friction surfacing can also be used as a promising tool for underwater applications.

14.2 ADVANTAGES AND DISADVANTAGES WITH FRICTION SURFACING

Friction surfacing offers wide advantages compared with other surface coating/depositing techniques as given below.

- No melting is occurred at the interface, and hence, the issues such as oxidation, formation of secondary phases, residual stresses and cracking/shrinkage due to solidification can be decreased or completely eliminated.
- Materials containing hard phases also can be easily deposited.
- Dissimilar materials which are not compatible with other liquid state methods can be deposited.
- Better microstructural features due to the lower heat inputs compared with liquid state methods and the material plastic deformation mechanisms similar to other friction-assisted processes promote improved properties.
- As the process eliminates the liquid state, the dilution issue at the interface also can be controlled.
- Wide range of similar and dissimilar materials including ferrous and non-ferrous metals can be successfully deposited.
- High reactive material systems such as aluminum and magnesium alloys are also easily coated.
- Direct alloy or composite coatings can be developed by using a single respective consumable rod.
- Friction surfacing does not result spatter, emission of radiation and fumes and therefore, environmentally classified as pollution free and green technology.

In spite of several advantages, friction surfacing suffers from several technical and productivity issues as given below which limit the sphere of this process in the current industrial applications.

- Poor bonding between the deposited layer and the substrate at the coating edges is the prime limitation that necessary requires post coating machining to removing the edges by appropriate machining process. This leads to decrease the width of the coating.

- Some material from the consumable rod forms a revolving flash at the tip which decreases the material transfer efficiency.
- The thickness and width of the coating are dependent of process parameters and result poor control over the thickness of the coating material.
- Unlike wire feeding or powder feeding, friction surfacing does not give continuous feeding of the coating material which decreases the productivity.
- Limited control over the coating layer cross-section geometry and post process machining is inevitable to bring the surface coating cross section as per the desired geometry.
- Quantification of material deposition rates and energy consumption characteristics of the process are not yet completely explored.
- Most of the investigations are done with linear depositions. Uneven contours of the substrates add further complex to the processing.

However, friction surfacing has been emerged as an alternate surface coating technique with its unique advantages in the manufacturing industry. Several industrial applications which can be easily addressed by friction surfacing and scope of the process for future applications are summarized in Chapter 18.

KEYWORDS

- **advantages with friction surfacing**
- **disadvantages with friction surfacing**
- **process development**

REFERENCES

1. Klopstock, H., & Neelands, A. R., (1941). An improved method of joining or welding metals, UK Patent No. GB572789.
2. Gandra J., Krohn H., Miranda R. M., Vilaca P., Quintino L., & dos Santos, J. F., (2014). Friction surfacing—A review, *Journal of Materials Processing Technology 214,* 1062–1093.

3. Thomas W. M. (1985).An introduction to friction surfacing. *Surface Engineering Conference*, Brighton, UK, *3*(49), 261–277.
4. Nicholas, E. D., & Thomas, W. M., (1986). Metal deposition by friction welding. *Welding Journal,* 17–27.
5. Bedford, G. M., & Richards, P. J., (1985). On the absence of dilution in friction surfacing and later friction welding. In: *1st International Conference on Surface Engineering,* Brighton, 279–290.
6. Bedford G. M., (1990).Friction surfacing for wear applications. *Metals Mater. 6*(11), 702–705.
7. Bedford, G. M., (1991).Friction surfacing a rotating hard metal facing material onto a substrate material with the benefit of positively cooling the substrate, US Patent No. 5,077,081 A.
8. Bedford, G. M., Sharp, R. P., Wilson, B. J., & Elias, L. G., (1994). Production of friction surfaced components using steel metal matrix composites produced by Osprey process. *Surface Engineering 10,* 118–122.
9. Nicholas, E. D., (1993). Friction surfacing. In: Olson, D., Siewert, T., Liu, S., Edwards, G. (Eds.), *ASM Handbook—Welding Brazing and Soldering.* ASM International, Ohio, United States of America, 321–323.
10. Nicholas, E. D., (2003). Friction processing technologies. *Welding in the World, 47,* 2–9.
11. Bedford, G. M., Vitanov, V. I., & Voutchkov, I. I., (2001). On the thermo-mechanical events during friction surfacing of high-speed steels. Surface and Coatings Technology, *141*, 34–39.
12. Khalid Rafi, H., Janaki Ram, G. D., Phanikumar, G., & Prasad, R. K., (2010). Friction surfaced tool steel (H13) coatings on low carbon steel—a study on the effects of process parameters on coating characteristics and integrity. *Surf Coat Technol205,* 232–242.
13. Liu, X. M., Zou, Z. D., Zhang, Y. H., Qu, S. Y., & Wang, X. H., (2008). Transferring mechanism of the coating rod in friction surfacing. *SurfCoat Technol202,* 1889–1894.
14. Liu, X., Yao, J., Wang, X., Zou, Z., & Shiyao Qu (2009). Finite difference modeling on the temperature field of consumable-rod in friction surfacing. *J Mater Process Technol209,* 1392–1399.
15. Voutchkov, I., Jaworski, B., Vitanov, V. I., & Bedford, G. M., (2001). An integrated approach to friction surfacing process optimization. *SurfCoatTechnol141,* 26–33.
16. Vitanov, V. I., Voutchkov, I. I., & Bedford, G. M., (2000). Decision support system to optimize the Frictec (friction surfacing) process. *J Mater Process Technol 107,* 236–242.
17. Vitanov, V. I., & Voutchkov, I. I., (2005). Process parameters selection for friction surfacing applications using intelligent decision support. *J Mater Process Technol159,* 27–32.
18. Chandrasekaran, M., Batchelor, A. W., & Jana, S., (1997). Friction surfacing of metal coatings on steel and aluminum substrate. *J Mater Process Technol72,* 446–452.
19. Chandrasekaran, M., Batchelor, A. W., & Jana, S., (1997). Study of the interfacial phenomena during friction surfacing of aluminum with steels. *J Mater Sci32,* 6055–6062.

20. Chandrasekaran, M., Batchelor, A. W., & Jana, S., (1998). Study of the interfacial phenomena during friction surfacing of mild steel with tool steel and Inconel. *J Mater Sci33,* 2709–2717.
21. Prasad Rao, K., Arun Sankar, Khalid Rafi, H., Janaki Ram, G. D., & Madhusudhan Reddy, G., (2013). Friction surfacing on nonferrous substrates: a feasibility study, *Int J Adv Manuf Technol 65,* 755–762.
22. Madhusudhan Reddy, G., Srinivasa Rao, K., & Mohandas, T., (2009).Friction surfacing: novel technique for metal matrix composite coating on aluminum-silicon alloy, *Surface Engineering, 25*(1), 25–30.
23. Baker, T. N., Xin, H., Hu, C., & Mridha, S. (1994). *Mater. Sci. Technol., 10*, 536–544.
24. Choo, S. H., Lee, S., & Kwon, S. J. (1999). *Metall. Mater. Trans. A, 30A*, 1211–1221.
25. Wang, Y. S., Zhang, X. Y., Zeng, G. T., & Li, F. C. (2001). *Composites, 32*, 281–286.
26. Gui, M. C. & Kang, S. B. (2000). *Mater. Lett., 46*, 296–302.
27. Deuis, R. L., Yellup, J. M., & Subramanian, C. (1998). *Compos. Sci.Technol., 58*, 299–309.
28. Madhusudhan Reddy, G., Satya Prasad, K., Rao, K. S., & Mohandas, T., (2011). Friction surfacing of titanium alloy with aluminum metal matrix composite, *Surface Engineering, 27*(2), 92–98.
29. Gandra, J., Vigarinho, P., Pereira, D., Miranda, R. M., Velhinho, A., & Vilaça, P., (2013). Wear characterization of functionally graded Al–SiC composite coatings produced by Friction Surfacing, *Materials and Design 52*, 373–383.
30. Batchelor, A. W., Jana, S., Koh, C. P., & Tan, C. S., (1996). The effect of metal type and multi-layering on friction surfacing, Journal of Materials Processing Technology *57,* 172–181.
31. Li, J. Q. & Shinoda, T., (2000).Underwater friction surfacing, *Surface Engineering, 16*(1) 31–35.
32. Nicholas, E. D. (1983). *Proc. Conf. Underwater Welding,* Trondheim Norway, June 1983. International Institute of Welding, IV-4.
33. Andrews, R. E., & Mitchell, J. S. (1990). *Met. Mater., Dec.* 796–797.
34. Tasaki, Y. et al. (1976). *Rep. Gov. Ind. Res. Inst., Nagoya, 25*(7), 183–187.
35. Tasaki, Y. et al. (1980). *Rep. Gov. Ind. Res. Inst., Nagoya, 19*(3), 63–71.

CHAPTER 15

Mechanism Behind Coating Formation in Friction Surfacing

Friction surfacing involves a combination of several material transfer mechanisms. The material transformation can be viewed as a combined effect of hot-working and welding at the processed zone. The processed region in friction surfacing remains in a solid-state similar to that of other friction-assisted processes. It is essential to understand the thermo-mechanical behavior during friction surfacing right from the first stage of material plastic deformation to the formation of a strong bond between the transferred material and the substrate in order to develop successful surface coatings. Heat is generated during friction surfacing by two reasons: (i) due to friction between the rotating consumable rod and the substrate, and (ii) due to the plastic deformation of the material from the consumable rod.

13.1 THERMO-MECHANICAL BEHAVIOR AND MATERIAL TRANSFER MECHANISMS

The phenomenon in material transfer during friction surfacing was explained by Thomas [1] and named as "third-body region" concept in which the material temperature reaches to a certain level above the recrystallization temperature and below the melting temperature. This leads to form a viscoplastic region at the tip of the consumable rod which does not have any constraints and three-dimensionally flow as a liquid to deposit on the substrate. The heat generated due to the friction raises the temperature of the tip of the depositing rod but less than the melting temperature of the material. The generated heat is conducted through the consumable rod (mechtrode) and results in a temperature gradient across the consumable rod and influence the deformation behavior of the consumable rod [2].

The material in the thermally gradient region of the consumable rod gradually softens and plastically deform due to the torsion and compression by the material of upper regions of the consumable rod which is in cold condition as explained by Bedford et al.[3] and schematically shown in Figure 15.1. On the other hand, Fukakusa [4] explained the material transfer phenomena with reference to the rotational contact plane along which slippage takes place between the consumable rod and the deposited layer. The difference in the speeds of the rotating consumable rod and the material which is deposited as a layer on the substrate brings viscous slipping, and the thickness of the deposited layer is governed by the distance of the contact plane from the substrate [5]. Additionally, heat is generated due to the viscous shearing at the surfacing interface as explained by Gandra et al. [6]. During this process when the consumable tool rotates and travels on the substrate, the material of the consumable rod at the interface will travel either towards the flash or towards the substrate to get deposited in a rolling pattern. Then the transferred material is cooled due to the heat dissipation through the consumable rod, substrate and due to convection through the air. Hence, a strong metallic bond is produced between the transferred layer of material and the substrate.

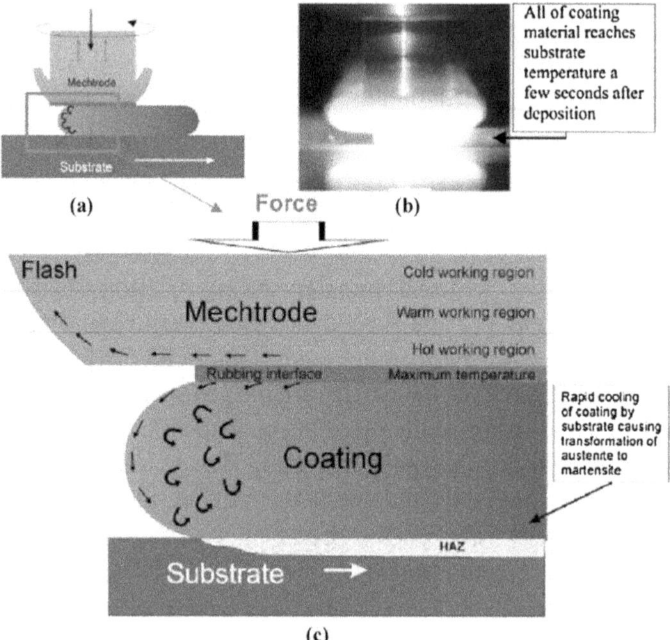

FIGURE 15.1 Thermo-mechanical events happened during the material transfer in friction surfacing of high-speed steels as explained by Bedford et al. (*Source:* Reprinted with permission from Bedford et al., [3]. © 2001 Elsevier.)

Furthermore, Prasad Rao et al., [7], reported significant information in understanding the thermo-mechanical behavior during friction surfacing. Three different combinations of materials:(1) D2 tool steel (high carbon and high chromium steel) on steel substrate, (2) commercial pure (Cp) copper on steel substrate, and (3) Cp rod on Cp substrate were used, and thermal profiles at the interface and within the consumable rod were investigated by using infrared camera during the process. A typical time-temperature (thermal profile) plot was generated as shown in Figure 15.2. Based on the experiments, Prasad Rao et al., [7] proposed four stages in the thermal profile as given below.

1. Stage 1 where the temperature is increased slowly and gradually for a given dwell time.
2. Stage 2 where the temperature is suddenly increased within a short time as shown in Figure 15.2 (b).
3. Stage 3 where the temperature is gradually increased.
4. Stage 4 where a steady state temperature is observed and also the formation of the surface coating.

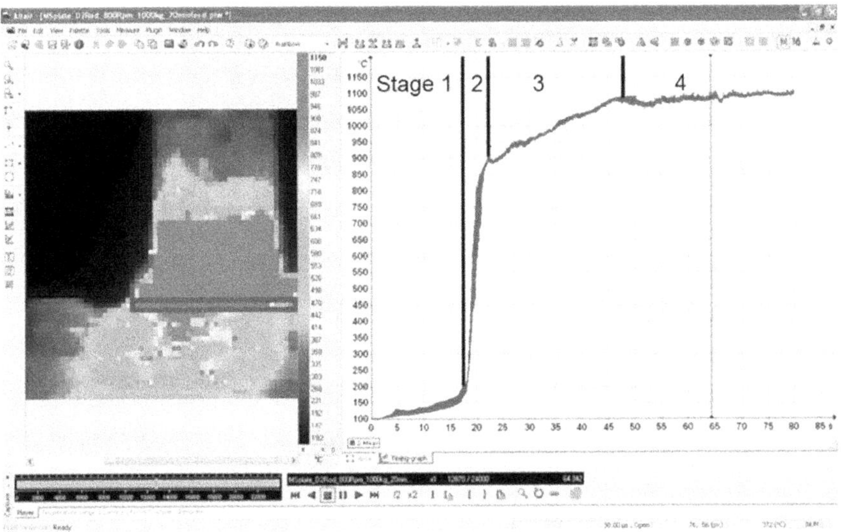

FIGURE 15.2 (See color insert.) Thermal profile obtained during friction surfacing in which tool steel (high carbon and high chromium) D2 rod was used as a consumable rod to develop surface coating on steel substrate. (*Source:* Reprinted with permission from Rao et al., [7]. © 2012 Elsevier.)

Figure 15.3 shows the rubbed surface of consumable rods after the experiment. Scanning electron microscope (SEM) image of the interface after producing D2 steel surface coating on a steel substrate indicates a strong perfect bonding (Figure 15.3(a)). The photograph of the used D2 rod after friction surfacing (Figure 15.3(b)) shows a rough surface that is an indication of material transfer. Interestingly, the rubbed surface of the copper rod was observed as flat as shown in Figure 15.3(c) where no coating was resulted. It was demonstrated in several other works that the flow stress of the material influence the success of the material deposition in friction surfacing. By referring to the thermal profile, the authors demonstrated a relationship between the study state temperature and the hot extrusion/hot forging temperatures of the material. This

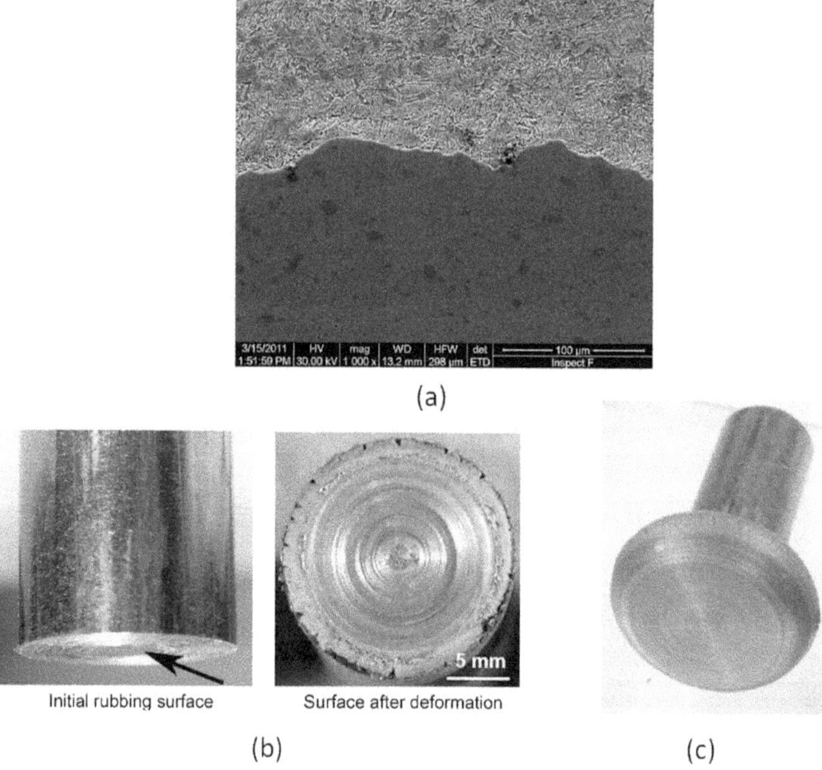

FIGURE 15.3 (a) Scanning electron microscope (SEM) image showing the interface of the D2 steel and coating on steel substrate, (b) used D2 steel consumable rod and (c) used copper consumable rod (*Source:* Reprinted with permission from Rao et al., [7]. © 2012 Elsevier.)

observation really helps to understand the potential of a particular material as a consumable rod to be used in friction surfacing. For example, in their work, a good surface bonding was observed for the tool steel (D2) on steel and Cp copper on Cp copper. However, no material deposition was observed when Cp copper was used as consumable rod to coat on steel substrate. Therefore it can be understood that by maintaining the steady state temperature (stage 3) close to that of consumable rod hot extrusion temperature, successful coatings can be developing in friction surfacing. Hence, maintaining the steady state temperature by means of aiding any additional care during friction surfacing can bring success to transfer the material to the substrate.

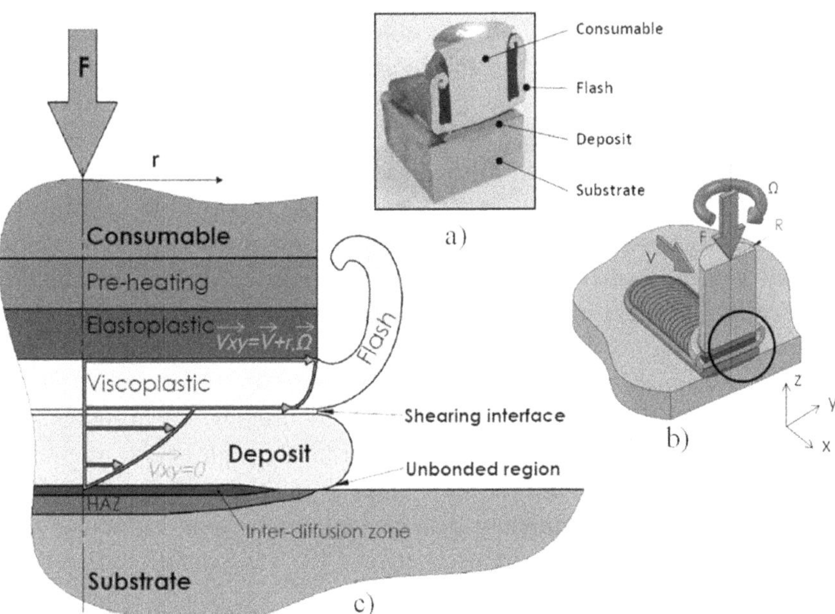

FIGURE 15.4 Important zones and influencing parameters of friction surfacing: (a) photograph of the cross-section, (b) different process parameters, and (c) mechanical features and speed profile at the cross-section. "(F-forging force (or plunging speed, Vz), Ω-rotation speed, Vx-ravel speed, Vxy-rod tangential speed in-plane xy given by composition of rotation and travel movements)"(*Source:* Reprinted with permission from Vilaça et al., [9]. © 2014 Springer Nature.)

15.2 FLASH FORMATION IN FRICTION SURFACING

During friction surfacing, the "third-body region" generated at the tip of the consumable rod is not totally constrained form its flow. During the process, a high amount of axial load is applied on the plastically deformed material at the tip of the consumable rod against the target. Due to no constraints, material is flowed outside of the diameter of the consumable rod. Then the excess material flow at the circumference of the tip of the consumable rod generated a revolving shaped plasticized material called "flash" [8]. Vilaça et al., [9] investigated the primary flash formation (PFF), and secondary flash formation (SFF) during friction surfacing and explained the influence of the flash formation at two different levels on successful formation of the coating. Figure 15.4 shows the schematic representation of important zones resulted during flash formation in friction surfacing. As reported in the literature, the material is transferred from the advancing side to the retreating side, and the material from the center of the consumable rod is usually coated to the substrate, and the material from the periphery of the consumable rod is expelled as flash [2, 4]. As shown in Figure 15.4 the difference in the speeds of the plastically deformed (viscoplastic) material that rotates along the consumable rod at Vxy, and the transferred material to the substrate (Vxy=0), results the plastically deformed material to detach from the consumable rod. Heat is generated due to the friction between the deposited coating and the rotating consumable rod.

In other friction-based techniques such as friction stir welding and processing, the plastically deformed material, i.e., third body region is mostly enclosed within the processed region and the flash is minimized, and in some cases completely disappeared. In friction surfacing, this "third-body region" is constrained between the consumable rod and the substrate similar to "open die forging condition" and results a flash. This can be termed as primary flash formation (PFF). As explained by Vilaça et al., [9], the rate of flash formation (RFF) is given in equation (1) can be assessed by calculating the difference between the volumetric rod consumption rate (*CR*), and volumetric deposition rate (*DR*).

$$RFF\ (m^3/s) = CR - DR \tag{1}$$

$$CR\ (m^3/s) = A_r V_z = \pi r^2 V_z \tag{2}$$

$$DR\ (m^3/s) = A_d V_x \tag{3}$$

where V_z is consumable rod plunging speed, A_r is the cross-sectional area of the consumable rod, and r is the radius. V_x is thetravel speed, and A_d is the deposited cross-section area. Then the deposition efficiency $\eta_{deposit}$ can be obtained as given below.

$$\eta_{\text{deposit}} = DR/CR \tag{4}$$

In friction surfacing, strong bonding is found in the middle, and no bonding is observed at the edges. Therefore, joining efficiency ($\eta_{joining}$)can be calculated from equation (5).

$$\eta_{\text{joining}} = W_b/W_d \tag{6}$$

where W_b is the width of the bonded coating, and W_d is maximum coating width. The effective coating efficiency ($\eta_{coating}$) is the fraction of total material transferred and bonded to the substrate as given below.

$$\eta_{\text{coating}} = \eta_{\text{deposition}} \text{x}\ \eta_{\text{joining}} = [A_d V_x/(\pi r^2 V_z)]\ W_b/W_d \tag{7}$$

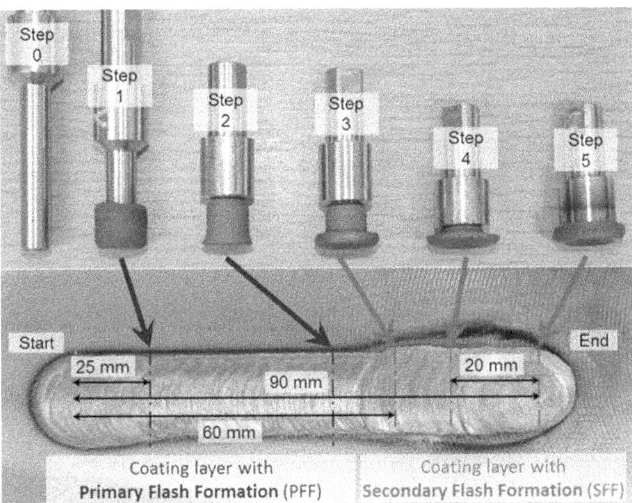

FIGURE 15.5 Photograph showing development of primary and secondary flash forming (PFF and SFF) and corresponding surface coating. (*Source:* Reprinted with permission from Vilaça et al., [9]. © 2014 Springer Nature.)

Due to the poor bonding between the coating and the substrate at the edge of the coating, the fraction of unwanted material as flash is increased. As the length of the coating layer is increased, the formation of flash further increases the necessity of more consumable rod. The increased size of the flash further increases the pressure distribution during the rubbing at the interface. This further leads to develop variations in the thickness and width of the coating because of the variation in the amount of heat generation and applied pressure. Flash cutters are suggested to use to prevent the adverse effects of formation of PFF. As demonstrated by Vilaça et al., [9], Figure 15.5 shows the sequence of flash formation during friction surfacing. After PFF, the formation of surface coating layer due to secondary flash formation (SFF) can also be seen as indicated with arrows. The sequence of steps during PFF and SFF during the coating formation was explained by Vilaça et al., [9] as given below.

- *Step 0:* With a given plunging speed, Vz (while Vx=0) the consumable rod is plunged into the substrate. The material at the end of the consumable rod undergoes viscoplastic state and formation of flash can be observed [Gandra et al., 2011].
- *Step 1:* Material from the consumable rod is transferred to the substrate, and the width of the rubbing area is increased due to the increased flash.
- *Step 2:* As the coating layer is increased, the primary flash is increased and touches the shoulder. As soon the PFF touches the shoulder, the pressure kinetics and the formation of the coating with initiation of SFF can be seen.
- *Step 3:* During this step, the flash is developed in the radial direction until the equilibrium pressure is reached in the viscoplastic region before the material transferred to the substrate.
- *Step 4:* At this stage, the radius of the flash is increased. The enlargement of the flash radius is usually observed between step 3 to step 4.
- *Step 5:* The formation of the coating is done by the SFF, and the coating layer experiences an additional thermo-mechanical processing which leads to form a coating layer with different geometrical, metallurgical and mechanical properties.

After completion of these steps which includes PFF and SFF a typical microstructure at the interface of the coating and substrate as obtained from

EBSD studies is shown in Figure 15.6 [9]. The waviness of the interface as usually observed due to the formation of diffusion bonding is clearly observed (Figure 15.6 9(c)). These findings demonstrate the formation of the interface is in the solid-state, and the gains are grown in the orientation of the substrate grains.

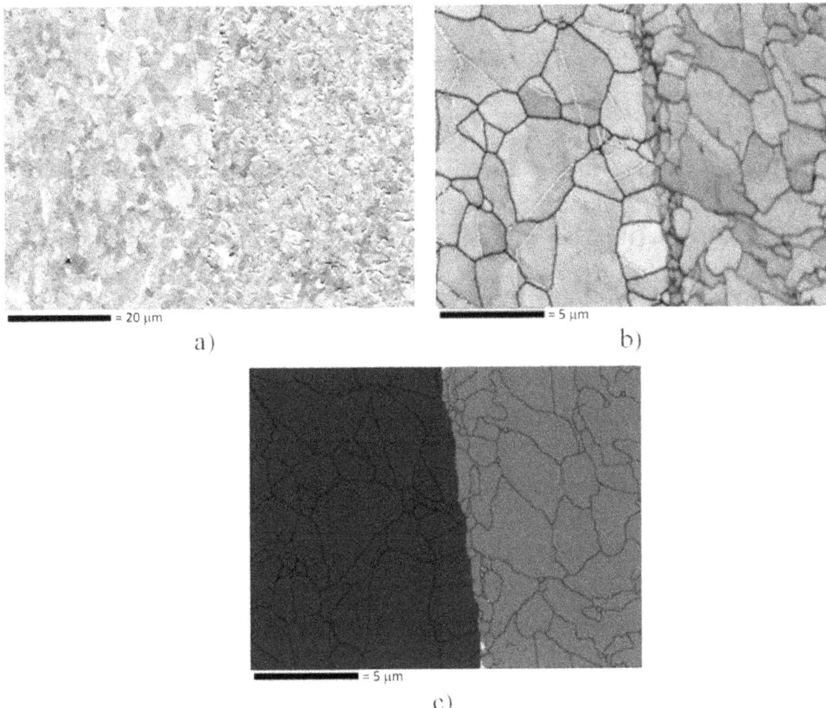

FIGURE 15.6 Microstructure obtained at the cross-section of the friction surfacing coating: (a) SEM image, (b) EDSD maps of grain boundaries, and (c) map of face-centered cubic (blue) and body-centered cubic (red) crystal structures (*Source:* Reprinted with permission from Vilaça et al., [9]. © 2014 Springer Nature.)

15.3 MATERIAL FLOW IN FRICTION SURFACING

From the works of Bendzsak et al., [10] and Schmidt et al., [11] the fundamental behavior of material flow during friction surfacing can be understood. Material flow in friction surfacing is resulted due to the combination of stirring, extrusion, and forging [2]. The coating process is completed in two stages: (i) dwell time at the beginning of the process, and (ii) material coating. During the first stage, the rotating consumable rod is allowed to contact with the substrate and undergoes plastic state. In the second stage, the plastically deformed material is transferred to the substrate. During the first stage, metal flow follows a circular path and leaves a circular shape to the deposited material. In later stage, the material flow is distributed linearly as the substrate is moved and a layer of coating is appeared. Along with these two factors, axial load also influence the material flow patterns, and therefore, the mechanisms of material flow are complex. However, a few investigations provide significant information to understand the material flow behavior during friction surfacing. In order to study the material flow, Rafi et al., [12] introduced tungsten powder into a consumable rod made of stainless steel AISI 304 and surface coatings were developed. Tungsten powder was filled in the holes provided in the consumable rods on the periphery so that the powder is distributed when coating is done.

The distribution of powder was studied by X-ray radiography and scanning electron microscopy which directly showed the material flow. Figure 15.7 shows the schematic representation of the cross-section view of the process consisting of the rotating consumable rod, the transferred material (coating) and the target (substrate). The flow patterns of the bottom layers of the coating material that is close to the moving substrate can be seen as close with the velocity of the moving substrate. Whereas the flow patterns of the coating layers on the top of the coating, i.e., close to the rotating consumable rod shows a direction that is governed by the movement of

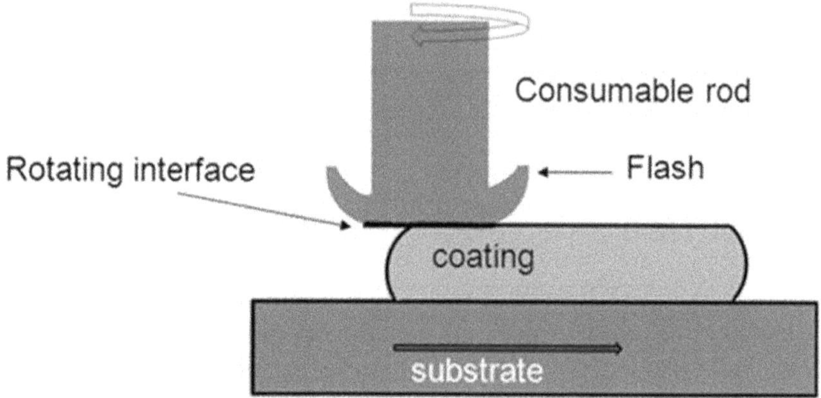

FIGURE 15.7 Schematic of the cross-section of friction surfacing process.

the consumable rod. This phenomenon was clearly observed from the X-ray radiographs of the surface coating. The top and bottom layers of the coating tend to move in opposite directions. As the tungsten powder is distributed along with the deposition of AISI 304 stainless steel, the flow patterns show a separation at the middle of the top and bottom layers of the coating which suggests that the material flow within the coating exhibit opposite direction. This phenomenon can be compared with a fluid flow between two parallel plates among them one plate is subjected to axial load [13]. In friction surfacing, the plasticized material that flows between the substrate and the rotating consumable rod can be visualized as a fluid that is constrained between the bottom substrate and the rotating consumable rod. Additionally, the rotation of the consumable rod adds more parameters to investigate the flow of the material, unlike fluid system. Furthermore, the authors found that the material flow occurs from the advancing side to the retreating side and then is terminated at the center of the coating layer [12]. From the available literature, the material flow behavior during friction surfacing is sufficiently understood. However, further research is needed to explore the possibilities with different combinations of the materials.

KEYWORDS

- **material flow**
- **primary flash formation**
- **rate of flash formation**
- **secondary flash formation**

REFERENCES

1. Thomas, W. M., (2009). *An Investigation and Study into Friction Stir Welding of Ferrous-Based Material.* University of Bolton, Bolton, United Kingdom.
2. Rafi, H. K., Balasubramaniam, K., Phanikumar, G., & Rao, K. P., (2011). Thermal profiling using infrared thermography in friction surfacing. *Metallurgical and Materials Transactions,A.*, *42*, 3425–3429.
3. Bedford, G. M., Vitanov, V. I., & Voutchkov, I. I., (2001). On the thermo-mechanical events during friction surfacing of high-speed steels. Surface and Coatings Technology, *141*, 34–39.

4. Fukakusa, K., (1996). On the characteristics of the rotational contact plane—a fundamental study of friction surfacing. Welding International, *10*, 524–529,.
5. Bedford, G. M., (1991). Friction surfacing a rotating hard metal facing material onto a substrate material with the benefit of positively cooling the substrate, US Patent No. 5,077,081 A.
6. Gandra, J., Miranda, R. M., & Vilac¸ A, P., (2012). Performance analysis of friction surfacing. *Journal of Materials Processing Technology,212*, 1676–1686.
7. Prasad Rao, K., Veera Sreenu, A., Khalid Rafi, H., Libin, M. N., & Krishnan Balasubramaniam, (2012). Tool steel and copper coatings by friction surfacing – A thermography study, *Journal of Materials Processing Technology, 212*, 402– 407.
8. Nicholas, E. D., Thomas, W. M., (1986). Metal deposition by friction welding. *Welding Journal,* 17–27.
9. Pedro Vilaça, Hannu Hänninen, Tapio Saukkonen, & Rosa Miranda, M., (2014). Differences between secondary and primary flash formation on coating of HSS with AISI 316 using friction surfacing, *Welding in the World, 58*, 661–671.
10. Bendzsak, G. J., North, T. H., & Li, Z. (1997). *Acta Mater.,45*, 1735–1745.
11. Schmidt, H. N. B., Dickerson, T. L. & Hattel, J. H. (2006). *Acta Mater.54*, 1199–1209.
12. Khalid Rafi, H., Phanikumar, G., & Prasad Rao, K., (2011). Material flow visualization during friction surfacing, *Metallurgical and Materials Transactions, A.,42a*, 937–939.
13. Marques, F., Sanchez, J., & Weidman, P. D. (1998). *J. Fluid Mech., 374*, 221–49.

CHAPTER 16

Influencing of Process Parameters

16.1 PROCESS PARAMETERS

Process parameters are the governing factors which influence the success of any manufacturing process. Additionally, the material properties of post-processed structures are also dependants of the processing parameters. In friction surfacing, there are several processing parameters which directly and indirectly influence the material flow, the formation of the coating, bonding strength of the coating to the substrate, formation of the flash, rate of consumable rod consumption, metallurgical and mechanical properties of the coating and the interface. This chapter presents different influencing processing parameters in friction surfacing. Usually, the selection of a surface coating process is made based on the thickness of the intended coating, width, bonding strength, the substrate, and the coating material. During friction surfacing, the type of substrate and the coating material gives different kinds of thermo-mechanical events. Heat is generated due to the combined effect of friction between the substrate and consumable rod and viscoplastic deformation in the coating and substrate materials. The difference in the relative speeds of the depositing material and the substrate leads to shear the bonding interface and disrupts the ongoing diffusion bonding. Therefore, higher rotational and transverse speeds may decrease the cross section bonding area as demonstrated by Shinoda et al., [1].

It was believed that the higher rotational speeds results in a higher amount of heat generation and promotes better coating in friction surfacing. On the contrary, lower rotational speeds and relatively lower speeds between the transferring deposit and the substrate increased the contact area and developed strong bond [2]. In addition, tilting the consumable rod to an appropriate angle (0-3°) also has shown to influence the bonding width as the transfer of the material from consumable rod is precisely confined to the target [3]. Along with processing parameters, the material combination is also another important influencing factor must be considered in friction surfacing. Understanding the effect of process parameters on the success of coating deposition is complex in friction surfacing. However, based on the available literature, relative trends can be observed, and the concerned effects can be understood. The consumable rod (mechtrode) force (*F*), the rotational speed of the consumable rod (*N*) and the transverse speed of the substrate (*Vx*) are the three crucial processing parameters which dictate the successful formation of the surface coating and the quality of the coating in friction surfacing. Thickness of the surface coating (*Ct*), width of the coating (*Cw*) and the bonding strength of the coating (*Cbs*) are the three important factors which are considered into account while selecting a set of process parameters [4].

Another important influencing factor in friction surfacing is the type of equipment and mode of operation. The main objective behind using different equipment and mode of controls is to bring the rotating consumable rod in contact with the substrate. The required pressure and heat generation at the interface can be achieved either by applying sufficient pressure on the consumable rod or by giving appropriate axial feed to the consumable rod. Maintaining a constant pressure on the consumable rod throughout the processing cycle is crucial to achieve uniformity in the surface coating. Therefore, computer numerical controlled (CNC) machines are required. Several alternate arrangements were also proposed/adopted to do friction surfacing using conventional milling machines. Usually, the vertical milling machines are attached with specially controlled work tables, or ram with the help of pneumatic or hydraulic controllers and the uniform pressure is maintained by moving the work table or ram against the rotating tool. The feed also is given by moving the work table or ram against the rotating consumable rod. Both the movements of the workpiece can be attained by giving appropriate axial and traverse movement to the work table or the ram as demonstrated by different authors [2, 5, 6].

As reported by Vitanov et al., [7, 8] using conventional CNC machines to provide constant load by giving a specific velocity with which the consumable rod is axially pushed against the substrate can be interesting strategy to control the loading issues. As explained by Vitanov et al. [7], a close relationship can be drawn between the rod feed rate and the axial force during friction surfacing. Therefore, equipment that is load controlled can be easily developed for friction surfacing or the existing machines can be altered with minimum modifications by providing additional attachments. Hence, industries can readily adopt friction surfacing technique without adding additional capital cost.

16.2 BONDING TIME

The bonding time is an important factor which is connected with the width of the coating during friction surfacing. Bonding time is defined as the time required for the heat generation area to completely pass over a specific point on the substrate. Usually, the heat generated area when the mechtrode touches the substrate is assumed as a circular in area.However, the area which comes in contact when the rotating mechtrode touches the substrate was observed as elliptical in shape rather than circular as shown in Figure 16.1 (a) as demonstrated by Vitanov and Voutchkov [4]. The bonding area is always less than the cross-sectional area of the consumable rod. The macroscopic observations at the interface of the coating reveal the ratio (R) of the bonding area to the mechtrode area which can be related as given in Eq. (1). The bonding area is a dependent of substrate speed.

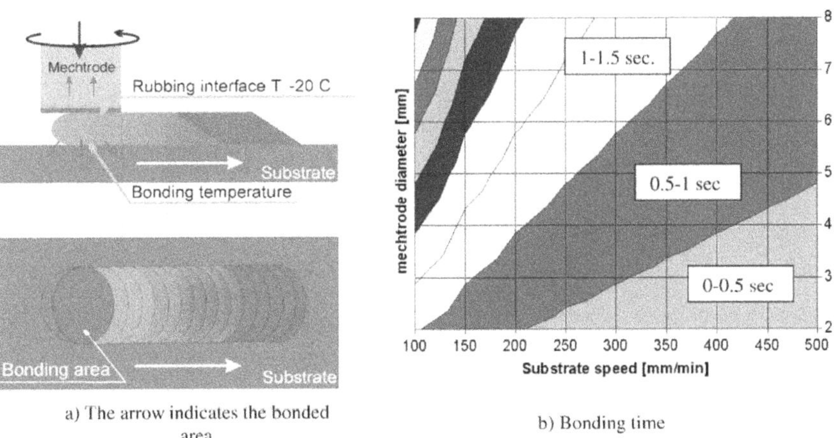

a) The arrow indicates the bonded area

b) Bonding time

FIGURE 16.1 (See color insert.) Bonding time influence on coating formation: (a) Schematic representation, and (b) relation between the mechtrode diameter and the substrate speed. (*Source:* Reprinted with permission from Vitanov & Voutchkov [4]. © 2005 Springer Nature.)

$$D_b = R.D_m \tag{1}$$

where D_b is bonding area, and D_m is the diameter of the mechtrode. Then the bonding time can be obtained from the Eq. (2).

$$T_b = (R.\ D_m)/V_x \tag{12}$$

where T_b is bonding time and V_x is the substrate speed.

Optimum bonding times are shown in Figure 16.1 (b) as reported by Vitanov and Voutchkov [4]. In general, longer bonding times produce better bonding between the coating material and the substrate as sufficient time is given to complete the material transfer mechanisms at the interface. Care must be taken to balance the heat flux generated at the interface by choosing appropriate combination of mechtrode diameter and processing speed. Figure 16.2 shows the microstructure of the interface of 316 stainless steel coating on mild steel. Figure 16 (a) shows the strong bond at the middle of the coating and Figure 16 (b) shows the undercut at the edge of the coating as usually observed in the coatings developed by friction surfacing. Figure 16.3 (c) schematically shows the bandwidth which is a fraction of the actual diameter of the consumable rod.

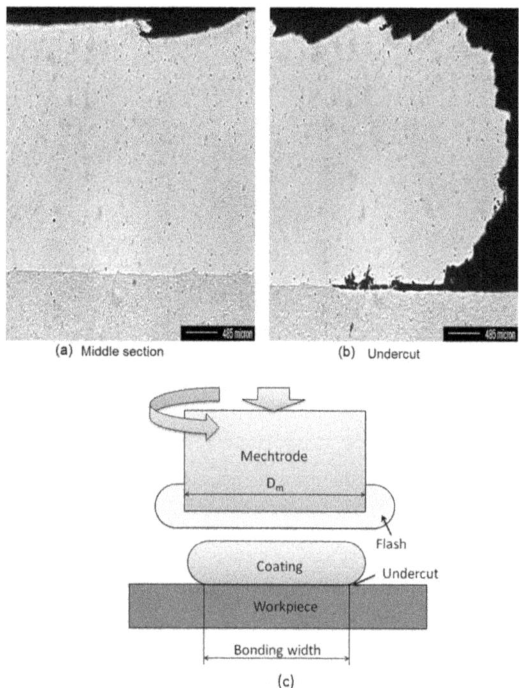

FIGURE 16.2 Cross section of 316 stainless steel coating on mild steel showing (a) middle region and (b) undercut at the edge of the coating, and (c) schematic representation of bonding width. (*Source:* Reprinted with permission from Vitanov & Voutchkov [4]. © 2005 Springer Nature.)

16.3 MECHTRODE FORCE (F)

Appropriate pressure is imposed upon the mechtrode to introduce strong bonding between the coating material and the substrate which is called as mechtrode force or axial force (F). Shinoda et al., [1] investigated the effect of axial force on the bond width and observed that the bond width was increased with thinner coating layers when the axial force was increased. Gandra et al., [3] investigated the effect of axial force on the coating formation and observed that the insufficient axial force decrease the bonding strength and excess axial force decreases the thickness of the coating. While depositing mild steel on mild steel substrate different axial force was applied, and the cross sections were analyzed. Figure 16.3 shows the microscope images obtained at the cross section after depositing coating material by applying 0.5, 2, 3, 4, 5 and 6 kN, respectively. It is evident from the observations that lower axial force results poor bonding as shown in Figure 16.3 (a), (b) and (c) and higher axial force results to develop thinner coating layers as shown in Figure 16.3 (d) (e) and (f).

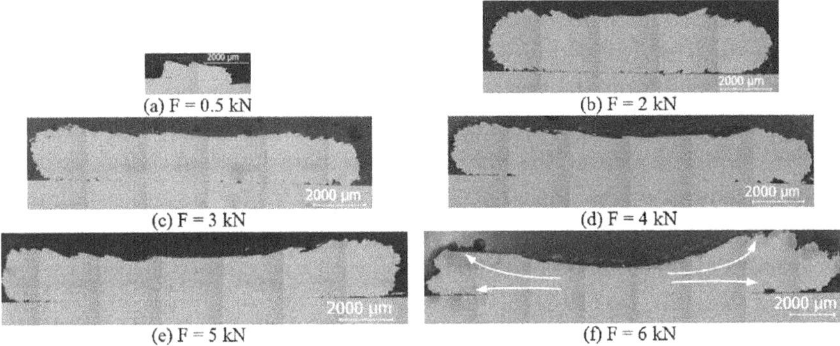

FIGURE 16.3 Effect of axial force on the bond formation at the interface: (a–c) poor bonding, and (d–f) coating thinning with good bonding obtained at tool rotational speed 2500 rpm with 4.2 mm/s feed. (*Source:* Reprinted with permission from Gandra et al., [3]. © 2012 Elsevier.)

A. M. Kalken [9] reported deeper heat-affected zones in the substrate after friction surfacing done at higher axial forces while coating stainless steel on mild steel. Similarly, Sakihama et al. [10] produced aluminum alloy coating on similar substrate and observed better mechanical properties for the surface coating layers when deposited at higher axial forces. Therefore, selecting appropriate axial force is crucial to eliminate the poor bonding issue and to reduce the material loss due to excess flash and to develop coatings with better mechanical properties.

16.4 MECHTRODE ROTATIONAL SPEED (N)

Mechtrode rotational speed (N) is another influencing parameter directly affects the quality of the bonding between coating and the substrate, width of the coating and the surface roughness of the coating. If the rotational speed is lower, the width of the coating layer is increased with a strong bonding with substrate. With higher rotational speeds the width of the bonded coating is decreased [10, 11].

Sakihama et al. [10] demonstrated decreased thickness and width of the coating when the mechtrode rotational speeds are increased while friction surfacing of 5052 aluminum alloy. Figure 16.4 shows the photographs of H13 steel deposited coatings over mild steel as reported by Rafi et al. [11]. With the same processing conditions(consumable rod of 18 mm diameter, 10 kN force and a travel speed of 4 mm/s) by increasing the tool rotational speed from 600 RPM to 2400 RPM, the width of the coating and the roughness were observed as decreased. Tokisueet al. [12] explained the reason behind the decreased surface roughness with higher rotational speeds while depositing AA5052 aluminum alloy on AA2017 plates. At higher rotational speeds, the interval of circularity pattern created by the material deposition becomes narrow, and therefore the surface roughness is decreased, and a smooth surface coating is resulted.

Gandra et al., [3] demonstrated the effect of tool rotational speed on the bonding strength. Figure 16.5 shows the cross-sectional microstructure of the coating-substrate interface obtained at different tool rotational speeds at a constant force (3 kN) and traveling speed (4.2 mm/s). As the tool rotational speed was increased to 1000 rpm to 2000 rpm, the width of the bonding region can be observed as increased. However, the thickness of the coating was observed as decreased with the increase in the tool rotational speed. Additionally, when the tool rotational speed is increased to 2500 and 3000 rpm, poor bonding between the coating layer and the substrate was observed. This is similar to what reported in their earlier work where lower tool rotational speeds were observed as beneficial [13]. Akram et al., [14] investigated the effect of process parameters on the dimensions of the coating layer by developing surface coatings in four combinations:

(i) steel coating on steel substrate, (ii) steel over stainless steel substrate, (iii) stainless steel coating on stainless steel substrate, and (iv) Inconel 718 coating on stainless steel substrate. Coatings were done at four different tool rotational speeds (1000, 1200, 1600 and 2000 rpm) and three different feeds (50, 100 and 150 mm/min) by applying three different loads (3000, 6000, 9000 and 12000 N). It was clearly demonstrated that the width of the coating was increased with higher mechtrode rotational speed and the thickness of the coating was increased with lower rotational speed.

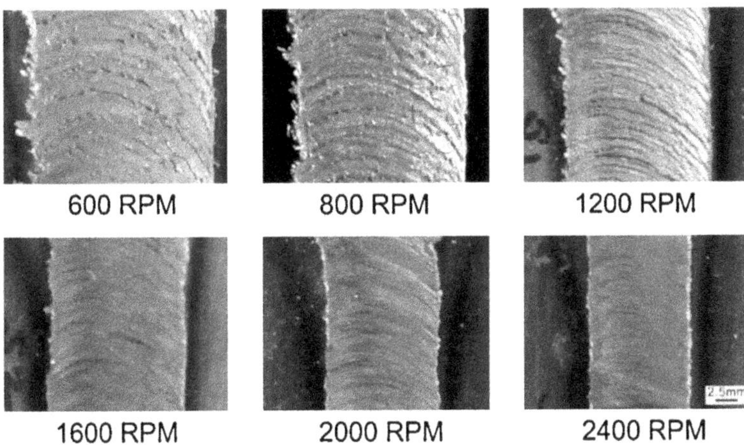

FIGURE 16.4 Photographs showing the effect of rotational speed on the width of the coating and surface roughness of the coating (*Source:* Reprinted with permission from Rafi et al., [11]. © 2010 Elsevier.)

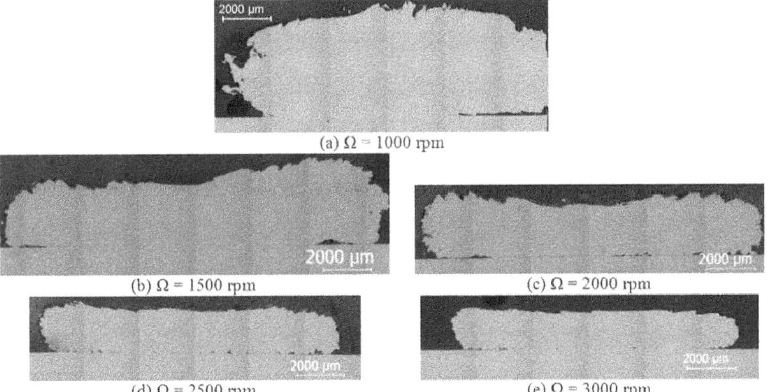

FIGURE 16.5 Photographs showing the effect of rotation speed on coating bonding at the interface (at 3 kN force and 4.2 mm/s traveling speed) (*Source:* Reprinted with permission from Gandra et al., [3]. © 2012 Elsevier.)

16.5 SUBSTRATE TRAVEL SPEED

Substrate travel speed is another influencing factor in friction surfacing which dictates the thickness and width of the coating layer. Higher substrate speeds decrease the thickness and the width of the coating [15]. It was observed that up to certain travel speed, the increment is favorable to increase the bonding strength as reported by Rafi et al., [15]. Depending on the thickness of the coating layer, the microstructure within the surface coating is varied from fine to coarse. Heat is quickly dissipated from the thinner surface coatings, and hence, the resulted grain size is very fine. Thin coatings are produced when the travel speed is higher. With lower substrate travel speeds, more amount of heat is concentrated, and grain growth is resulted. Rafi et al., [11] developed AISI H13 steel on mild steel at different process parameters, and the coatings were subjected different tests to investigate the mechanical behavior. The shear failure at the interface of the coating and the substrate was observed as quickly in the coatings produced at lower travel speeds. On the other hand, the coatings produced at the higher substrate speeds exhibited better shear resistance. Bending strength was also observed as higher for the surface coatings produced at the higher travel speed compared with the coatings produced at lower speeds.

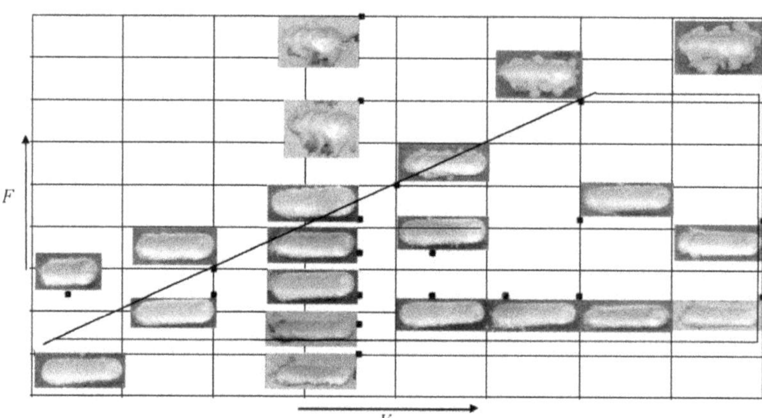

FIGURE 16.6 Combination of different axial force – substrate speed on the friction surfacing. (*Source:* Reprinted with permission from Vitnov et al., [16]. © 2000 Elsevier.)

From the works of Vitanov et al., [16], the increased travel speed was observed as affected the bonding strength between the coating and the substrate. Figure 16.6 shows the appearance of the surface coating layer developed at different combinations of axial force and substrate travel speed. The trapezium on the chart shows the promising combination of process parameters. Due to insufficient time allowed at higher travel speeds the metallurgical bonding at the interface is poor and may lead to fail the transferred coating material to strongly adhere to the substrate. Recently, Gandra et al., [3] also observed similar kind of coating formation while developing different grades of steels on a mild steel substrate.

16.6 MECHTRODE DIAMETER

The diameter of the consumable rod is another important factor must be considered as it influences the amount of heat that is produced and the width of the coating layer. Bedford et al., [17] demonstrated the influence of the mechtrode diameter on thermal aspects and further on the coating formation. Figure 16.7 shows the outcome of their research work in the form of a graph relating the substrate travel speed, temperature and the time with the diameter of the mechtrode.

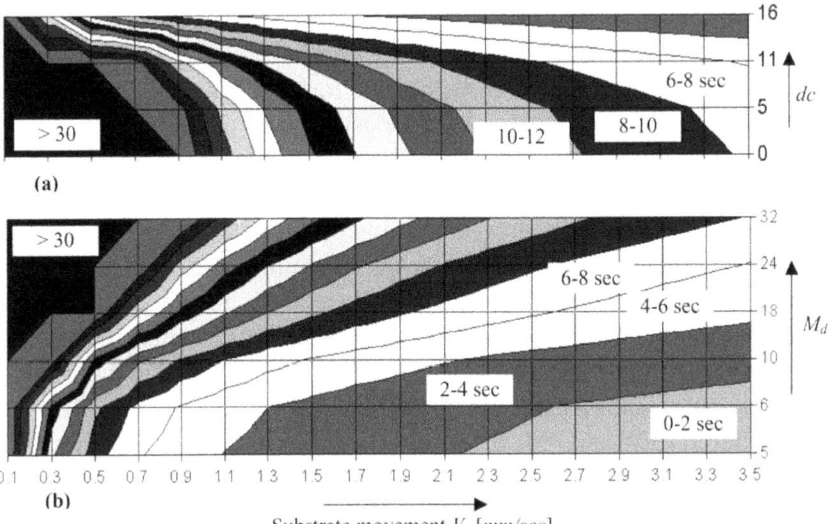

FIGURE 16.7 "Time spent at elevated temperatures in austenitizing region as a function of traverse speed (Vx): (a) radial positions for a 32-mm diameter mechtrode and (b) time spent by central region of coating for given mechtrode diameters (Md)." (*Source:* Reprinted with permission from Bedford et al., [17]. © 2001 Elsevier.)

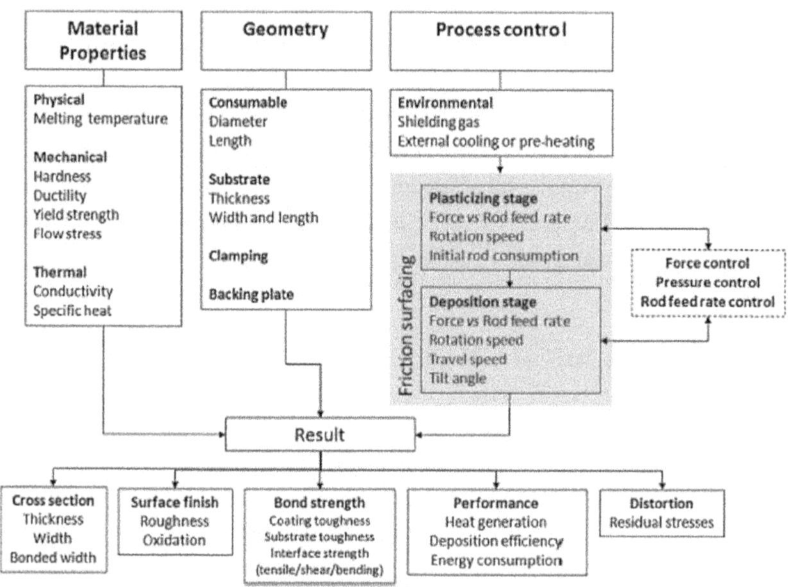

FIGURE 16.8 Influencing process parameters and variables in friction surfacing. (*Source:* Reprinted with permission from Gandra et al., [18]. © 2014 Elsevier.)

The variation in the heat loss at the cross-section of a mechtrode is higher due to the difference in the heat dissipation from the center of the mechtrode to the circumference of the mechtrode. Therefore, as shown in Figure 16.7, the temperature was observed as different when the diameter of the mechtrode is changed. Gandra et al., [18] summarized and presented the influencing process parameters in successfully developing surface coatings by friction surfacing as shown in Figure 16.8. Obtaining appropriate thickness and width of surface coating with good surface finish, coating with high bonding strength and lower residual stresses are the important objectives must be considered before designing the friction surfacing process for a given set of material systems.

Similar to FSW and FSP, the success of friction surfacing also depends on selection of appropriate process parameters. It is true that several dissimilar surface layers can be developed by friction surfacing which is very difficult with conventional liquid state methods, but the information on several combinations of non-ferrous and ferrous metals is still insufficient. Future research in this area may provide better understanding and may widen the application of friction surfacing for different industries.

KEYWORDS

- **computer numerical controlled**
- **mechtrode diameter**
- **mechtrode rotational speed**
- **substrate travel speed**

REFERENCES

1. Shinoda, T., Li, J. Q., Katoh, Y., & Yashiro, T., (1998). Effect of process parameters during friction coating on properties of non-dilution coating layers. Surface Engineering, *14*, 211–216.
2. Kramer de Macedo, M. L., Pinheiro, G. A., dos Santos, J. F., & Strohaecker, T. R., (2010). Deposit by friction surfacing and its applications. *Welding International 24*, 422–431.
3. Gandra, J., Miranda, R. M., & Vilac¸A, P., (2012). Performance analysis of friction surfacing. *Journal of Materials Processing Technology, 212,* 1676–1686.
4. Vitanov, V. I., & Voutchkov, I. I., (2005). Process parameters selection for friction surfacing applications using intelligent decision support, *Journal of Materials Processing Technology, 159*, 27–32.
5. Chandrasekaran, M., Batchelor, A. W., & Jana, S., Study of the interfacial phenomena during friction surfacing of mild steel with tool steel and Inconel. *Journal of Materials Science, 33,* 2709–2717,.
6. Chandrasekaran, M., William Batchelor, A., & Jana, S., (1998). Friction surfacing of metal coatings on steel and aluminum substrate. Journal of Materials Processing Technology, (1997). 72, 446–452,.
7. Vitanov, V. I., & Javaid, N., (2010). Investigation of the thermal field in micro-friction surfacing. *Surface and Coatings Technology, 204*, 2624–2631.
8. Vitanov, V. I., Javaid, N., & Stephenson, D. J., (2010). Application of response surface methodology for the optimization of micro-friction surfacing process. *Surface and Coatings Technology, 204*, 3501–3508.
9. Kalken, A. M., (2001). (Master Thesis) *Friction Surfacing of Stainless Steel on Mild Steel With a Robot*. Delft University of Technology.
10. Sakihama, H., Tokisue, H., & Katoh, K., (2003). Mechanical properties of friction surfaced 5052 aluminum alloys. *Materials Transactions, 44*, 2688–2694.
11. Rafi, H. K., Ram, G. D. J., Phanikumar, G., & Rao, K. P., (2010). Friction surfaced tool steel (H13) coatings on low carbon steel: a study on the effects of process parameters

on coating characteristics and integrity. *Surface and Coatings Technology,205*, 232–242.

12. Tokisue, H., Katoh, K., Asahina, T., Usiyama1 T., (2006). Mechanical Properties of 5052/2017 Dissimilar Aluminum Alloys Deposit by Friction Surfacing, *Materials Transactions,47*, 874–882.
13. Gandraa J., Pereira D., Miranda R. M., Vilaçac P., (2013). Influence of process parameters in the friction surfacing of AA 6082-T6 over AA 2024-T3, *Procedia CIRP, 7*, 341 – 346.
14. Javed Akram, Prasad Rao Kalvala, & Mano Misra, (2014). Effect of Process Parameters on Friction Surfaced Coating Dimensions, Advanced Materials Research, *922*, 280-285.
15. Rafi, H. K., Ram, G. D. J., Phanikumar, G., & Rao, K. P., (2010). Friction surfacing of austenitic stainless steel on low carbon steel: studies on the effects of traverse speed. In: *World Congress on Engineering,* London.
16. Vitanov, V. I., Voutchkov, I. I., & Bedford, G. M., (2000). Decision support system to optimize the Frictec (friction surfacing) process. *Journal of Materials Processing Technology, 107,* 236–242.
17. Bedford, G. M., Vitanov, V. I., & Voutchkov, I. I., (2001). On the thermo-mechanical events during friction surfacing of high-speed steels. *Surface and Coatings Technology,141*, 34–39.
18. Gandra J., Krohn H., Miranda R. M., Vilac P., Quintinoa L., & dos Santos, J. F., (2014). Friction surfacing—A review, *Journal of Materials Processing Technology 214,* 1062– 1093.

CHAPTER 17

Material Systems Processed by Friction Surfacing

Several materials such as steels, aluminum alloys, magnesium alloys, titanium alloys were used as substrates to deposit different coating materials by friction surfacing. This chapter summarizes and presents the work done using different material systems in developing surface coatings by friction surfacing.

17.1 STEELS

Steels are the most widely used substrates and coating materials to develop surface coatings by friction surfacing. During coating, phase transformation from BCC to FCC is happened as the temperature is reached beyond the austenitizing temperature. Similar to the heat treatment of steels, the mechanical behavior of the coating depends on the phase transformation after quenching, level of grain refinement, presence of alloying elements, etc. Bedford et al., [1] produced high-speed steels coating on mild steel substrate, and the temperature was observed as reached to austenite zone and the carbide was dissolved. The bond between coating and substrate was demonstrated as a result of diffusion bonding. Due to the cooling rate around 400°C/s, martensite transformation was resulted. The coating contained fine and homogeneous grains in hardened state. Rafi et al., [2] also explained the mechanism of formation of fine carbides while developing AISI H13 hot work tool steel on mild steel substrates. Figure 17.1 shows the photograph of the coating and the perfect interface at different regions of the coating. In their work, it was a clear observation that the carbide particles appeared in the H13 tool steel before the coating was disappeared after developing the coating. This strongly suggests the dissolution of carbide into the matrix. After coating, re-precipitation was arrested due to the fast cooling, and fine equiaxed grains with homogeneous distribution of martensite was noticed within the coating. The grain size of the consumable rod was refined during the process from 50–60 μm to 2–10 μm. From the hardness measurements, more hardness was measured on the coating which is 190% higher than the starting value. Chandrasekaran et al., [3] produced different surface coatings on steel and aluminum substrate to investigate the feasibility of friction surfacing to develop solid-state coatings. On steel substrate, different coatings such as tool steel, Inconel, aluminum, titanium coatings were developed. Additionally, an aluminum substrate, stainless steel, Inconel and mild steel coatings were developed by friction surfacing. From their work, significant information was obtained. Success of coating formation was observed as dependant of material properties. Mild steel has shown better coating formation compared with stainless steel on aluminum substrate. Materials with lower hardness were developed surface coatings on both mild steel and, aluminum substrate at lower pressures. Similarly, Chandrasekaran et al., [4] also developed surface coatings of AISI01 tool steel and Inconel 600 on mild steel 1020 substrate. A sharp boundary was observed at the interface of tool steal coating and the mild steel substrate. Formation of interfacial compound was observed at the interface of the Inconel coating and the mild steel substrate. Usually, if the joining metals are dissimilar, mechanical interlocking is one mechanism by which bonding is developed in material processing. However, the authors did not see significant mechanical interlocking, but the bonding at the interface was observed as a solid-state bonding.

Rafi et al., [5] demonstrated the effect of process parameters on the coating formation while developing surface coatings of H13 tool steel on mild steel substrate. Seven mechtrode rotational speeds (350, 650, 800, 1200, 1600, 2000 and 2400 rpm), nine substrate travel speeds starting from 1.2 to 4.4 with 0.4 mm/s increment with a constant axial force (10 kN) were adopted. Figure 17.2 (a) shows a typical macroscopic view at the interface of the coating and substrate usually observed in friction surface coating which is a characteristic with unbounded regions at the ends and well-bonded coating in the middle. The bond strength was observed as influenced by the process parameters as shown in Figure 17. 2 (b) and (c). Coating width was highly influenced by the mechtrode speed, and the coating thickness was influenced by the substrate travel speed. Among the all combinations, 800–1200 rpm with 4 mm/s at 10 kN axial force was observed as optimum combination. Other works of Rafi et al., [6, 7] also demonstrate the advantage of friction surfacing in developing coatings of stainless steel, mild steel, austenitic stainless steel on low carbon steel.

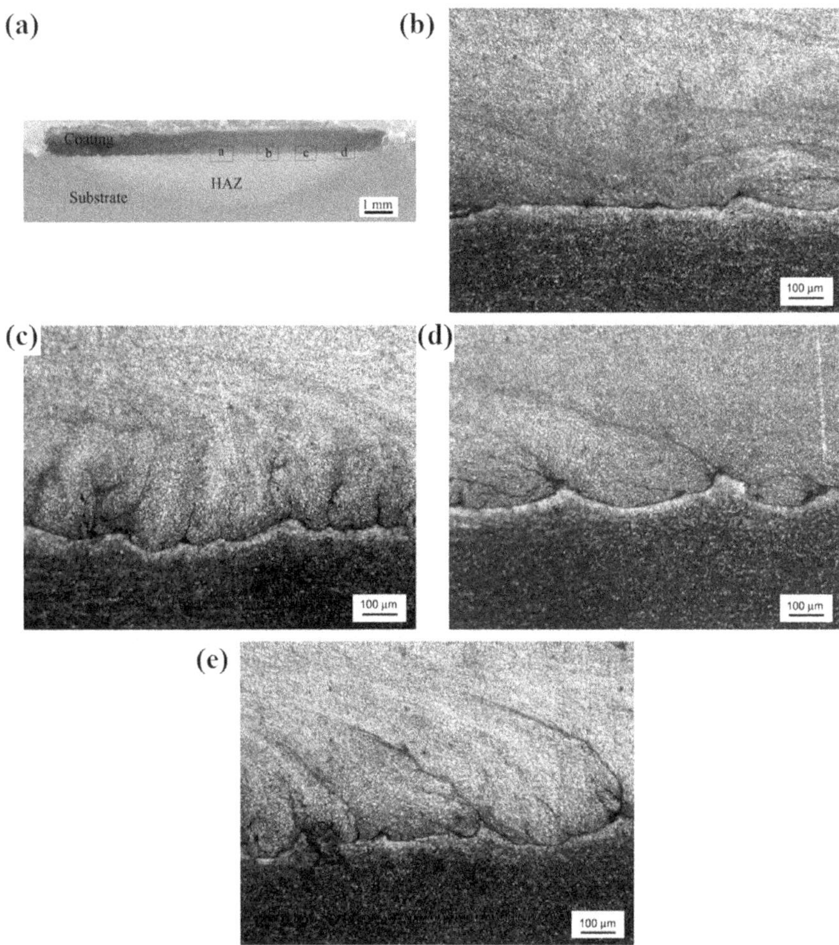

FIGURE 17.1 (a) Macrostructure of the coating on the substrate (transverse section) and (b–e) microstructure at the interface obtained at different points as shown in (a). (*Source:* Reprinted with permission from Rafi et al., [2]. © 2011 Elsevier.)

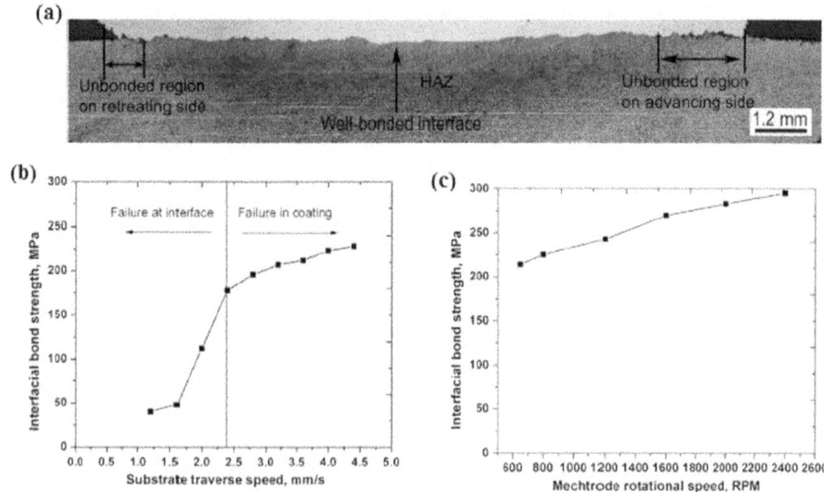

FIGURE 17.2 (a) Macroscope view of the coating-substrate interface, (b) interfacial bond strength with respect to increased substrate travel speed (mm/s) and (c) interfacial bond strength with respect to increased mechtrode rotational speed (RPM). (*Source:* Reprinted with permission from Rafi et al., [5]. © 2010 Elsevier.)

Sekharbabu et al., [7] developed D2 steel on low carbon steel and grain refined coating was observed due to dynamic recrystallization which increased the surface hardness and wear resistance by promoting martensitic transformation and refining the primary carbides. Similarly, Govardhan et al., [7] also selected stainless steel to coat on low carbon steel to investigate the efficacy of friction surfacing in developing dissimilar surface coatings. Different combinations of process parameters were adopted such as two rotational speeds (1500 and 2400 rpm), two values of applied pressure (29, 47 MPa) and two different tool travel speeds (78, 190 mm/min). Figure 17.3 (a) shows the photographs of the coating layer obtained at different process parameters. From their study, they observed the requirement of higher loads during the initial stage of the process due to the dry friction. Further by optimizing the process parameters, a strong bond with no defects and excellent coating integrity was achieved. Mechanical properties of the coating surface were observed with excellent toughness. From the bending studies, the coatings were exhibited excellent adhesion as observed from the side bending samples as shown in Figure 17. 3 (b), (c), and (d). Enhanced corrosion properties were also observed for the coatings. The authors proposed to adopt friction surfacing to develop surfaces with improved mechanical and corrosion properties for petrochemical vessels, pumps for chemical industry and equipment for operating aggressive chemical solutions.

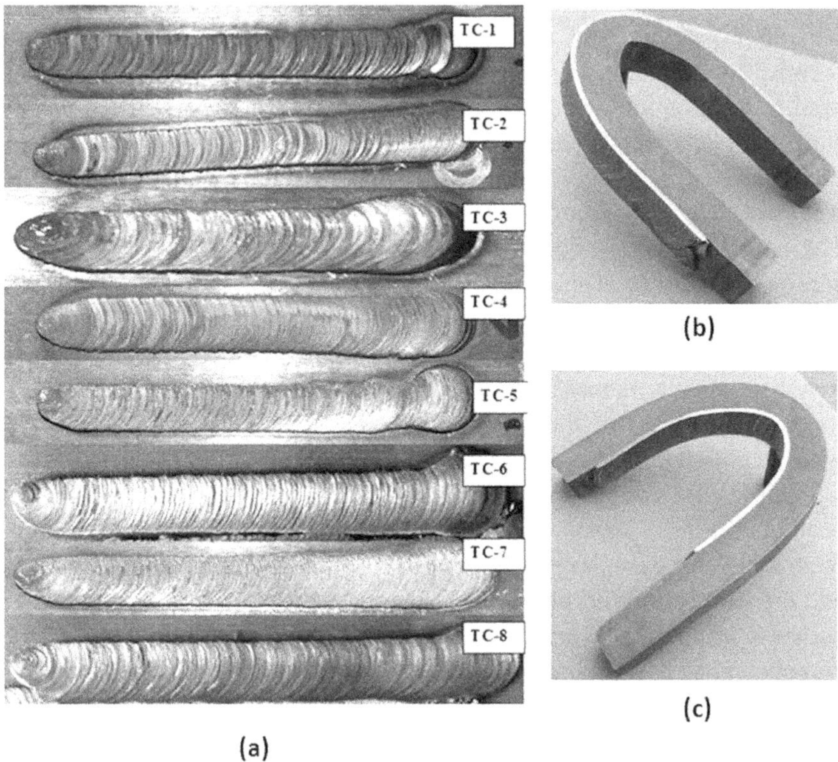

FIGURE 17.3 (a) Photographs of friction surface coating layers obtained at different process parameters, b), (c) and (d) photographs showing bending of outer, inner and side coated layers. (*Source:* Reprinted with permission from Govardhan et al., [7]. © 2012 Elsevier.)

Nixon et al., [10] produced AISI 304 stainless steel over graphite iron substrate by friction surfacing. Due to the combined effect of forging pressure and shear action imposed during friction surfacing resulted excellent interfacial bonding between the dissimilar steels. Several other alloys such as AISI 4140, AISI 8620, AISI 310, AISI 316L, AISI 321, AISI 410, AISI 416, AISI 431, AISI 440 and high speed steels (BM2, BT15 and ASP 30) were used as coating materials to develop solid-state surface coating on mild steel [11–17].

17.2 ALUMINUM ALLOYS

Aluminum alloys are the most widely used nonferrous metals in the automobile, aerospace and marine applications. Friction surfacing has been used to develop various surface coatings of aluminum alloys. Sakihama et al., [18] developed coatings of 5052 aluminum alloys on a substrate of same material by friction surface.The microstructure of coating material was relatively finer compared with that of mechtrode and the substrate. The hardness of the coating material was observed as close to that of annealed condition of the alloy. The effect of process parameters on the formation of coating revealed the influence of mechtrode rotational and travel speed on the width and the thickness of the coating. Tensile strength was observed as increased with increased axial pressure and mechtrode rotational speed. But the tensile strength was significantly decreased with increased substrate travel speed. Tokisue et al., [19] selected 2017 aluminum alloys as coating material to develop monolayer and multilayer coatings on 5052 substrates. In the monolayer surfaced coating, incomplete welded regions were noticed at the edge of the deposited material and the substrate. With multi-layer coating, better bonding was observed between the coating layer and the substrate. Material deposition efficiency was also reported as higher with the multilayer compared with monolayer. Mechanical behavior and the softening region in the coating was observed as similar to what reported by Sahikama et al., [18]. Suhuddin et al. [20] developed surface coating of an AA6082-T6 aluminum alloy over AA2024-T3 substrate. From the Electron backscatter diffraction studies, the authors demonstrated the evidence of shearing deformation behind the occurrence of dynamic recrystallization. Vilaca et al., [21] also developed AA6082-T6 aluminum coatings on AA2024-T3 alloy and microstructural studies were carried out to investigate the transformation of the consumable material during coating. Heat affected zone and thermomechanical affected zones were clearly differentiated. Uday Bhat et al., [22] produced pure aluminum surface coatings of 40–50 μm thickness on mild steel substrates by friction surfacing and follow up heat treatment was done at 700°C for 2 h. Formation of intermetallic phases such as Fe_2A_{15} and

Fe_4Al_{13} was observed. Further, FeAl and Fe_3Al phases were also observed in minor fraction. The formation of these phases was reported due to the reactive diffusion mechanism. Similarly, Vijay Kumar et al., [23] also developed 6063 aluminum alloy coatings on mild steel and demonstrated that the feasibility is limited to very narrow set of process parameters with the formation of intermetallic phases.

Producing surface coatings of ammonium-based composites is another development in which the coating layer is composed of composite deposited by using a consumable composite mechtrode. Gandra et al., [24] developed functionally graded composite layers by drilling holes in AA6082-T6 aluminum rods and used as mechtrode rods to coat on AA2024-T3 aluminum alloy substrate. Except at the edges of the coating layers, a strong bonding was noticed at the substrate and coating interface as well as between the each coating layers. Improved wear properties for the composite coatings were obtained compared with base aluminum alloy. Similarly, Madhusudhan Reddy et al., [25] also developed aluminum metal matrix coating on Ti6Al4V alloy substrate. From the interdiffusion of the alloying elements at the interface of the MMC coating and the substrate, formation of metallurgical bond was reported. Better wear characteristics were noticed for the MMCs coating due to the decreased coefficient of friction. The interface was also noticed with the presence of Ti_3Al particles. Nakama et al., [26] used 6061 aluminum pipes filled with alumina particles to develop composite coatings on 5052 aluminum substrate. Consumable rods with different inner diameters such as 5, 7, 9 and 10 mm were used, and better wear resistance was demonstrated.

17.3 TITANIUM ALLOYS

Initially, Beyer et al., [27] demonstrated producing surface coatings of titanium alloy and presence of martensite acicular α phase was observed in the coating. Increased hardness up to 400 Hv was measured in the surface coating. Brittle fracture was noticed due to the presence of porosity at the interface. Prasad Rao et al., [28] have done a detailed study to investigate the feasibility of coating formation with different combinations of material systems. Pure titanium coating was produced on Ti6Al4V substrate. However, the coatings were found to be discontinuous. Later on, Fitseva et al., [29] demonstrated developing Ti6Al4V coatings in the presence of

Argon gas on Ti6Al4V substrate. Initially, the coatings were observed as discontinuous. The deposition of the coating material was observed as proper when the axial force was 6 kN. However, when the axial force was decreased to 4 kN, the coating layer was observed as discontinuous. From a minimum rotational speed from 300 rpm to 6000 rpm with a constant substrate travel speed 16mm/min and material consumption rate of 1.6 mm/s. From their studies, it was clearly demonstrated that maintaining constant mechtrode consumption rate is crucial to develop titanium alloy coatings by friction surfacing. Increase in the mechtrode rotational speed increased the width of the coating. Figure 17.4 shows the photographs of the surface coatings obtained at different rotational speeds with a constant mechtrode transfer rate and travel speed. The flash formation was also observed as dependent of the combination of the process parameters. Furthermore, Fitseva et al., [30] studied the temperature distribution during developing surface coating and a temperature distribution of 1021°C to 1299°C was measured at different processing parameters which are above the temperature of β phase formation and therefore presence of β phase was reported from the microstructural observations. Better mechanical properties were observed for the coating compared with the as-received wrought alloy due to the development of new microstructure.

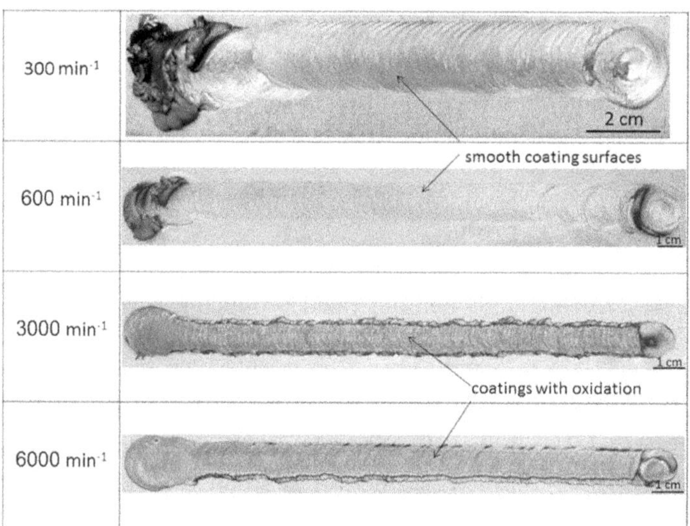

FIGURE 17.4 Photographs showing appearances of Ti6Al4V coatings at different speeds with a travel speed of 16mm/s and a fixed material consumption rate of 1.6mm/s. (*Source:* Reprinted with permission from Fitseva et al., [29]. 2015 Elsevier.)

17.4 NIAl-BRONZE

Hanke et al., [31] produced NiAl-Bronze coating layer on the same substrate using friction surfacing. The microstructure studies revealed the presence of homogeneous and globular α phase surrounded with β phase. Resistance against the cavitation erosion was found to be higher for the coatings compared with the substrate. Cracks were initiated from the boundaries of the secondary phase in both the base and the coating material. The coating material exhibited ductile behavior and therefore, better wear behavior was observed compared with the base material. On the other hand, the base material has undergone severe attack at the grain boundaries which lead to increase the wear in the base material compared with the surface coating.

17.5 MAGNESIUM ALLOYS

The information on developing magnesium based surface coatings by friction surfacing is insufficient in the literature. Nakama et al., [32] used AZ91 Mg alloy as mechtrode and AZ31 Mg alloy as substrate to develop surface coating. The coating layer was appeared as circular at all mechtrode rotational speeds. These circular patterns appeared after surfacing was similar to that of aluminum coatings. The microstructure was appeared as very fine in the coating compared with the base microstructure. At the interface of the coating and the substrate, sharp interface was noticed without metallurgical bonding. Highest temperature was noticed 75 mm after the coating was started. From the hardness and wear studies, the coating exhibited higher hardness and better wear characteristics.

Even though the process was proven as promising in developing surface coatings within the solid-state itself, the material systems which were processed by friction surface were limited to a few groups of alloys. The necessity of research about other material systems is in a great need. With possible future developments, friction surfacing may become more potential route for addressing several issues in the surface engineering.

KEYWORDS

- **aluminum alloys**
- **magnesium alloys**
- **NiAl-bronze**
- **steels**
- **titanium alloys**

REFERENCES

1. Bedford, G. M., Vitanov, V. I., &Voutchkov, I. I., (2001). On the thermo-mechanical events during friction surfacing of high-speed steels. *Surface and Coatings Technology, 141,* 34–39.
2. Rafi, H. K., Ram, G. D. J., Phanikumar, G., & Rao, K. P., (2011). Microstructural evolution during friction surfacing of tool steel H13. *Materials and Design, 32,* 82–87.
3. Margam Chandrasekaran, Andrew William Batchelor, & Sukumar Jana, (1997). Friction surfacing of metal coatings on steel and aluminum substrate, *Journal of Materials Processing Technology, 72,* 446–452.
4. Chandrasekaran M., Batchelor A. W., & Jana S., (1998). Study of the interfacial phenomena during friction surfacing of mild steel with tool steel and Inconel, *Journal of Materials Science, 33,* 2709–2717.
5. Khalid Rafi, H., Janaki Ram, G. D., Phanikumar, G., & Prasad Rao, K., (2010). Friction surfaced tool steel (H13) coatings on low carbon steel: A study on the effects of process parameters on coating characteristics and integrity, *Surface & Coatings Technology 205,* 232–242.
6. Khalid Rafi, H., Janaki Ram, G. D., Phanikumar, G. & Prasad Rao, K. (2010). Microstructure and properties of friction surfaced stainless steel and tool steel coatings, *Materials Science Forum, 638–642,* 864–869.
7. Khalid Rafi, H., Kishore Babu, N., Phanikumar, G., Prasad Rao, K., Microstructural Evolution During Friction Surfacing of Austenitic Stainless Steel AISI 304 on Low Carbon Steel, Metallurgical and Materials Transactions, A., (2013). 345–350.
8. Sekharbabu, R., Khalid Rafi, H., & Prasad Rao, K., (2013). Characterization of D2 tool steel friction surfaced coatings over low carbon steel, *Materials and Design 50,* 543–550.
9. Govardhan, D., Kumar, A. C. S., Murti, K. G. K., & Madhusudhan Reddy, G., (2012). Characterization of austenitic stainless steel friction surfaced deposit over low carbon steel, *Materials and Design 36,* 206–214.
10. George Sahaya Nixon, R. & Mohanty, B. S., (2013). Friction surfacing of metal coatings on stainless steel Aisi 304 over spheroidal graphite iron substrate, *Advanced Materials Research,816–817,* 271–275.

11. Kramer de Macedo, M. L., Pinheiro, G. A., dos Santos, J. F., & Strohaecker, T. R., (2010). Deposit by friction surfacing and its applications. *Welding International 24*, 422–431. http://dx.doi.org/10.1080/09507110902844535.
12. Lambrineas, P., Jenkins, B. M., & Doyle, E. D., (1990). Low-pressure friction surfacing:adhesion of stainless steel coatings on mild steel. In: *International Tribology Conference*, Brisbane, 12–15.
13. Lambrineas, P., & Jewsbury, P., (1992). Areal coverage using friction surfacing. *Journal of Ship Production 8*, 131–136.
14. Puli, R., Kumar, E. N., & Ram, G. D. J., (2011). Characterization of friction surfaced martensitic stainless steel (AISI 410) coatings. *Transactions of the Indian Institute of Metals 64*, 41–45.
15. Katayama, Y., Takahashi, M., Shinoda, T., & Nanbu, K., (2009). New friction surfacing application for stainless steel pipe. *Welding in the World 53*, 272–280.
16. Puli, R., & Janaki Ram, G. D., (2012a). Corrosion performance of AISI316L friction surfaced coatings. *Corrosion Science 62*, 95–103, http://dx.doi.org/10.1016/j.corsci.2012.04.050.
17. Puli, R., & Janaki Ram, G. D., (2012b). Microstructures and properties of friction surface coatings in AISI 440C martensitic stainless steel. *Surface and Coatings Technology 207*, 310–318, http://dx.doi.org/10.1016/j.surfcoat.2012.07.001.
18. Hidekazu Sakihama, Hiroshi Tokisue, & Kazuyoshi Katoh, (2003). Mechanical Properties of Friction Surfaced 5052 Aluminum Alloy, *Materials Transactions*, *44*(12) 2688–2694.
19. Tokisue, H., Katoh, K., Asahina, T., & Uchiyama, T., (2006). Mechanical properties of5052/2017 dissimilar aluminum alloys deposit by friction surfacing. *Materials Transactions 47*, 874–882.
20. Suhuddin, U., Mironov, S., Krohn, H., Beyer, M., & Dos Santos, J. F., (2012). Microstructural evolution during friction surfacing of dissimilar aluminum alloys. *Metallurgical and Materials Transactions A: Physical Metallurgy and Materials Science, 43A*(13), 5224–5231. http://dx.doi.org/10.1007/s11661–012–1345–8.
21. Vilaç A, P., Gandra, J., & Vidal, C., (2012). Linear friction based processing technologies aluminum alloys: surfacing, stir welding and stir channeling. In: Ahmad, Z. (Ed.), *Aluminum Alloys—New Trends in Fabrication and Applications*. InTech, Rijeka, Croatia.
22. Udaya Bhat K., Nithin, Suma Bhat, & Sudeendran, (2015). Heat Treatment of Friction Surfaced Steel-Aluminum Couple, *Materials Science Forum, 830–831*, 135–138.
23. Vijaya Kumar, B., Madhusudhan Reddy, G., & Mohandas, T., (2014). Identification of suitable process parameters for friction surfacing of mild steel with AA6063 aluminum alloy, *Int J Adv Manuf Technol*, DOI 10.1007/s00170-014-5964-7.
24. Gandra, J., Vigarinho, P., Pereira, D., Miranda, R. M., Velhinho, A., & Vilaça, P., (2013). Wear characterization of functionally graded Al–SiC composite coatings produced by Friction Surfacing, *Materials and Design 52*, 373–383.
25. Madhusudhan Reddy, G., Satya Prasad, K., Rao, K. S., & Mohandas, T., (2009). Friction surfacing of titanium alloy with aluminum metal matrix composite, *Surface Engineering*, *25*(1), 25–30.

26. Dai Nakama, Kazuyoshi Katoh, & Hiroshi Tokisue, (2009). Effect of filling condition of alumina particle on dispersibility of particle by friction surfacing, *Journal of Japan Institute of Light Metals*, *59*(3), 114–120.
27. Beyer, M., Resende, A., & Santos, J. F.dos, (2003). *Friction surfacing for multi-sectorial applications – FRICSURF,* Institute for Materials Research, GKSS Forschungszentrum Geesthacht GmbH, Technical report.
28. Prasad Rao, K., Sankar, A., Rafi, H. K., Janaki Ram, G. D., & Reddy, G. M., (2012). Friction surfacing on nonferrous substrates: a feasibility study, *Int. J. Adv. Manuf. Technol. 65,* 755–762.
29. Fitseva, V., Krohn, H., Hanke, S., & dos Santos, J. F., (2015). Friction surfacing of Ti–6Al–4V: Process characteristics and deposition behaviour at various rotational speeds, *Surface & Coatings Technology 278,* 56–63.
30. Fitseva, V., Hanke, S., dos Santos, J. F., Stemmer, P., & Greising, B., (2016). The role of process temperature and rotational speed in the microstructure evolution of Ti-6Al-4V friction surfacing coatings, *Materials and Design 110,* 112–123.
31. Stefanie Hanke, Alfons Fischer, Matthias Beyer, & Jorge dos Santos, (2011). Cavitation erosion of NiAl-bronze layers generated by friction surfacing. 273(1), 32–37.
32. Nakama, D., Katoh, K., & Tokisue, H., (2008). Some characteristics of AZ31/AZ91 dis-similar magnesium alloy deposit by friction surfacing. *Materials Transactions, 49,* 1137–1141.

CHAPTER 18

Applications, Challenges, and Future Scope

18.1 APPLICATIONS OF FRICTION SURFACING

Friction surfacing facilitates to develop structures with tailored chemical composition which is not possible with conventional manufacturing methods. Material can be added at selected regions which exhibit completely different mechanical, wear, corrosion and degradation properties. Mechanical assembling and fastening of different components can be eliminated. Repairing the worn out surfaces, shafts and agriculture parts by adding additional layers is one of important application reported in the literature [1–4]. In the area of machine tools and cutting tools applications, friction surfacing was used to develop hard surfaces in cutting tool and punches [5]. By friction surfacing, stellite 6 and high-speed tool steel materials were deposited on mild steel and stainless steel substrate. They demonstrated the application of friction surfacing in producing cutting edges for blades and knives for components used in the consumer goods processing in the packing industry. Other workshop tools such as screwdrivers and chisels also can be produced by developing them from the multilayered materials and follow up processes. Treating turbine blades for extending the life span is another important application in the surface engineering. Friction surfacing can also be used to treat wear-resisting turbine blade tips. Bedford et al., [6] produced Stellite 12 single and multi-layered coatings on narrow substrates of stainless steel AISI316 to investigate the suitability of friction surfacing for turbine blade applications. Additionally, repairing the surface locally through friction surfacing was also demonstrated.

In the year 1984, Dunkerton and Thomas [7] reported the promising application of friction surfacing to deposit material on pipe flange contact faces and brake discs. Further, in the year 1993, Amos [8] proposed this technique to develop hard facing steam turbine blade trailing edges. Later on, Foster et al., [9] presented a method of producing material deposition at the circumference of the round drums and discs and suggested to produce different turbine blades from the deposits. The potential of friction surfacing to engineer the surfaces of the components and structures used in the petrochemical industries such as pressure vessels, pipes and pumps was also suggested by Govardhan et al., [10] while demonstrating the improved corrosion performance of austenitic stainless steel coatings over mild steel substrate. Surface recovery in extreme environmental conditions such as in the presence of moisture, solvents and underwater circumstances, offshore pipes and structures can also be achieved as demonstrated by Li and Shinoda [11]. Yamashita and Fujita [12] reported using friction surfacing for in-situ repair of damaged structures and components due to stress corrosion cracking in nuclear power stations. Friction Surfacing resulted lower level of heat input at the processing zone and decreased the residual tensile stresses which are usually observed in fusion welding techniques. The surface cracks can also be closed by developing a layer of coating through friction surfacing.

Treating the worn railway trails by friction surfacing is another area as investigated and reported by Doughty et al., [13] in demonstrating a specially designed portable system for repairs in the railway industry. The set up is directly fixed on the rail where repair is required. The special design facilitates to control the consumable rod rotational speed and travel speed along with by applying suitable axial load. Angular adjustments are also possible so that the consumable rod can be operated to deposit material in different regions on the rail. Layer-by-layer coatings can be developed by friction surfacing which opened new area of applications in additive manufacturing engineering. The process can also be easily controlled by robotics and automation as demonstrated by Beyer et al. [14]. 3D deposition of material on linear horizontal paths was demonstrated by using a robot in repairing a cylindrical part successfully.

The use of friction surfacing is not limited to the above-mentioned areas as it offers several advantages compared with several liquid based surface coating or treatment methods. As the substrate dilution, development of residual stresses and distortion of structures are minimum or in some cases completely eliminated, friction surfacing can be an alternate technique in the manufacturing industry. Instead of replacing the entire structure or component due to the degraded surface or presence of crack, localized healing by applying friction surfacing at selected regions can decrease the cost in manufacturing.

18.2 CHALLENGES INVOLVED IN ADOPTING FRICTION SURFACING PROCESS

The important challenges involved in friction surfacing process to become as a potential tool in surface engineering are studied by several research groups and reported. Based on the overall observations, the challenges and limitations which restrict the method from wide usage in the manufacturing industry are summarized and given below.

- Process parameters play crucial role in friction surfacing. The important process parameters such as consumable rod rotational speed, substrate travel speed, axial load, material type and preheating of the substrate are the crucial process parameters required to be optimized which involves rigorous study through different experiments.
- The optimization of process parameters must be done for every type of material system.
- The material transfer mechanisms and formation of flash associated with complex mechanisms involving thermo-mechanical events.
- The region which is significant in friction surfacing required to study is very thin at the interface of the mechtrode and the substrate
- The surface geometry is another factor based on which the design of the equipment is altered. For example, highly automated equipment is required to treat the surface of curves or inclined compared with flat surfaces.
- Formation of secondary phases due to the chemical reaction between the coating material and the substrate as usually observed in several dissimilar material combinations is one common issue needs close study to develop coatings with desired phases.
- The thickness of the coating layer is usually higher compared with thin film coating techniques and hence, friction surfacing is not suitable for developing coatings with lower thickness.
- Poor bonding or in some case no bonding at the edges of the coating material is an important limitation in friction surfacing. Designing the process to achieve a strong bond between the coating and the substrate is required to decrease the poor bonding.
- Formation of excess flash by which consumption of more material is unproductive which is another factor must be considered in friction surfacing.

- Friction surfacing is done in open atmosphere which may lead to develop oxide layers after completion of the coating deposition. When layer-by-layer is deposited by friction surfacing to develop 3D structures, presence of oxides at the interfaces of the layers is an important limitation.

18.3 SCOPE FOR THE FUTURE DEVELOPMENTS

The process is at its infant stage compared with other surface coatings techniques. However, it is appreciated that many authors have significantly contributed in developing friction surfacing as a promising alternative in solid-state methods to produce surface coatings. There are several areas of this novel technique yet to be investigated to sufficiently understand the process to make it suitable for wide variety of surface engineering applications as briefly given below.

- Material flow mechanisms and patterns during friction surfacing were sufficiently reported in the literature while processing several material combinations. However, further investigations are essentially required to understand the material flow mechanisms and patterns particularly in some of difficult to process materials such as magnesium alloys, titanium and other alloys.
- Understanding the material flow mechanisms by using modeling and simulation studies certainly helps the engineers along with the experiments
- Additional studies on improving the bonding strength and decreasing the edge effect with poor bonding are crucial issues yet to be investigated in detail.
- Developing the processing equipment best suitable for processing all kinds of surfaces including irregular counter surfaces or regions of curved geometry makes the process is liable to be readily adopted by the manufacturing industry.
- As friction surfacing allows deposition of different dissimilar materials layer-by-layer, multilayers surfaces can be developed which needs rigorous study as it opens new areas in connection with additive manufacturing field.
- Developing surface composites by friction surfacing is another promising area in friction surfacing. However, the information

available in this area is limited. By developing surface composites using friction surfacing, functionally gradient 3D structures with different chemical composition can be produced.

As the area of the surface layer produced in one cycle after friction surfacing is smaller, developing specially designed equipments to process more area of the surface is essentially required to make the process more attractive to adopt for bulk applications.

KEYWORDS

- **applications of friction surfacing**
- **scope for the future developments**

REFERENCES

1. Logik, N. V., (1970). Friction hardfacing steel with Stellite V3D. *Svarochnoe Proizvod-Stvo 8*, 16–17.
2. Thomas, W. M., (1988). Solid phase cladding by friction surfacing. In: *TWI International Symposium,* Cambridge, p. 18.
3. Tyayar, K. H. A., (1959). Friction welding in the reconditioning of worn components. *Svarochnoe Proizvodstvo 1*, 3–24.
4. Zakson, R. I., & Turukin, F. G., (1965). Friction welding and hardfacing of agriculturalmachine parts. *Avesta Svarka 3,* 48–50.
5. Bedford, G. M., & Richards, P. J., (1990). *Method of Forming Hard Facings on Materials,* US Patent No. 4,930,675 A.
6. Bedford, G. M., Sharp, R. J., & Davis, A. J., (1995). Micro-friction surfacing in the manufacture and repair of gas turbine blades. In: *3rd International Charles Parsons Turbine Conference: Materials Engineering in Turbines and Compressors, Newcastle Upon Tyne*, pp. 689–698.
7. Dunkerton, S. B., & Thomas, W. M., (1984). *Repair by Friction Welding*. Repair and Reclamation, London.
8. Amos, D. R., (1993). *Method of Forming a Trailing Edge on a Steam Turbine Blade and the Blade Made Thereby,* US Patent No. 5,183,390 A.
9. Foster, D. J., Gillbanks, P. J., & Moloney, K. C., (1996). *Integrally Bladed Disks or Drums,* US Patent No. 5,556,257 A.
10. Govardhan, D., Kumar, A. C. S., Murti, K. G. K., & Madhusudhan Reddy, G., (2012). Characterization of austenitic stainless steel friction surfaced deposit over low carbon steel. *Materials and Design 36*, 206–214, http://dx.doi.org/10.1016/j.matdes.2011.07.040.

11. Li, J. Q., & Shinoda, T., (2000). Underwater friction surfacing. *Surface Engineering 16,* 31–35, http://dx.doi.org/10.1179/026708400322911483.
12. Yamashita, H., & Fujita, K., (2001). Newly developed repairs on welded area of LWRstainless steel by friction surfacing. *Journal of Nuclear Science and Technology, 38*, 896–900.
13. Doughty, R. W., Shaw, D. J., & Gibson, D. E., (2009). *Friction Stir Surfacing Process and Device for Treating Rails,* Patent No. WO 2009030960 A1.
14. Beyer, M., Resende, A., & Santos, J. F. D., (2003). *Friction Surfacing for Multi-Sectorial Applications – FRICSURF,* Institute for Materials Research, GKSS Forschungszentrum Geesthacht GmbH, Technical report.

PART IV

New Developments in Friction-Assisted Processes

CHAPTER 19

New Process Variants in Friction-Assisted Processes

Development of friction welding and friction stir welding opened a new group of solid-state processing techniques in the past two decades. The present chapter briefly presents two important variants developed in friction-assisted processes.

19.1 ADDITIVE MANUFACTURING BY FRICTION SURFACING

Adding material, layer-by-layer with the help of computer-aided design (CAD) to develop three dimensional (3D) components is the basic observation in all additive manufacturing processes. Figure 19.1 shows schematic representation of different steps involved in additive manufacturing processes. Initially, the component is modeled using a CAD, and then an STL file is generated. The file is then transferred to the equipment, and the part is produced. Post-processing operations are carried out to use the component in the targeted application. Other similar terms based on the technology used for this kind of manufacturing in which 3D structures are produced by layer-by-layer deposition are given below.

- Automated fabrication
- Solid freeform fabrication
- Layer based manufacturing
- Stereolithography
- 3D printing
- Rapid prototyping

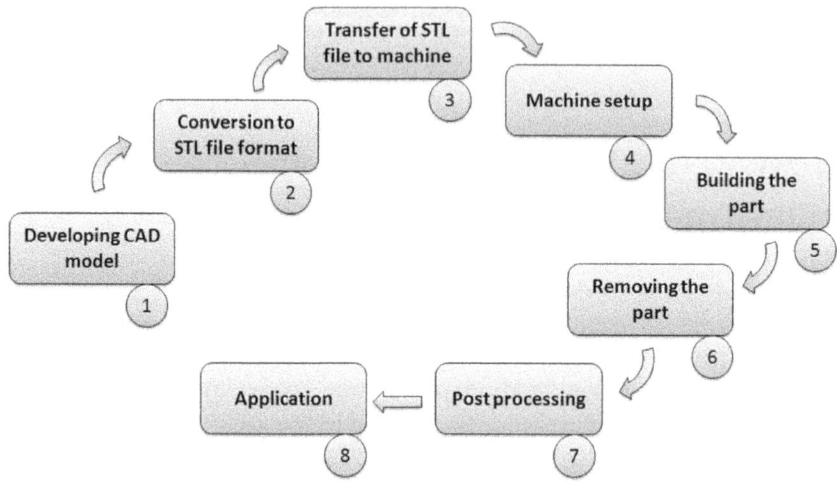

FIGURE 19.1 Steps involved in additive manufacturing process.

Very recently, material deposition by means of friction has been realized as a new method in additive manufacturing. A few reports demonstrated developing 3D components by friction-assisted processes. In 2011, John Samuel Dilip and his group [1] demonstrated *friction deposition*, a novel additive manufacturing technique which involves depositing material from a rotating consumable rod over a substrate by applying certain load to plastically soften the material at the end of the consumable rod and to transfer to the substrate. Similar to the other friction-assisted processes, generation of heat due to the friction at the interface of the rotating consumable rod and the substrate is the important factor in friction deposition process. After completion of a layer deposition, another layer is produced on the previous layer to build the thickness of the coated layer. Friction deposition is similar to friction surfacing, but the entire deposition is achieved at once when the rotating consumable rod is retreated in friction deposition. However, in friction surfacing, the mechtrode is moved along the direction on the substrate in a single track to complete the layer deposition [2–4]. Figure 19.2 schematically shows friction deposition and a typical photograph of friction deposited AISI 304 austenitic stainless steel.

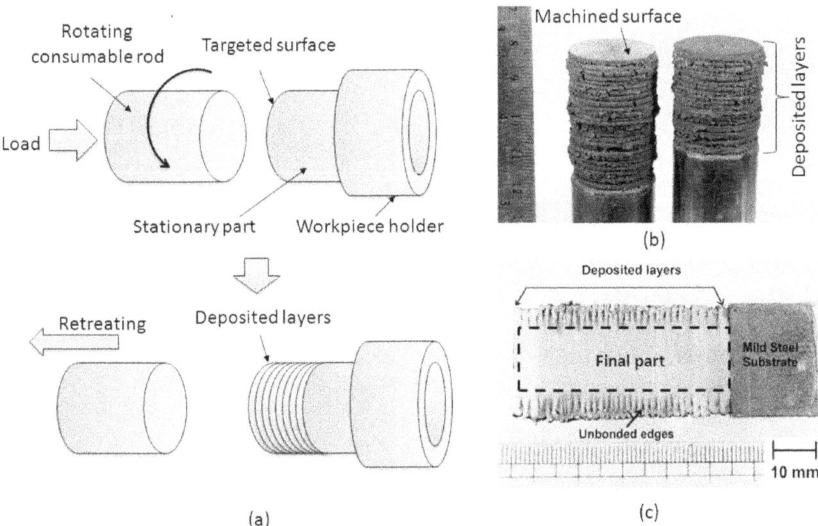

FIGURE 19.2 (a) Schematic diagram of friction deposition, (b) photograph showing friction deposited AISI 304 austenitic stainless steel and (c) cross section showing perfect metallurgical continuity at the center. (*Source:* Reprinted with permission from Dilip et al., [13]. © 2011 Springer Nature.)

FIGURE 19.3 (a) Macrograph of the friction deposit layers, (b) cross section showing 70 layers of deposition, (c) SEM image showing the edges of the deposited layers, and (d) optical microscope image at the cross section shows perfect metallurgical continuity. (*Source:* Reprinted with permission from Dilip et al., [15]. © 2013 Springer Nature.)

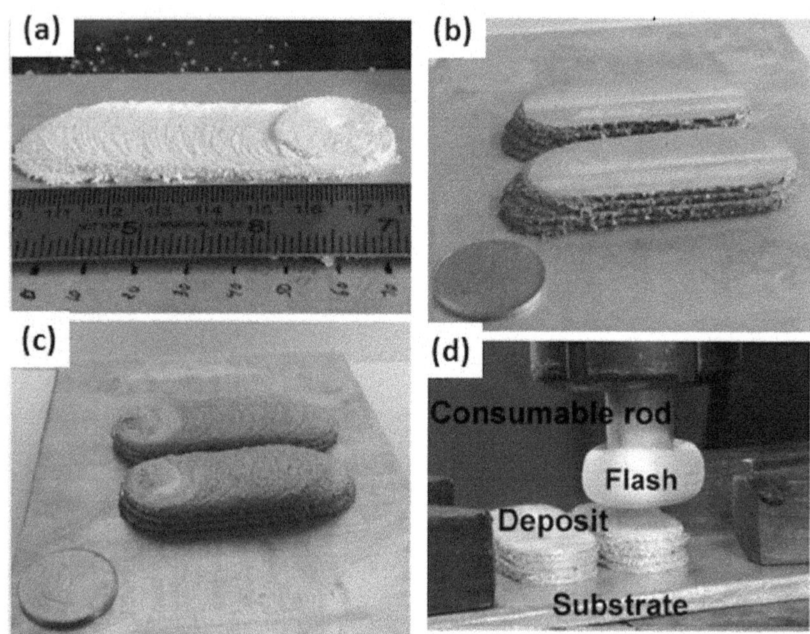

FIGURE 19.4 Friction deposition of aluminum alloy 2014: (a) single layer, (b) three layers, (c) fourth layer and (d) fifth layer deposition (*Source:* Reprinted with permission from Dilip et al., [16]. © 2013 Elsevier.)

The metallurgical continuity with perfect bonding of friction deposited layers was perfect without any defects as examined by microstructural studies. Figure 19.3 shows the photographs of the friction deposited layers and corresponding cross-sectional microstructure. In friction deposition, similar to the other solid-state mechanical processes, the consumable rod material undergoes severe plastic deformation at the high homogeneous temperatures (less than melting temperature) and strain rates. Due to the mode of material transfer under high strains, if any large particles or secondary phases present in the consumable rod will be broken into smaller pieces and new microstructure with finely distributed particles is resulted. However, the applicability of friction deposition to produce large components is still required detailed investigation as economical issues also play crucial role when the component size is increased. Overall, emerging new additive manufacturing processes based on friction-heat generation principle is really a great step must be recognized in the advanced manufacturing industry. Figure 19.4 shows aluminum alloy 2014 layered structure developed by friction deposition method. In the year of 2013, John Samuel Dilip and his group [7] demonstrated for the first time, another variant of friction surfacing in depositing material layer-by-layer on mild steel substrate using consumable mild steel and named the process as *friction-free form*. It is interesting to learn that the development of the multi-layered 3D structure by friction free form from the basic principles of friction surfacing. Figure 19.5 (a) schematically explains the procedure of friction-free form. A multi-track and multilayer deposition of mild steel on a 6 mm thick mild steel sheet was done by friction surfacing using a 19 mm mild steel rod. Initially, 5 tracks were deposited and then four, and three tracks were deposited layer-by-layer. While placing the tracks on the first layer, processing was done such a way that the second layer tracks were placed in the middle of the tracks of the first player to bring uniformity with respect to the center of the track and edge of the track after completing all the tracks and layers. Figure 219.5 (b) and (c) shows the photographs of the surface layers deposited on the mild steel plate.

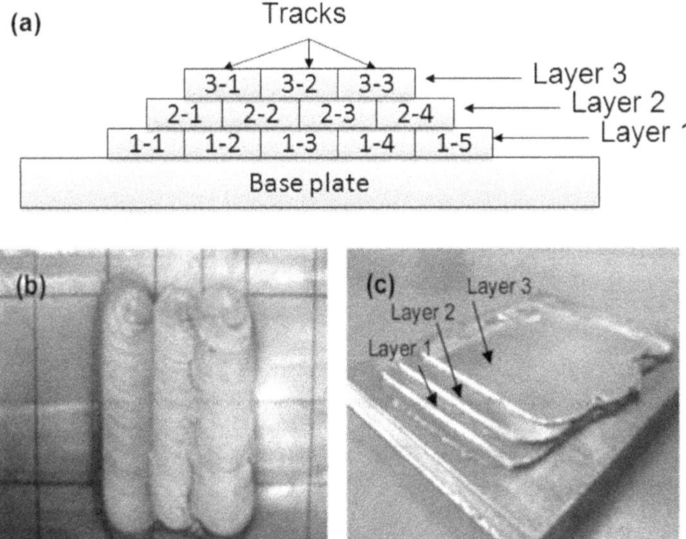

FIGURE 19.5 (a) schematic diagram showing deposition of multi-tracks and multilayers (the first number denotes track and the second number denotes layer); (b) Photograph showing multi-track friction surfaced deposit and (c) photograph showing a multi-track multi-layer friction surfaced deposit. (*Source:* Reprinted with permission from Dilip et al., [7]. © 2013 Elsevier.)

In addition to fabricating 3D components by this method, the same research group demonstrated developing components with internal closed cavities/features as shown in Figure 19.6 in the shape of English alphabets IIT produced within the material. In order to achieve this, initially, the single track was produced as shown in Figure 19.6 (a) and subsequently, holes were produced in the shape of alphabets I, I and T (Figure 19.6 (b) and (c)). The next layers were deposited to close the holes as shown in Figure 19.6 (d) and (e). The X-ray radiograph shows the cavities in the shape of IIT alphabets.

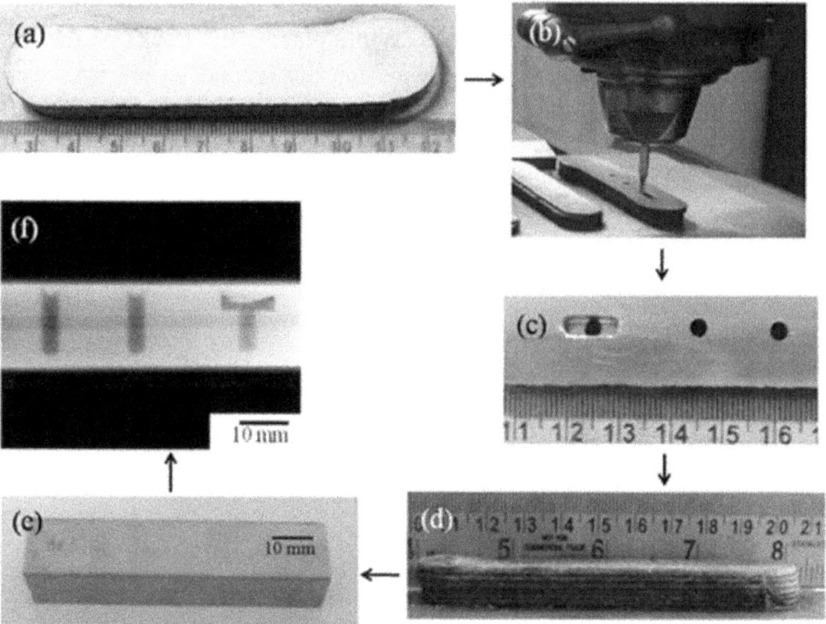

FIGURE 19.6 (See color insert.) Photographs of a 3D part produced by friction deposition containing internal cavities: (a) initial surface layer, (b) producing holes by drilling, (c) top vies of the cavities and (d) surface layers deposition, (e) machined structure and (f) X-ray radiograph of the developed 3D structure. (*Source:* Reprinted with permission from Dilip et al., [7]. © 2013 Elsevier.)

Furthermore, friction-based additive manufacturing processes can be used to develop components with dissimilar layers of metals. For example, Figure 19.7 shows the alternate layers of alloy 316 and alloy 410 deposited using friction surfacing technique to develop 3D structure and corresponding microstructure of the interface [7]. Obtaining a layered structure of soft and ductile alloy 316 and strong and hard alloy 410 can be easily achieved. Additionally, high corrosion resistance from alloy 316 and wear resistance form alloy 410 can be coupled by developing alternate layered material. From the microstructural observation, good bonding with excellent metallurgical features can be observed. Producing these kinds of samples is very difficult using other additive processes without disturbing the properties of the base materials. However, research on friction-assisted additive manufacturing methods is in its early stage. The preliminary findings reported in the literature are promising with a positive future scope.

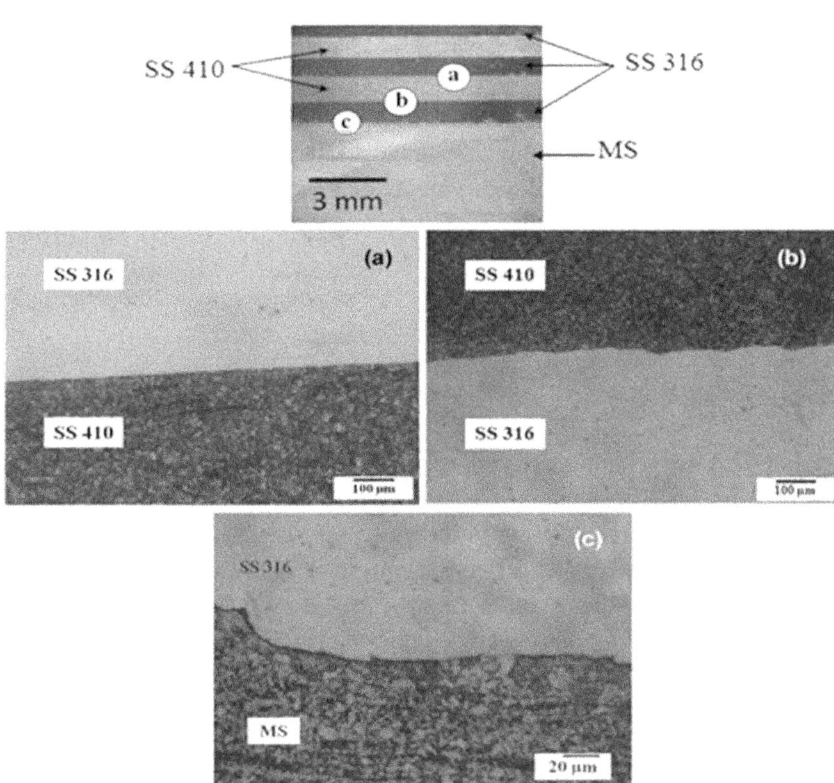

FIGURE 19.7 Optical microscope images showing alternate layers of alloy 316 and alloy 410: (a), (b) and (c) interface of the layers (*Source:* Reprinted with permission from Dilip et al., [7]. © 2013 Elsevier.)

19.2 FRICTION STIR SPOT WELDING

Another novel friction based method developed to address issues in the welding industry is friction stir spot welding. In lightweight applications, metallic sheets and panels are joined by resistance spot welding, self-piercing rivets and clinching. High electric energy, electrode dressing, and cost of consumables are the limitations with these methods. Friction stir spot welding (FSSW) is a solid-state method developed by Mazda Motor Corporation [8]. Initially, FSSW was developed to process aluminum alloy sheets and later adopted to process several similar and dissimilar metals. It was also observed that the cost of spot welding was decreased to a great extent by adopting FSSW compared with resistance spot welding. The overall FSSW equipment cost was also lower compared with resistance spot welding equipment which is economically attracting factor in the context of the technology transfer. Due to these attractive factors, several automobile and aviation companies have adopted FSSW process. For the past decade, based on the developments that happened in using FSSW in the industry, three different FSSW processes can be observed as given below:

i) Basic FSSW
ii) Refill FSSW
iii) Swing FSSW

Figure 19.8 schematically shows the Basic FSSW process. Initially, the rotating FSW tool is plunged into the sheets by applying a suitable load. Due to the stirring action of the tool, material from both the sheets is metallurgically mixed, and a perfect joint is formed [9]. However, in the basic FSSW, a keyhole remains after the process which was addressed by the next variant developed by GKSS (Germany).

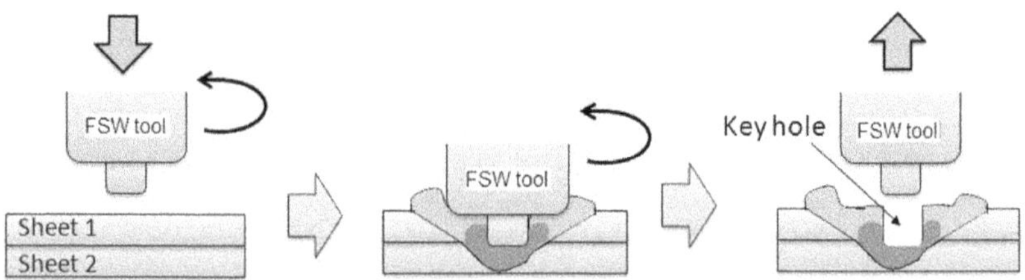

FIGURE 19.8 Schematic representation of friction stir spot welding process.

In the Refill FSSW process, the sheets are placed in lap configuration, and a specially designed rotating FSW tool is plunged into the sheets. Figure 19.9 schematically shows the process of Refill FSSW. The tool used in Refill FSSW consisting of three-piece tool system includes clamp ring, an outer shoulder and inner pin [10, 11]. Each of these three elements of the tool can move independently. The shoulder and the pin of the tool rotate in the same direction with the same number of revolutions per minute. The stationary clamping element holds the workpiece during the spot welding. Initially, the rotating tool comes into contact with the surface of the workpiece and due to friction, heat is generated. Once the sufficient amount of heat is generated, the inner pin is penetrated into the sheets and does the material to deform and flush out plastically. The outer shoulder acts as a reservoir to hold the excess material comes out form the weld zone as flash. Then the inner pin is retreated, and the outer shoulder is forwarded towards the workpiece and causes extrusion of the excess flush back into the weld zone. At the end of the process, no material loss is recorded, and no indentation on the surface also observed. Issues such as sheet lifting, separation, and expulsion and spitting of material from the weld zone can be addressed by the clamping element during the welding.

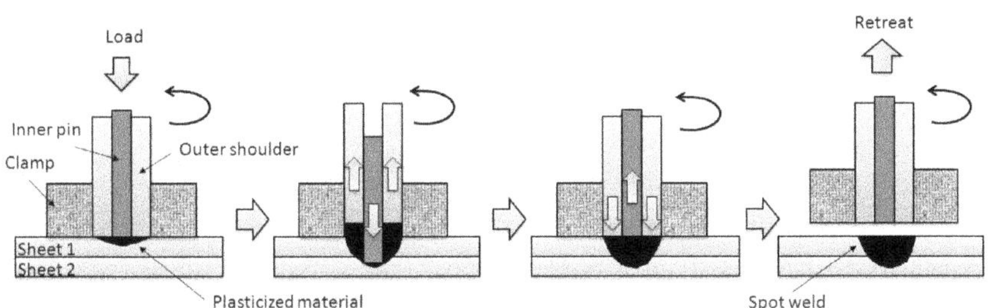

FIGURE 19.9 Schematic representation of refill friction stir spot welding process.

Swing FSSW is another variation developed by The Welding Institute (TWI), UK in which tool is plunged into the sheets to be joined similar to that of Basic FSSW but during the stirring step, tool is made to move in a linear motion by giving a large radius and a small tilt angle [12]. Therefore, as schematically shown in Figure 19.9, the tool travels in a predefined path which includes three sub-paths as a number from 1 to 3. Stirring in path 1 and three are starting and ending of swing respectively and path 2 gives one complete circular path. During stirring, the tool is tilted with a small tilt angle. During retreating, similar to Basic FSSW the rotating tool is retreated from the weld zone. Swing FSSW results higher joint area compared with Basic FSSW and gives higher joint strength. Figure 19.10 compares the starting and ending of the process and the tool travel path in Basic FSSW and Swing FSSW.

There are other variants in FSSW processes reported in recent years. Pinless FSSW is one of such methods in which tool without the pin and a groove on its shoulder are used [13]. This process is simple and can be completed within shorter dwell time. This process leaves no impression or keyhole and gives high strength. Another variation as proposed by Sun et al., [14], involves welding in a two-step. In the first step, a specially designed backplate is used to do Basic FSSW. After the first step, a keyhole and a protuberance in the lower sheet are formed. In the second step, a tool, without a pin is used to remove both the keyhole and the protuberance and leaves no indication of the keyhole. The developments in joining sheets and panels through these variants in FSSW demonstrated the opening of promising areas in the welding industry.

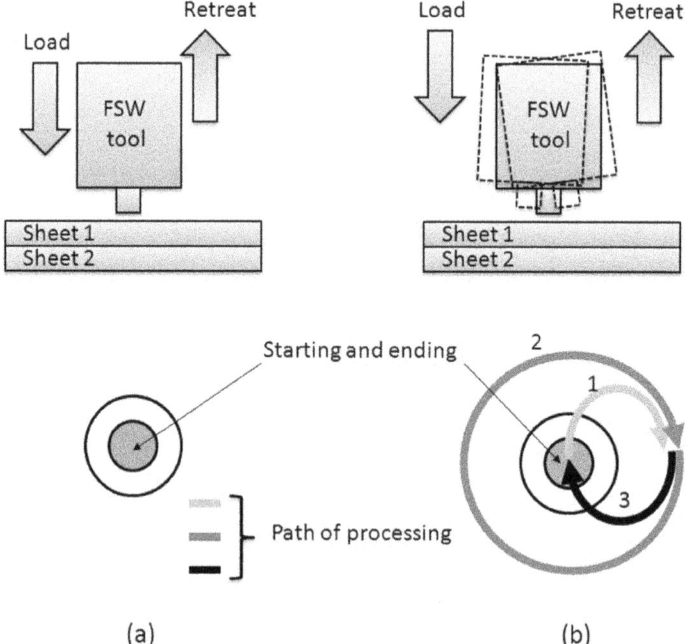

FIGURE 19.10 Schematic comparison of tool path in (a) Basic FSSW and (b) Swing FSSW process.

19.3 FRICTION STIR RIVETING

Joining at one side of aluminum alloys by friction stir riveting (FSR) is another new joining method which eliminates the necessity of drilling a hole to insert a self-piercing rivet. A rotating blind rivet is brought into contact with the workpiece. Due to the generated heat, the force required to penetrate the rivet is decreased and sufficiently inserted into the work-piece material. Then the upsetting of the rivet is done by using the internal mandrel similarly as observed in the conventional riveting. The material that is mixed around the rivet helps to develop a strong bond between the workpieces. The rotational speed of the rivet, feed rate, penetration depth and preheating time are the important process parameters which influence the success of the process [15, 16]. Ma [17] demonstrated friction riveting by adopting rotational speeds from 500 to 3000 rpm with a feed rate from 0.05 in per min and by increasing the feed rate in steps of 0.05 in per min. From the microstructural observations, mixing of material from the sheets around the rivet was reported. The softened material due to the rotation of the rivet allowed proper mixing of the material for the development of a good joint. The plastically deformed material is then confined around the rivet. Due to the gap which is generated between the cap of the rivet and the top of the processed zone in FSR, the joint strength is a concern compared with the conventional riveting process. A wide variety of aluminum alloy sheets can be joined by FSR. The process can be adopted by simply altering the existing equipment which does not required high capital cost. The rate of joint formation is high, and therefore, high production rate can be achieved [18,19]. Figure 19.11 schematically illustrates the FSR process.

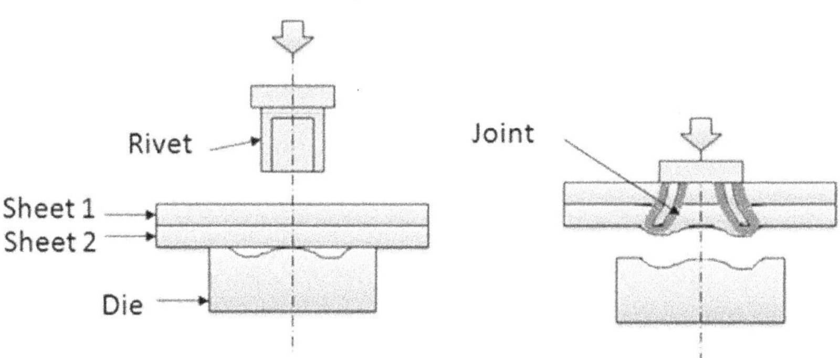

FIGURE 19.11 Schematic representation of friction stir riveting process.

19.4 FRICTION STIR CLADDING

Cladding is a surface coating process to develop thin surface coating layer on a substrate. Friction stir cladding is a solid-state process in which thin coating layer is deposited on a substrate by using a modified friction stir processing tool which is similar to a hollow rod. The hollow rod allows filler material to be coated on the substrate [20]. Similar to FSP, tool rotational speed, substrate travel speed, and axial force are the important process parameters which influence the success of the process. Limited amount of work has been carried out in friction stir cladding. Form the preliminary results, improved wear and corrosion properties were observed.

19.5 FRICTION STIR CHANNELING

In friction stir channeling (FSC), a non-consumable rotating tool is used similar to that of friction stir processing. The removed material due to the stirring of specially designed probe provided at the end of the shoulder of FSC tool facilitates to develop a channel underneath the surface of the processed zone as schematically shown in Figure 19.12. [21]. There is a clear difference between FSP and FSC in the material flow and consolidation mechanisms as explained by Balasubramanian et al., [22]. In FSP or FSW, the gap between the shoulder surface to the workpiece surface is crucial. The surface of the tool shoulder touches the workpiece and imposes a forging pressure to confine the plastically deformed material under the shoulder itself. This step is crucial in order to achieve defect-free processed regions in FSW and FSP. Whereas in FSC, an upward force is generated by using specially designed tool which rotates in clockwise direction (with right hand threaded probe) or anticlockwise direction (with left hand threaded probe) to initiate the material flow towards upward direction. A clear channel is produced when the plastically deformed material is flowed around the pin and moved towards the shoulder. The flow of the plasticized material towards the surface of the workpiece is achieved by the orientation of the threads on the probe. The material which is moved towards the shoulder of the FSC tool is deposited at the top of the nugget zone.

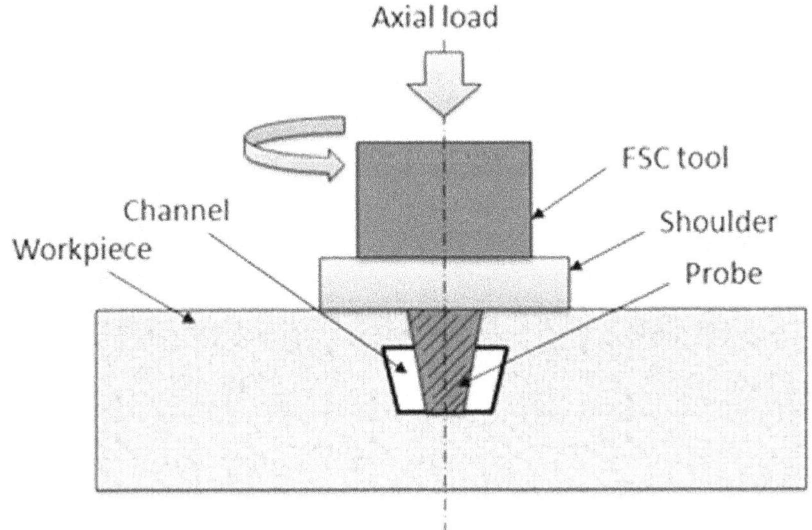

FIGURE 19.12 Schematic representation of FSC process.

The shape and the size of the channel is highly influenced by the tool design and process parameters such as tool rotational speed, travel speed and the design of the shoulder. Figure 19.13 (a) shows typical channel produced by FSC in commercially available 6061 aluminum alloy obtained by processing with the direction of tool counterclockwise. The regions A and B show the stir zone after FSC. Unprocessed region is denoted with C, and the channel is indicated with D. Region E represents the material deposition on the surface after FSC. Figure 19.13 (b) and (c) shows the influence of the process parameters on the shape and size of channel as demonstrated by Balasubramanian et al., [22]. Vidal et al., [23] also produced a channel in AA7178-T6 aluminum alloy plate through FSC and demonstrated the potential of FSC to be used as a promising tool to develop channels. FSC can be used in several cooling systems or heat exchangers used in automobile and aerospace industry as proposed by Balasubramanian et al., [24].

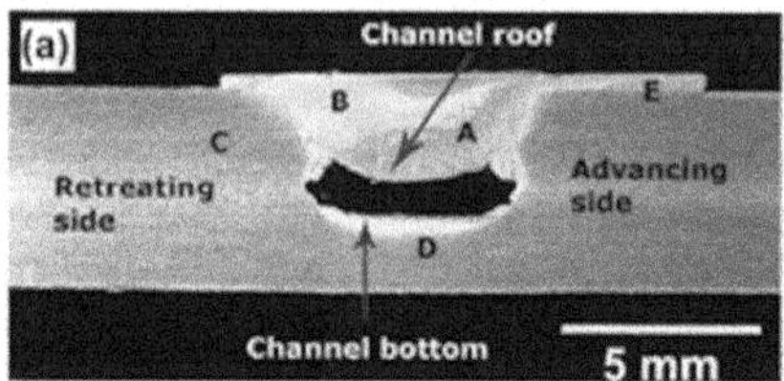

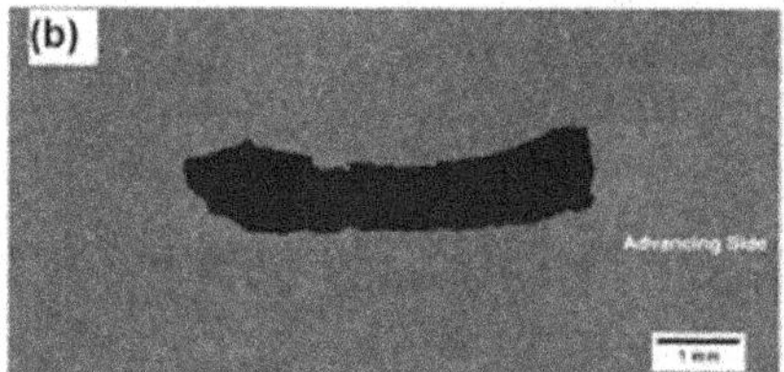

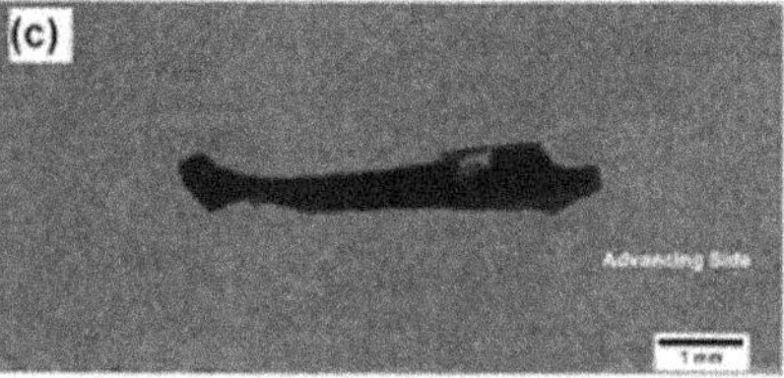

FIGURE 19.13 Optical microscope images of the cross-section after FSC of 6061 alloy: (a) different regions, (b) processed at 1100 rpm with 2.11 mm/s, and (c) processed at 1100 rpm with 2.96 mm/s. (*Source:* Reprinted with permission from Balasubramanian et al., [22]. © 2011 Elsevier.)

The development of new friction based processes enhanced the scope of the solid-state processing techniques to address critical issues faced by the manufacturing engineers particularly in fabricating non-ferrous light metals. Friction based additive manufacturing, friction stir spot welding, friction stir cladding, friction stir riveting, and friction stir channeling techniques are at the infant stage in the process of evolution. However, these processes demonstrated promising preliminary results as potential routes in producing energy efficient structures for the development of manufacturing industry.

KEYWORDS

- **friction deposition**
- **friction stir channeling**
- **friction stir cladding**
- **friction stir riveting**
- **friction stir spot welding**
- **friction-free form**

REFERENCES

1. Dilip Samuel John, J., Khalid Rafi, H., & Ram Janaki, G. D., (2011). A new additive manufacturing process based on friction deposition, *Transactions of the Indian Institute of Metals, 64*(1 & 2), 27–30.
2. Ratna Sunil, B., Reddy Pradeep Kumar, G., Mounika, A. S. N., Navya Sree, P., Rama Pinneswari, P., Ambica, I., Babu Ajay, R., & Amarnath, P., (2015). Joining of AZ31 and AZ91 Mg alloys by friction stir welding, *Journal of Magnesium and Alloys, 3*(4), 330–334.
3. Mishra, R. S., Ma, Z. Y., & Charit, I. (2003). Friction stir processing: a novel technique for fabrication of surface composite. *Mater Sci Eng A.341,* 307–310.
4. Bedford, G. M., Vitanov, V. I., & Voutchkov, I. I., (2001). On the thermo-mechanical events during friction surfacing of high-speed steels, *Surface and Coatings Technology, 141,*34.
5. Dilip, J. J. S., & Janaki Ram, G. D., *Microstructures and properties of friction freeform fabricated borated stainless steel.*
6. Dilip, J. J. S., & Janaki Ram, G. D., (2013). *Microstructures and properties of friction freeform fabricated borated stainless steel. Journal of Materials Engineering and Performance, 22* (10) 3034–3042.
7. Dilip, J. J. S., & Janaki Ram, G. D., (2013). Microstructure evolution in aluminum alloy AA 2014 during multi-layer friction deposition, *Materials Characterization, 86* 146–151.
8. Dilip, J. J. S., Babu, S., Varadha Rajan, S., Rafi, K. H., Janaki Ram, G. D., & Stucker, B. E., (2013). Use of Friction Surfacing for Additive Manufacturing, *Materials and Manufacturing Processes, 28*(2), 189–194.
9. Sakano, R., Murakami, K., Yamashita, K. et al., (2004). Development of spot FSW robot system for automobile body members, in *Proceedings of the 3rd International Symposium on Friction Stir Welding,* Kobe, Japan.
10. Nguyen, T. N., Kim, Y. D., & Kim, H. Y., (2011). Assessment of the failure load for an AA6061-T6 friction stir spot welding joint, Proceedings of the Institution of Mechanical Engineers B: *Journal of Engineering Manufacture, 225*(10), 1746–1756.
11. Rosendo, T., Parra, B., Tier, M. A. D. et al., (2011). Mechanical and microstructural investigation of friction spot welded AA6181-T4 aluminum alloy, *Materials, and Design, 32*(3), 1094–1100.
12. Prangnell, P. B., & Bakavos, D., (2010). Novel approaches to friction spot welding thin aluminum automotive sheet, *Materials Science Forum,* 638–642, 1237–1242.
13. Fang, Y., (2009). *The Research on the Processes and Properties of Pinless Friction Stir Spot Welding,* Jiangsu University of Science and Technology, Jiangsu, China.
14. Sun, Y. F., Fujii, H., Takaki, N., & Okitsu, Y., (2012). Microstructure and mechanical properties of mild steel joints prepared by a flat friction stir spot welding technique, *Materials Design, 37,* 384–392.
15. Sun, Y. F., Fujii, H., Takaki, N., & Okitsu, Y., (2011). Novel spot friction stir welding of 6061 and 5052 Al alloys, *Science and Technology of Welding and Joining, 16*(7), 605–612.
16. Rodelas, J. M., Lippold, J. C., Rule, J. R., & Livingston, J., (2011): Friction stir processing as a base metal preparation technique for modification of fusion weld

microstructures. In: Mishra, R. S., Mahoney, W. W., & Lienert, T. J., (Ed) Friction Stir Welding, and Processing VI, *Minerals, Metals & Materials Society (TMS),* pp. 21–36.

17. Li, Y. B., Wei, Z. Y., Wang, Z. Z., & Li, Y. T., (2013). Friction self-piercing riveting of aluminum alloy AA6061-T6 to magnesium alloy AZ31B. *J Manuf Sci Eng135,* 1–7. doi:10.1115/1.4025421.
18. Ma, G., (2012). A thesis-friction-stir riveting, characteristics of friction-stir riveted joints. The University of Toledo, Toledo.
19. Luo, H., (2008). *New Joining Techniques for Magnesium Alloy Sheets,* MS thesis. Institute of Metal Research, Chinese Academy of Sciences, China, pp. 48–63.
20. Yang, X. W., Fu, T., & Li, W. Y., (2014). Friction stir spot welding: a review on joint macro- and microstructure, property, and process modeling. *Adv Mater Sci Eng* 1–11.
21. Rodelas, J. M., Lippold, J. C., Rule, J. R., & Livingston, J., (2011). Friction stir processing as a base metal preparation technique for modification of fusion weld microstructures. In: Mishra, R. S., Mahoney, W. W., Lienert, T. J., (Ed) Friction Stir Welding, and Processing, V. I., *Minerals, Metals & Materials Society* (TMS), pp. 21–36.
22. Vilaça, P., & Vidal, C., (2011). Modular adjustable tool and correspondent process for opening continuous internal channels in solid components. National patent pending N.105628T.
23. Balasubramanian, N., Mishra, R. S., & Krishnamurthy, K., (2011). Process forces during friction stir channeling in an aluminum alloy. *Int J Mater Prod Technol211,* 305–311.
24. Vidal, C., Infante, V., & Vilaça, P., (2012). Mechanical characterization of friction stir channels under internal pressure and in-plane bending. *Key Eng Mater 488–489,* 105–108.
25. Balasubramanian, N., Mishra, R. S., & Krishnamurthy, K., (2009). Friction stir channeling: characterization of the channels. *J Mater Process Technol209,* 3696–3704.

CHAPTER 20

Conclusions

Development of friction-based processes is a significant event that revolutionized the solid-state processing routes in the manufacturing industry. These processes can offer several advantages compared with their conventional counterparts and can be adopted in a wide variety of applications in automobile, aerospace and marine industries. Particularly surfaces of the structures can be engineered to increase mechanical, corrosion, wear and bioproperties by modifying the microstructure. Superplasticity can be introduced in metals through friction stir processing (FSP). The increased formability after FSP is attributed to the fine and uniform grain size produced during the process. Enhanced mechanical properties such as hardness, fracture strength, tensile strength, and bending strength can be achieved by FSP. Corrosion resistance of a few materials can be increased by modifying the microstructure by using FSP. However, some studies also demonstrated the decreased corrosion resistance due to the grain refinement. A wide variety of material systems can be successfully processed by FSP such as different steels, aluminum alloys, magnesium alloys, titanium, and its alloys. The level of grain refinement and uniformity in the grain size can be altered by increasing the number of passes. Repairing the defect surfaces at certain local sites such as tiny cracks, pores, casting defects by carrying FSP at site-specific areas can reduce the manufacturing cost and the life of the structure.

The basic principle of FSP can be further used to develop surface composites without melting which opened another interesting area in surface engineering. The intended secondary phase particles are introduced into the surface during FSP and the stirring action of the tool distribute the particles into the substrate. The level of distribution of the particles can be altered by increasing the number of FSP passes. Different methods such as grooves, holes, surface coating and sandwich methods can be used to incorporate the distributed phase. Another interesting development to embed the secondary phase particles is the use of direct friction stir processing (DFSP) tool which contains a through hole within the tool that facilitates a continuous supply of the secondary phase powder during FSP. Modifying the porous surfaces is another area recently FSP has shown the significant effect to modify the mechanical properties at the surface. Development of friction stir channeling is another variant recently demonstrated to produce through blind cavity which can be used for internal circulation of fluids for coolant or in the applications of heat exchangers.

Development of friction surfacing to produce coating layers on the surface within the solid-state is another important development in friction-assisted processes. By using a consumable rod, strong and sound joints with excellent coating integrity with the substrate can be produced by friction surfacing. Wow to its potential to develop a wide variety of surface coatings without melting. Difficult to coat alloys and composites can be used as consumable rod and coating can be successfully developed. However, the poor bonding at the edges of the coating is a prime concern must be considered while designing surface coatings by friction surfacing. From the successful evolution of friction surfacing, development of 3D structures was also demonstrated by depositing layer-by-layer coatings. Friction deposition or with a specific name "friction free form" is such a good example that explains how friction surfacing leads to additive manufacturing. Development of cladding is another variation where the coating layer is deposited by consumable rods supplied through the tool. These processes have opened a new area in the solid-state processing and engineering which really offer several advantages compared with conventional processes. The research on some of these methods is in its preliminary stage, but however, the promising characteristics of these processes certainly revolutionize the manufacturing industry.

20.1 CRITICAL ISSUES AND FUTURE SCOPE

Friction-assisted surface engineering processes have undoubtedly demonstrated their efficacy as advanced manufacturing processes. However, some issues still needed to be understood to make these processes more viable for the industry.

Friction-assisted processes are new to the industry and are at present mostly at the interest of academic research level. A few large scale industries have already adopted these processes in developing different structures. But the small and medium scale industries are not yet ready to adopt. The scientific knowledge that is available in not sufficiently understood by the small and medium scale industries. Usual resistance to adopt new technology in the place of existing conventional technologies by the manufacturers is another main limitation. Specifications and standards are crucial for the wide acceptance and adaptation of any technology. Several technical committees have already been specified certain standards in connection with friction stir welding and friction stir processing. The aerospace industry is the first industry which has seen the reliable fruits of these techniques. However, generalized specifications and standards have not been streamlined which is the important limitation why the widespread of using these techniques is not yet seen in the manufacturing industry. Importance of appropriate and optimum design in developing tools is another area that plays a crucial role in the success of these techniques.

Some industries have developed their own standards in addition to coupling the technology with automation and robotics. Precise control over the process can be achieved by designing the process with the help of automation as process parameters can be altered with the help of programmable controllers. In addition, a higher level of production rate with lower defects and better repeatability can be attained with the automated technology.

With the advancement of computational methods, the use of computer modeling and simulation routes to understand the process and the material deformation mechanisms is the new area in which several research groups are working. The development of commercially available numerical codes for the other design and manufacturing applications can also be used to analyze these techniques. A few dedicated codes have been developed but which are applications. For the past two decades, the interest on adopting numerical methods to analyze different aspects of friction-assisted processes has been considerably grown. In particular, thermal stresses and distortion of the structures, material flow mechanisms during the process were widely investigated by using finite element analysis methods. These kinds of studies explore more scientific knowledge and enhance the level of understanding of the processes. In the coming years manufacturing industry witness, the active role of these friction-assisted processes in surface engineering.

KEYWORDS

- **direct friction stir processing**
- **friction stir processing**
- **future scopes**

Index

F

G

V

W

X

Y

Z

Milton Keynes UK
Ingram Content Group UK Ltd.
UKHW022059141024
449569UK00031B/1696